MW00355517

ELLAVUT

Ann Fienup-Riordan *&* Alice Rearden

ELLAVUT

Our Yup'ik World *&* Weather

//

CONTINUITY AND CHANGE ON THE BERING SEA COAST

UNIVERSITY OF WASHINGTON PRESS Seattle *&* London

in association with

CALISTA ELDERS COUNCIL Anchorage, Alaska

© 2012 by the University of Washington Press

16 15 14 13 12   5 4 3 2 1

All rights reserved. No part of this publication may be reproduced or transmitted in any form or by any means, electronic or mechanical, including photocopy, recording, or any information storage or retrieval system, without permission in writing from the publisher.

University of Washington Press

PO Box 50096, Seattle, WA 98145, USA

www.washington.edu/uwpress

Library of Congress Cataloging-in-Publication Data

Fienup-Riordan, Ann.

Ellavut / our Yupik world and weather : continuity and change on the Bering Sea coast / Ann Fienup-Riordan and Alice Rearden.

    p. cm.

Includes bibliographical references and index.

ISBN 978-0-295-99161-0 (pbk. : alk. paper)

1. Yupik Eskimos—Science—Alaska—Bering Sea Coast. 2. Yupik Eskimos—Alaska—Bering Sea Coast—Social conditions. 3. Indigenous peoples—Ecology—Alaska—Bering Sea Coast. 4. Climatic changes—Alaska—Bering Sea Coast. 5. Global environmental change—Alaska—Bering Sea Coast. 6. Bering Sea Coast (Alaska)—Environmental conditions. I. Rearden, Alice. II. Title.

E99.E7F45 2012

305.897'14016451—dc23   2011041461

Printed and bound in the United States of America

Designed by Ashley Saleeba

Composed in Minion Pro and ITC Franklin Gothic

The paper used in this publication meets the minimum requirements of American National Standard for Information Sciences—Permanence of Paper for Printed Library Materials, ANSI Z39.48-1984.∞

Frontispiece: A young Kwigillingok hunter holding up a spotted seal taken along the ice edge sometime in the late 1940s. The kayak is covered in canvas. *Warren Petersen, Petersen Family Collection*

CONTENTS

ACKNOWLEDGMENTS

Our greatest thanks to the dozens of Yup'ik men and women from throughout the region who so generously shared their wisdom and knowledge. Their hope is that readers can learn from their experiences and live better—and safer—lives. We are especially grateful to Simeon and Anna Agnus of Nightmute, Frank Andrew and Roland Phillip of Kwigillingok, Nick Andrew Sr. of Marshall, George Billy of Napakiak, Lizzie Chimiugak and Martina John of Toksook Bay, John Eric and Paul Tunuchuk of Chefornak, Michael John and Peter John of Newtok, Paul Kiunya Sr. of Kipnuk, and John Phillip Sr. of Kongiganak. We also thank Paul John of Toksook Bay, the Nelson Island leader whom John Eric referred to as *akagarcailkutiit* (one who prevents them from rolling). Throughout our discussions, he kept us on the right track.

The Calista Elders Council (CEC), which is the primary heritage organization in southwest Alaska representing the region's 1,300 elders sixty-five years and older, guided our work. CEC's executive director, Mark John, and its board of directors, including Bob Aloysius of Kalskag, Nick Andrew Sr., Peter Elachik of Kotlik, Peter Jacobs Sr. of Bethel, Paul John, Paul Kiunya Sr., Martin Moore Sr. of Emmonak, Moses Paukan Sr. of St. Marys, John Phillip Sr., and Moses White Sr. of Kasigluk, provided invaluable guidance and support throughout.

The National Science Foundation, Office of Polar Programs, funded our work under two grants. The first (Grant 9909945) supported a traditional knowledge project that was carried out between 2000 and 2005 and enabled the CEC to host small gatherings with elders from throughout the region, during which staff documented Yup'ik views of the environment. The second was carried out under NSF's BEST (Bering Ecosystem Study) Program (Grant 0611978, Publication BEST BSIERP 14). BEST is a multiyear, multiproject study of the dynamics of change in the eastern Bering Sea involving more than a hundred researchers. Our Nelson Island Natural and Cultural History Project was the one BEST project that dealt with sociocultural issues. The others all focused on the physical science and biology of the Bering Sea. Being part of such a large, integrated research initiative has been rewarding. CEC staff and board members have learned a great deal from other BEST researchers about the *imarpik* (ocean). We have also tried to share with our partners a nearshore Yup'ik perspective on the Bering Sea, based on decades of personal experience. Special thanks to our NSF program officers, Anna Kerttula de Echave and William Wiseman, for setting these collaborations in motion.

Other organizations have contributed to our work, especially during the three-week circumnavigation of Nelson Island in July 2007 that was a major component of our Nelson Island project. Calista Corporation provided the CEC with office space in Anchorage, financial support, and the expertise of their vice president of Land and Natural Resources, June McAtee. The Association of Village Council Presidents provided office space in Bethel as well as the services of archaeologist Steve Street. And the US Fish and Wildlife Service loaned us newly hired biologist Tom Doolittle, who spent his first weeks on the delta learning from Nelson Island elders and sharing his experiences in turn. The Nelson Island project also received endorsement from the International Polar Year organizing committee as part of the larger SIKU (Sea Ice Knowledge and Use) initiative, providing linkages to other projects, primarily ELOKA (Exchange for Local Observations and Knowledge of the Arctic). Special thanks also to the Alaska Humanities Forum and the Rasmuson Foundation, whose generous assistance made this publication possible.

This book is the ninth produced by the CEC in our efforts to document Yup'ik oral traditions, not as arcane facts but as knowledge systems with continuing relevance in our rapidly changing world. The transformation of the social landscape during the last fifty years has been matched by dramatic changes in the lowland delta and coastal environments of southwest Alaska. The first CEC publications— *Wise Words of the Yup'ik People: We Talk to You because We Love You* and *Yupiit Qanruyutait/Yup'ik Words of Wisdom*—were written to share the foundational values of Yup'ik people embodied in their *qanruyutet* (words of wisdom), which continue to guide their relations with one another. This book presents the values and *qanruyutet* that guide Yup'ik people in their relations with the land and sea. That one book should follow the other is no accident, as how one should act on the ocean and the land and how one should treat one's fellow humans are closely bound.

Yup'ik views of *ellavut*—our world and its weather—are multitudinous, and we cannot claim that our treatment is comprehensive. Although we have talked with elders from throughout southwest Alaska, we have worked most closely with men and women from the Canineq (lower Kuskokwim coastal) region and Nelson Island. In regional gatherings elders often stated that what they heard from others applied in their areas as well. Likewise, I am often struck by how much their words apply to my own *kass'aq* life.

Elders always shared information in the Yup'ik language, and my friend and colleague Alice Rearden did the lion's share of the transcription and translation, ably assisted by David Chanar and Marie Meade. Alice's commitment to accuracy and clarity shines through these pages. Moreover, Alice and I were partners at elders' gatherings during which this information was recorded. I came armed with questions and word lists, but it was Alice's sharp mind and boundless curiosity that encouraged elders to share so much. As I worked on these chapters, Alice was always my first and best reader—answering questions, spotting errors, making suggestions. I shaped the words she provided. Alice taught me as much as any elder, and always with humor and kindness. She made collaboration a real joy. After working with Alice for over ten years, how can I claim sole authorship in the old pith-helmet-wearing anthropological tradition? I can't.

Alice and I were also fortunate to have the assistance of friends and colleagues with special expertise who reviewed particular sections, including Jim Brader, Tom Doolittle, Hajo Eicken, Torre Jorgenson, Igor Krupnik, June McAtee, Travis Rector, Ned Rozell, Steve Street, Matthew Sturm, Francisca Yanez, and NSF program officers Anna Kerttula de Echave and William Wiseman. We are both deeply indebted to Steven Jacobson and his colleagues at the Alaska Native Language Center for the hard work and dedication that produced the *Yup'ik Eskimo Dictionary*, which guided our work. The clarity of the English text benefited from the sharp eyes and light touch of my longtime editor and friend Judith Meidinger.

Many individuals and organizations provided illustrations for this book. Nick Therchik Jr., Mark John, and Simeon John photographed sea ice on the Nelson Island coast during spring 2008, 2009, and 2010, and George Smith photographed the results of a fall storm near Scammon Bay. Jeff Foley, June McAtee, and Rob Retherford of Calista, Tom Doolittle and Josh Spice of the US Fish and Wildlife Service, Yukon Delta National Wildlife Refuge, Warren Jones of Quinhagak, and friends Matt O'Leary and Janet Klein shared stunning photographs of features of the coastal environment, allowing readers to see what elders describe. We thank Jim Barker for his powerful black-and-white photographs of contemporary harvesting activities. Historical photographs were also supplied by the Anchorage Museum, the Moravian Archives, the National

Museum of the American Indian, and the family of Warren Petersen. Cameron Byrnes photographed Yup'ik artist Milo Minock's drawing from the Kline family collection, and Barry McWayne contributed his photograph of the North Wind mask from the National Museum of the American Indian. Once again, my friends and colleagues Patrick Jankanish and Matt O'Leary collaborated on a map of the region. Michelle Pearson and Steve Street generously contributed their expertise to produce our Nelson Island map. And thanks to Nick Riordan for making our illustration of land and water forms presentable.

Thanks are also due to the University of Washington Press and their fine staff who helped us through the publication process, including Pamela Bruton, Beth Fuget, Lorri Hagman, Ashley Saleeba, Pat Soden, Marilyn Trueblood, and others.

Last but not least, thanks to our families, especially our children, Kyle, Kayla, Christopher, Frances, Jimmy, and Nicky. They make our work worthwhile.

YUP'IK CONTRIBUTORS

	Residence	Birthplace	Birth Year
Nick Andrew Sr. / *Apirtaq*	Marshall	Ohagamute	1933
Tim Myers / *Uparquq*	Pilot Station	Pitka's Point	1926
Marie Myers / *Luqipataaq*	Pilot Station		
Mary Mike / *Arrsauyaq*	St. Marys	Uksuqalleq	1912
Jasper Louis / *Kaligtuq*	St. Marys	Anagciq	1916
Johnny Thompson / *Cakitelleq*	St. Marys	Pilot Station	1923
Matthew Beans	Mountain Village		
Simon Harpak / *Agqerralria*	Mountain Village	Nanvaruk	1935
Willie Kamkoff / *Uankaaq*	Kotlik	Nunapiggluugaq	1923
Peter Elachik / *Ilacik*	Kotlik		
Alex Bird / *Apaliq*	Emmonak	Caniliaq	1921
Teddy Sundown / *Canaar*	Scammon Bay		

Mike Utteryuk / *Uteryuk*	Scammon Bay	Scammon Bay	1926
Neva Rivers / *Aluskaamutaq*	Hooper Bay	Hooper Bay	1920
Louise Tall / *Atsaruaq*	Chevak	Upaucugmiut	1902
Elsie Tommy / *Nanugaq*	Newtok	Kaviarmiut	1922
Michael John / *Qukailnguq*	Newtok	Cevtaq	1931
Mark Tom / *Nuyarralek*	Newtok		
Peter John / *Miisaq*	Newtok	Kayalivik	1936
Joseph Patrick / *Agiyangaq*	Newtok	Nerevkartuli	1937
Mary George / *Nanurniralria*	Newtok	Cevtaq	1942
Nicholas Tommy / *Mikcuar*	Newtok		
Paul Charles / *Ayurun*	Newtok	Kayalivik	1948
John Roy John / *Aliurtuq*	Newtok		1964
Susie Angaiak / *Uliggaq*	Tununak	Tununak	1923
Edward Hooper / *Maklak*	Tununak	Tununak	1925
Lucy James / *Kakgailnguq*	Tununak		1928
Tommy Hooper / *Cuk'ayaq*	Tununak	Tununak	1931
Rita Angaiak / *Nanurrualek*	Tununak	Chefornak	1932
John Walter / *Cungauyar*	Tununak	Chefornak	1939
James James / *Cagmaran*	Tununak		
Frances Usugan / *Piiyuuk*	Toksook Bay	Up'nerkillermiut	1915
Phillip Moses / *Nurataaq*	Toksook Bay	Nightmute	1925
Theresa Moses / *Ilanaq*	Toksook Bay	Cevv'arneq	1926
Paul John / *Kangrilnguq*	Toksook Bay	Cevv'arneq	1928
Sophie Agimuk / *Avegyaq*	Toksook Bay	Cevv'arneq	1928
John Alirkar / *Allirkar*	Toksook Bay	Cevv'arneq	1929
Lizzie Chimiugak / *Neng'uryar*	Toksook Bay		1930
Joe Asuluk / *Atrilnguq*	Toksook Bay	Kayalivik	
Martina John / *Anguyaluk*	Toksook Bay	Nightmute	1936
Ruth Jimmie / *Angalgaq*	Toksook Bay	Nightmute	
Nick Therchik Jr. / *Taneksak*	Toksook Bay		1970
Albertina Dull / *Cingyukan*	Nightmute		1918
Dick Anthony / *Minegtuli*	Nightmute	Cevv'arneq	1922
Simeon Agnus / *Unangik*	Nightmute	Nightmute	1930
Anna Agnus / *Avegyaq*	Nightmute	Nightmute	1930

Camilius Tulik / *Cirmirraq*	Nightmute	Umkumiut	1935
Peter Dull / *Arcunan*	Nightmute	Qunguq	1939
Thomas Jumbo / *Arcunaq*	Nightmute	Qiingssaq	1940
Theresa Anthony / *Ceturngalria*	Nightmute		
Hilary Kairaiuak / *Kairaiyuaq*	Chefornak	Calit'lleq	1922
Martina Wasili / *Cuyanguyak*	Chefornak		
Maria Eric / *Qamulria*	Chefornak	Cevv'arneq	
Jobe Abraham Sr. / *Kumak*	Chefornak	Nightmute	1933
David Jimmie Sr. / *Akak'aq*	Chefornak	Kavirlirralek	1936
Pauline Jimmie / *Kangrilnguq*	Chefornak	Cevv'arneq	1937
Theresa Abraham / *Paniliar*	Chefornak	Cevv'arneq	1941
John Eric / *Cungauyar*	Chefornak	Cevv'arneq	1942
Paul Tunuchuk / *Qerrataralria*	Chefornak	Arayakcaarmiut	
Joe Maklak / *Acivaralria*	Chefornak	Cevv'arneq	
Peter Matthew / *Angutekayak*	Chefornak	Arayakcaarmiut	
John Jimmie / *Yang'aq*	Chefornak	Kavirlirralek	
Walter Tirchik / *Tutmalria*	Chefornak	Bethel	1952
Edward Kinegak / *Sak'aq*	Chefornak	Akiachak	
David Martin / *Negaryaq*	Kipnuk	Cal'itmiut	1914
Paul Kiunya Sr. / *Kayungiar*	Kipnuk	Pengurpagmiut	1930
Frank Andrew / *Miisaq*	Kwigillingok	Kwigillingok	1917
Lena Atti / *Kayungiar*	Kwigillingok	Kipnuk	1925
Roland Phillip / *Angutekaar*	Kwigillingok		1927
John Phillip Sr. / *Ayagina'ar*	Kongiganak	Anuurarmiut	1925
Paul Andrew / *Pug'uralria*	Tuntutuliak		
Nick Mark / *Uyaquq*	Quinhagak		
Annie Blue / *Cungauyar*	Togiak	Qissayaarmiut	1916
Irvin Brink / *Cagluaq*	Kasigluk	Nunacuaq	1925
Herman Neck / *Nug'aralria*	Nunapitchuk		
Paul Jenkins / *Pugtall'er*	Nunapitchuk	Cuukvagtuli	1925
Joseph Jenkins / *Pac'aq*	Nunapitchuk	Cuukvagtuli	
Katie Kernak / *Tutmaralria*	Napakiak	Eek	
George Billy / *Nacailnguq*	Napakiak		
Nick Charles Sr. / *Ayagina'ar*	Bethel	Nelson Island	1912

Elena Charles / *Nengqerralria*	Bethel	Nunacuaq	1918
Lucy Beaver / *Urutaq*	Bethel	Qinarmiut	
Peter Jacobs Sr. / *Paniguaq*	Bethel	Cuukvagtuli	1923
James Guy / *Qugg'aq*	Kwethluk		
Fannie Nicori	Kwethluk		
John Andrew	Kwethluk		
Joe Lomack / *Uyaquq*	Akiachak	Akiachak	1924
Wassilie Evan / *Misngalria*	Akiak	Napaskiak	1930
Joshua Phillip / *Maqista*	Tuluksak	Akiachak	1909
Elizabeth Andrew	Tuluksak		
Golga Effemka / *Ungagpak*	Sleetmute		1934
Jack Egnatty	Sleetmute		
Nick Mellick Jr.	Sleetmute		
Bob Aloysius / *Sliksuuyar*	Kalskag	Paimiut	1935
Theresa Alexie	Upper Kalskag		
Lucy Sparck / *Utuan*	Anchorage	Chevak	1940
David Chanar / *Cingurruk*	Anchorage	Umkumiut	1946
Marie Meade / *Arnaq*	Anchorage	Nunapitchuk	1947
Mark John / *Miisaq*	Anchorage	Nightmute	1954

NOTE: This list reflects Yup'ik protocol. Names are ordered by community, running north to south along the Bering Sea coast and upriver to Bethel. Within each community, individuals are listed by age (eldest to youngest); their Yup'ik names are in italics. In the text the first occurrence of each speaker's name is followed by place of residence, for example, Paul John (November 2000:81) of Toksook Bay. The gathering date and transcript page number of the statement follow the elder's name in parentheses.

ELLAVUT

Elders and youth discuss place-names during a CEC gathering in the Chefornak community hall, March 2007. *From left to right*, Mark John, David Jimmie Sr., David Chanar, and John Eric. *Ann Fienup-Riordan*

INTRODUCTION

O N A BLIZZARDY MARCH AFTERNOON IN 2008, ALICE REARDEN
and I met with a dozen elders and young people gathered in the Chefornak community hall to document discussion of their way of life.
They represented the Calista Elders Council (CEC), the primary heritage organization for elders in southwest Alaska, which sponsored our visit. We brought US
Geological Survey maps, pens, paper, and my thirty-year-old Sony tape recorder.
John Eric, age sixty-six, began by comparing what his grandfather taught him to
a college education:

> We experienced that [way of living]. I listened closely to people in the past, includ-
> ing my late grandfather and his peers, when they spoke in the *qasgi* [communal
> men's house]. My grandfather also brought me with him when he traveled by dog
> team. He constantly taught me about places around our village, including rivers and
> lakes or other landmarks. He'd tell me, "I'm revealing these places to you because
> I want you to use them in the future when you are alone." When I think about what
> he taught me, I realize now that he evidently gave me a college education about the
> Yup'ik way of life.

John then spoke about the form his grandfather's teachings took—pithy *qanruyutet* (wise words or instructions) covering every situation he would meet in

life—and regretted that today these instructions are rarely heard: "Since those people are now all gone, that way [of teaching] is also gone. Today I no longer hear many things that I heard in the *qasgi*. And although we have become instructors, that [form of teaching] is now nonexistent, since there are so many things that are distracting to our way of life today." John Eric then noted the gratitude he felt when CEC director Mark John spoke to him about holding a gathering in Chefornak: "I'm grateful that he wanted these people here to reveal things before they are all gone." John also supported CEC's past and future efforts to share elders' words in written form:

> Sometimes when I read the book that this person [Ann Riordan] sent me, I really like reading the things [Frank Andrew] from Kwigillingok said. Things that I heard and things that I will use are in that book. And Oscar Wasilie, who is part of the younger generation, also liked [the book]. Indeed, people who don't have anyone to instruct them are able to learn through [books] or through the work that we will do all day today. Even though a person is not in attendance here, by looking at that book [they can learn].

The book that John Eric spoke of was *Yupiit Qanruyutait/Yup'ik Words of Wisdom*, a bilingual compilation of *qanruyutet* that elders from throughout southwest Alaska had shared during CEC gatherings like the one that day in Chefornak. The CEC hosted dozens of gatherings in Anchorage, Bethel, and surrounding villages between 2000 and 2005. The CEC's board of elders worked with younger staff members to choose the topics. Like John Eric, these nine Yup'ik-speaking men, representing villages throughout the Yukon-Kuskokwim region, actively support the documentation and sharing of traditional knowledge, which all view as possessing continued value in the world today.

Over the years CEC staff have found that small gatherings like that in Chefornak—working with groups of elder experts and younger community members on a specific set of questions for two or three days—were effective in both documenting past traditions and addressing contemporary community concerns. Unlike interviews, during which elders answer questions posed by those who often do not already hold the knowledge they seek, gatherings (like academic symposia) encourage elders to speak among their peers at the highest level. John Phillip (October 2006:284) of Kongiganak observed during one gathering: "Hearing the story you just told, I learned what I didn't know. As we are sitting here one explains what we did not know, while another one explains something else. It is like we are still learning."

CEC gatherings always take place in the Yup'ik language, as the form in which information is shared is as important as the content. Translators Alice Rearden, Marie Meade, David Chanar, and others then create detailed transcriptions and translations of each gathering, and we work together to turn these transcripts into both bilingual publications, like *Yupiit Qanruyutait*, and accompanying English texts. To date the CEC has produced three sets of "paired" books—one in English for general audiences and the other bilingual for community use.[1]

In gatherings elders speak their past selectively, not comprehensively, and what is not said is often as significant as what is said. Long and careful listening to these conversations provides unique perspectives on Yup'ik knowledge.[2] In these forums elders teach more than facts; they teach listeners how to learn. They share not only what they know but how they know it and why they believe it is important to remember.

CEC staff and Yup'ik community members value topic-specific gatherings not merely as tools for documentation but as contexts of cultural transmission. The gatherings themselves are meaningful events that enrich lives locally at the same time they have the potential to increase cross-cultural understanding globally. Camilius Tulik (March 2007:596) of Nightmute declared: "Your work here causes one to be grateful. It seems like for the sake of our children and grandchildren, we have revealed those things that were out of their reach." Simeon Agnus, also of Nightmute, agreed: "Truly what you are working on is good because you have helped us remember these oral instructions we had forgotten. Through this work our oral traditions have been renewed." ·

Simeon Agnus points out a land feature near Arayakcaaq at the mouth of the Qalvinraaq River, July 2007. Michael John sits to his right and Theresa Abraham to his left. *Ann Fienup-Riordan*

Upper: Rita Angaiak, Anna Agnus, Martina John, and Joe Felix rest at Nacecararmiut, a reindeer herding camp on the upper Cakcaaq River, with Jackie Lincoln and Linda Joe in the background, July 2007. *Ann Fienup-Riordan*

Lower: Paul John documents the history of Kaviaq, on the southern shore of Toksook Bay, with a group of elders and students, July 2007. *Ann Fienup-Riordan*

Although touching on serious topics, gatherings are always enjoyable. Mark Tom (March 2007:1011) of Newtok shared: "When I tell stories briefly, I try to make people laugh, as people won't forget and will learn things from laughing. The *qanruyutet* are wonderful. Although it's a small piece of advice, it's significant."

Paul John (September 2003:217) of Toksook Bay is another passionate supporter of the CEC's efforts: "I am thankful that you are helping us by learning these things. It seems as if you are saying to us, about the things that we know, 'Don't keep them to yourselves.' Even when we are gone, we will be heard through the tapes and books that you work on, and you will use these and learn." Paul John then offered a short *qanruyun* on the importance of sharing knowledge: "It is a saying that when an elder is knowledgeable, he cannot say that he does not know anything. He is to quickly tell how it is done so his descendants can learn."

Many feel strongly that following the *qanruyutet* is the key to leading a good life. Frank Andrew (June 1995:17) observed: "If we follow teachings, everything that we work on is finished in a good way. That is also what a person's disposition is like. They will not become beautiful all by themselves, but the instructions will cause them to become beautiful. Even though their clothing is dirty, they will be attractive by the way they behave." Conversely, not following *qanruyutet* can bring one to ruin. Frank Andrew concluded: "These instructions are especially important for our younger generation. A child who is not given instructions cannot grow healthy. A plant cannot grow if it is not watered. That is what people are like."

Michael John rides up the Cakcaaq River toward the end of the Nelson Island circumnavigation.
Ann Fienup-Riordan

Linda Joe, Ryan Abraham, Ben Angaiak, and Anna Agnus enjoy a meal of raw pike dipped in seal oil while camping on the Urumangnaq River, July 2007. *Ann Fienup-Riordan*

Michael John (March 2007:1242) of Newtok stated: "Those of us who caught the oral instructions they gave in the past are the last generation to hear them [before times changed]. When our speakers are all gone, how are we going to bring up oral instructions? Before these oral instructions are lost, we must mention the few things that we know, the *inerquutet* [prohibitions] and *alerquutet* [laws]. If we don't keep those things hidden but speak of them, it will greatly improve the situation in our villages." This mandate to document and share *qanruyutet* pertaining to proper social relations was in part realized in the CEC's first two major publications: *Yupiit Qanruyutait/Yup'ik Words of Wisdom* and *Wise Words of the Yup'ik People: We Talk to You because We Love You*, both published in 2005. Yet, as elders emphasized from the beginning, everything has rules: no one book can hold all that a person needs to know.

Elders' primary concern has always been that young people understand how to treat one another—their parents, children, spouses, relatives, even their non-Yup'ik *aipait* (counterparts, partners)—and rules guiding these relationships fill both books. Yet elders also suffer over the fact that many young people lack knowledge of *ella*—translated variously as "weather," "world," or "universe"—which many continue to view as responsive to human thought and deed. Tommy Hooper (March 2007:1358) of Tununak exclaimed in frustration: "Our young people are completely clueless since they aren't following our way of living. They don't know the ways that we followed. When one can no longer travel in the wilderness and it gets dark, he could build a shelter, and it would be warm like going inside a home. These days they no longer do that because they don't know about it.

Jackie Lincoln and Linda Joe pick salmonberries along the Cakcaaq River, July 2007.
Ann Fienup-Riordan

They [get on the snowmobile] wearing a small coat and thin pants, and they take off at a fast speed."

John Eric (January 2007) observed, "The ocean cannot be learned," and Stanley Anthony (January 2007:147) of Nightmute stated: "One must not say, 'The ocean isn't daunting.' The ocean is aware and knowing. A person should not approach the ocean with confidence." Simeon Agnus (January 2007:3) added: "We must also teach our children about the land." These statements emphasize the tremendous knowledge hunting requires and how retaining this knowledge is critical to safe travel for contemporary young people. Insofar as our work could record and share this knowledge, John Eric (January 2007:2) enthused: "These things that you mentioned that you will be doing are wonderful."

Just as *qanruyutet* embodied time-tested rules for interacting with ones' fellow humans, *qanruyutet* likewise guided one's interaction with one's environment. In the past, elders shared these *qanruyutet* with their young people because, as they said, they loved them, and today they try to do the same. Edward Hooper (March 2007:1444) of Tununak said: "This work is good. We share with others around us, like sharing food." Anna Agnus (July 2007:685) of Nightmute declared: "The one who instructed us said, 'We advise our children, our descendants, so that they will learn about every type of work, even though you are women, men, and children, because we love you so.' They said that everyone hears the *qaneryarat* [words of advice]. They said a person speaks only to those whom he loves without censoring what they have to say."

Paul Tunuchuk (March 2007:216) of Chefornak also emphasized the importance of teaching young people about the world around them—not merely its physical features but the ways in which one's actions elicit reactions in a sentient and responsive world:

> These days, we aren't being dutiful and attentive to everything. It's as though we are sleepwalking. Everything on earth has customary teachings and instructions attached to them—the air, land, and water. How will our young people deal with potential dangers?
>
> They mention that we must treat it with care and respect, but these days, it's as though the land has become a dumping ground. What will become of us if we don't treat *ella* [the world] with care?

Many observe the sobering answer to Paul Tunuchuk's question. Paul Kiunya (October 2005:125) of Kipnuk remarked on the physical consequences of moral decline:

> *Ella* truly has changed since the time I became aware. Nowadays there is more wind and stormy weather because we people are not careful anymore. Our ancestors treated the environment and land with respect because of their food. People referred to as *kenegnarqenrilnguut* [ones who experienced death in their family] had admonitions. They instructed those people to be careful, that *ella* would be aware of them. Because they aren't following their traditional instructions, our environment has become windier and there is more stormy weather. Because [the weather] will change with the people, it has probably followed us since we have become bad.

Raised on Nelson Island, Mark John (April 2001:78) of Anchorage eloquently testified that the reason for educating young people with *qanruyutet* is far from academic. Such knowledge, he believes, can change their lives. He, too, spoke from personal experience, recalling how his grandmother's words had given him strength later in life. Frank Andrew (September 2000:13) agreed that *qanruyutet* are the tools that can change their world: "*Qanruyutet* change people's behavior. All our material belongings can be fixed with our arms. But a person's nature can be fixed only by *qanruyutet*." John Phillip (March 2000:17) echoed this view of the instrumental value of *qanruyutet*: "The teachings are like something that pushes one to a good life. Those who listen and apply the teachings will live good lives."

The Yup'ik Homeland

The CEC is grounded in the Yukon-Kuskokwim region, a lowland delta the size of Kansas and the traditional homeland of the Yupiit, or Yup'ik Eskimos. The region's current population of more than 23,000 (the largest Native population in

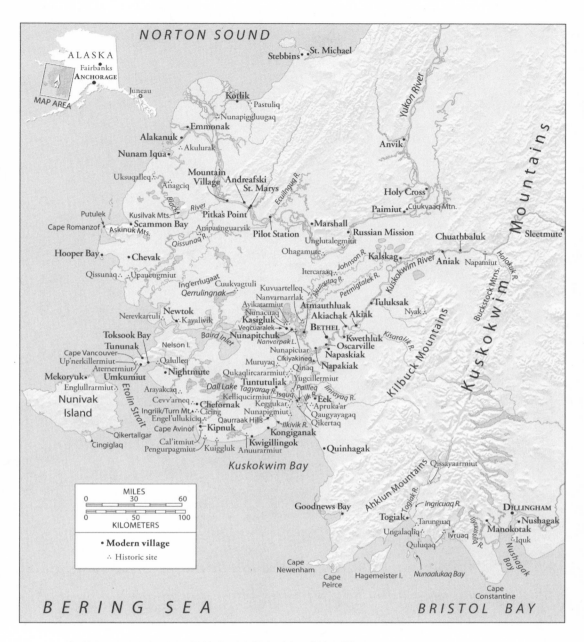

The Yukon-Kuskokwim delta, 2009. *Patrick Jankanish and Matt O'Leary*

Alaska) lives scattered among fifty-six villages ranging from about 200 to 1,000 persons each. Bethel, the regional center and home to nearly 6,000, is an hour's plane ride (four hundred miles) west of Anchorage. Today this huge region is cross-cut by historical and administrative boundaries, including two dialect groups, three major Christian denominations, five school districts, two census

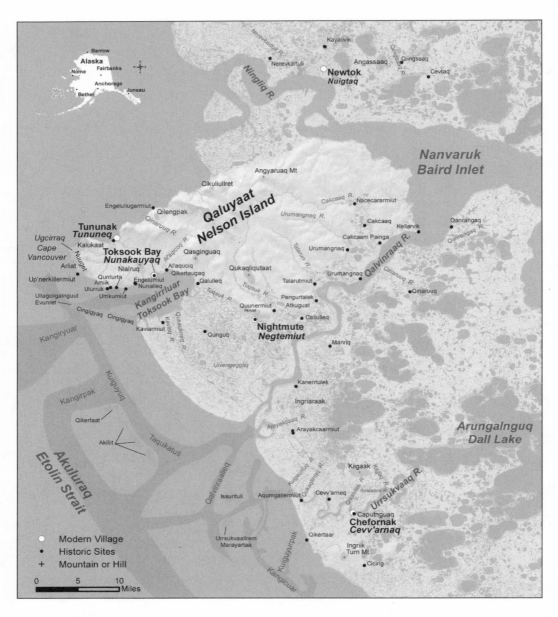

Qaluyaat (Nelson Island). *Michelle Pearson, Alice Rearden, and Steve Street*

areas, and three Native regional corporations established by the Alaska Native Claims Settlement Act (ANCSA) of 1971. Each village has an elementary and a secondary school, city government or traditional council, health clinic, church or churches, airstrip, and electricity. Some have indoor plumbing.

Weather on the delta can be harsh, and wind chills can drive temperatures to the equivalent of a brutal -80°F in winter. At the other extreme, the temperature on a sunny, windless day in July can reach 90°F. Precipitation averages no more than twenty inches a year, including fifty inches of snow.

The subarctic tundra environment nourishes a rich array of vegetation, including numerous edible greens and berries. Shrubs and trees, mostly willow and alder, crowd stream banks. River drainages sustain a mixture of spruce and birch upriver from Aniak on the Kuskokwim River and Pilot Station on the Yukon River. Though the region is beyond the reach of the dense forests of the Interior, breakup releases an ample drift of logs downriver every spring. The Yupiit are well supplied with wood, which they used in the past to build their homes, boats, tools, and elaborate ceremonial accoutrements.

The Bering Sea region supports not only a rich fishery but millions of seabirds and tens of thousands of marine mammals. Its productivity is fueled by nutrients annually replenished from oceanic waters across the broad continental shelf. This impressive variety of animals provides an annual cycle on which Yup'ik people focus both thought and deed. Far from seeing their environment as the insentient provider of resources available for the taking, many continue to this day to view it as responsive to their own careful action and attention or lack thereof.

Although the major gathering and butchering tasks of the year end with the first frosts in early October, men continue regular hunting and fishing trips away from the village. Tomcod run upstream and can be taken through the ice with barbless hooks. Men also travel out to nearby sloughs and streams to set wooden or wire traps for the small but tenacious blackfish and the more substantial burbot. These cone-shaped, spirally bound traps are set with their mouths upstream and rely on underwater barriers to channel the fish. If checked once or twice a week, they can provide hundreds of pounds of fresh fish through fall and early winter. Men also net sticklebacks, or "needlefish," through the ice with dip nets. Later in winter many set traps within snowmobile distance from the village for mink, muskrat, and beaver.

December brings temperatures as low as -20°F, too severe for ocean hunting. Yet families enjoy daily meals of dried fish either dipped or aged in seal oil, frozen fish, *egamaarrluk* (fish partially dried and then boiled), and moose and caribou stew from the previous season's bounty. Fresh food continues scarce through February. In the past, famine was an all-too-common antecedent to spring renewal.

As leads begin to open in the nearshore ice in spring, coastal hunters, protected from the treacherous open sea by shifting floes, set out to hunt seals. The first to come are the bearded seals (some as large as 750 pounds) on their way north. The smaller ringed seals, spotted seals, and ribbon seals follow as the shore ice begins to dissipate. Hunters also occasionally take sea lions, walrus, and beluga whales.

While coastal families enjoy fresh meat and oil, upriver residents concentrate on other tasty resources. By the end of March white-plumed rock ptarmigan begin to flock in the still-brown willow and alder, making them easy targets. Women and children hook tomcod and northern pike through the ice. Young men and boys travel farther afield, hunting Arctic fox and snowshoe and Arctic hare. On Nunivak and Nelson Islands they may harvest an occasional musk-ox from the herd introduced in 1963.

One Yup'ik name for April is Tengmiirvik, or "time that birds come." More than a million ducks and half a million geese annually make the long journey north to nest and breed in the ample wetlands of southwest Alaska.[3] The delta and its adjacent coastal waters host 216 species of birds, including more than 30 species of waterfowl, 30 species of seabirds, 30 species of sea ducks, and 33 species of shorebirds.

Families both work hard and eat well during this season of plenty. While men focus on harvesting activities, coastal women adjourn to their storehouses to prepare for drying the thousands of pounds of meat that have accumulated between the beginning of hunting and the time when the weather warms enough to defrost the frozen carcasses. They will also cut the seals' blubber and store it in a cool place, where it gradually renders into the oil that accompanies every meal.

Women and children in both coastal and riverine communities also venture out to the nearby tundra to gather a variety of edible greens, including the greens of marsh marigold and wild celery and the roots and shoots of wild parsnips. Although egg gathering from the thousands of nearby nests made by gulls,

Martina John smiles while using her *uluaq* (semilunar knife) to remove the blubber from the skin of an *amirkaq* (young bearded seal), while Anna Agnus works alongside, July 2007. *Ann Fienup-Riordan*

cranes, geese, and ducks is today restricted by law, coastal residents continue to harvest fresh clams and mussels from the tidal flats.

The summer fishing season begins in earnest by early June, and modern villages bustle with activity around the clock. Along the coast the first arrivals are herring, which residents harvest in abundance and dry for winter use. The summer harvest along the Yukon, Kuskokwim, and Nushagak Rivers focuses on the regular succession of five major species of salmon: king (chinook), red (sockeye), coho (silver), pink (humpback), and chum (dog). From June through August, millions of these fish struggle up the major waterways to spawn. As they begin their ascent, thousands are taken in gill nets and either sold commercially or dried, smoked, and stored away as winter food for both people and dogs.

Throughout summer, boats come and go with the tides. Fishermen harvest halibut and flounder with longlines and four-pronged jigging hooks. As summer comes to a close, men also use dip nets, spears, and traps to harvest a variety of freshwater fish, including sheefish, northern pike, Dolly Varden (char), burbot, and blackfish. Various species of whitefish also populate streams and lakes.

Fishing slows by the first of August as the berries ripen, and families disperse over the soggy tundra to harvest salmonberries, crowberries, blueberries, and cranberries. Later in the season women whip the stored berries together with seal oil or shortening and sugar to make the festive *akutaq*, or "Eskimo ice cream," which remains a requisite for hospitality during winter entertaining. Seals return to the river mouths, where hunters pursue them in open water into December if the winter is mild. Schools of beluga whales also enter river mouths and are sometimes beached and shared throughout the village.

Men stay busy before freeze-up, checking nets and traps on day trips away from the village or leaving in pairs on overnight hunting ventures. Some travel upriver to hunt for moose and caribou. Women tend to the last of their annual harvesting activities. On clear days they search the tundra for caches of "mouse food"—the stock of small roots and underground stems of sedges and cotton grass gathered by industrious tundra lemmings. They also gather rye grass blade by blade and dry it to use for weaving baskets during the long winter. This gathering used to be a major fall task, as winter boot insoles, mats, twined bags, and floor coverings required a huge supply. Though less important today, gathering grass remains part of the fall routine.

Fall is also a time for bird hunting, though less so today than in years past due to declining bird populations and increasing state and federal restrictions. From mid-August through mid-September, geese, cranes, and swans begin their journey south after summer nesting. By mid-October the ducks are gone as well, and the berry-fattened ptarmigan remain the only fresh fowl available for the table. Except for processing the daily catch of fish and birds, the major gathering and butchering tasks end with the frost in early October. Women retire into their homes for the winter, while men continue regular hunting and fishing forays away from the village.

Yup'ik people built on their rich resource base, developing a complex cultural tradition prior to the arrival of the first Euro-Americans in the early 1800s. They are members of the larger family of Inuit cultures extending from Prince William Sound on the Pacific coast of Alaska to both sides of Bering Strait and from there thousands of miles north and east along Canada's Arctic coast and into Labrador and Greenland. Within that extended family, they are members of the Yup'ik-speaking, not Inuit/Iñupiaq-speaking, branch. All claim a common origin in eastern Siberia and Asia. Before 2,400 years ago the ancestors of modern Yupiit were probably present in small numbers at headlands and other prime coastal locations. At that time the number of large coastal villages increased, triggered by the adoption of net-fishing technology and the shift toward salmon, which allowed use of previously marginal locations.[4] Their dependence on sea mammals decreased and was replaced by the mixed reliance on salmon, migratory waterfowl, caribou, and small game that characterizes the region to this day. As coastal populations increased, gradual encroachment eastward occurred along large river and lake systems at the expense of Interior residents.

The abundance of plants and animals in southwest Alaska allowed for a more settled life there than in other parts of the subarctic. Hundreds of seasonal camps and dozens of winter settlements lined river highways that still link communities. Like the northern Inuit, the coastal Yupiit were nomadic, yet their rich environment allowed them to remain within a relatively fixed range. Each of at least a dozen regional groups demarcated a largely self-sufficient area, within which people moved freely throughout the year in their quest for food.

Interregional relations were not always friendly. Intermittent skirmishes regularly interrupted delta life prior to the arrival of the Russians in the early 1800s. Ironically, death itself brought this killing to an end. Diseases that accompanied contact with Euro-Americans decimated the population. Although few Russians settled in southwest Alaska, the larger Russian trade network to the south introduced smallpox into the region, devastating the Native population. Entire villages disappeared. As much as 60 percent of the Yup'ik population in Bristol Bay and along the Kuskokwim River had died by June 1838.[5]

The smallpox epidemic of 1838–39 and epidemics of influenza in 1852–53 and 1861 resulted in both a decline and a shift in the population, undercutting inter-regional social distinctions. Although the introduction of communicable diseases damaged traditional social groups and patterns of intergroup relations, it left intact the routines of daily life throughout the remainder of the nineteenth century. Small bands of extended family groups continued to move over the landscape, seeking the animals they needed to support life and gathering in winter villages for an elaborate annual ceremonial round.

The Yupiit were left largely to their own devices into the early decades of the twentieth century. The bilateral extended family, numbering up to thirty persons, was the basic social unit. Spanning two to four generations, including parents, offspring, and parents' parents, the group might also encompass siblings of parents or their children. An overlapping network of family ties joined people in a single community. In larger villages most marriage partners came from within the group, though regional recruitment also occurred.

Extended family groups lived together most of the year, but normally not in family compounds. Rather, each winter village was divided residentially between a *qasgi* and smaller *enet* (sod houses), the latter occupied by women and young children. Married couples or groups of hunters often moved to outlying camps for fishing and trapping during spring and fall. Families gathered when temperatures dropped below freezing and sometimes during spring seal-hunting and summer fishing seasons. Winter villages ranged in size from a single extended family to a few hundred people. In these larger settlements there might be as many as three men's houses with up to fifty men and boys living in each.

All men and boys older than five years ate their meals and slept in the *qasgi*, the social and ceremonial center of village life. In winter men rose early in the morning to begin their day's work and returned home by sundown. They spent their spare time together talking and carving in the *qasgi*. Their daughters and wives brought their meals, waiting by their sides while each man emptied his personal bowl. The *qasgi* was also the location of the ubiquitous fire bath, during which participants socialized and cleansed themselves in the intense heat.

Every man's place in the *qasgi* reflected his social position, and the men's house framed a number of internal distinctions, including that between young and old, married and unmarried, and host and guest. The social structure of the *qasgi* mirrored that of the natural world. The Yupiit believed that sea mammals lived in huge underwater *qasgit* (plural), from which they observed, and chose whether to give themselves to, human hunters. Hunters who extended the most thought and care toward the animals they sought were richly rewarded, both socially and materially. The *nukalpiaq*, or good provider, was a man of considerable importance in village life. Not only did he contribute wood for the communal fire bath and oil to keep the lamps lit; he also figured prominently in midwinter ceremonial distributions. As youths, contemporary elders had listened in the *qasgi* to their fathers and grandfathers discuss the hunt and talk about the rules for right living.

Southwest Alaska lacked significant amounts of the commercially valuable resources that first drew non-Natives to other parts of the state, minimizing contact with the outside world. Missionaries were among the first outsiders to interact with the scattered and isolated peoples of the Yukon-Kuskokwim delta — Russian Orthodox in the 1840s, followed by Moravians, who settled in Bethel in 1885, and Jesuits, who established a mission on Nelson Island three years later.[6] Both Moravians and Catholics were less tolerant of pre-Christian ritual acts, especially the masked ceremonies, than their Russian Orthodox predecessors, referring to them as "heathen idol worship" and the "devil's frolic."[7]

A major demographic marker occurred in 1900 when an influenza epidemic arrived with annual supply vessels, halving the Native population in just three months. Although coastal communities were not as severely affected, many winter villages on the Yukon and Kuskokwim Rivers were abandoned. A sharp increase in the region's white population paralleled the decline of the Native population. The Yup'ik people supplied fish and cordwood to miners and steamship captains and participated in an expanding fur market, substantially changing their domestic economy.

Though the people of southwest Alaska increasingly spoke English, lived within four walls, worked for wages, and attended church, the region's isolation meant that they remained largely independent, their lives focused on extended family relations and the pursuit of animals on which they had relied for centuries. Animal husbandry and tilling the soil were embraced by an energetic minority, but the people ultimately ignored these eminently "civilized" activities when they conflicted with indigenous subsistence and settlement patterns, including their annual round of hunting and fishing. Though much changed, much also remained of their pre-Christian view of the world, especially their belief in the essential personhood of animals and in the responsiveness of the natural world to human thought and deed.

During the early decades of the twentieth century both the Jesuits and Moravians sought to replace Yup'ik animism, as expressed in their traditional cycle of ceremonies, with the form and content of the Christian faith. The discipline and order they preached appealed to the Yupiit as a novel spiritual solution for an unprecedented social crisis brought on in part by the epidemic diseases and social stresses of the early decades of the twentieth century. Combining their traditional sensitivity to the spirit world with the discipline of the Protestant work ethic, the majority found it possible to become fervent Christians without ceasing to be Yup'ik people.[8]

By the time Alaska became a state in 1959, continuities between past and present remained as significant as innovations. People continued to speak the Central Alaska Yup'ik language, enjoyed a rich oral tradition, participated in large ritual distributions, and focused their lives on extended family relations bound to the harvesting of fish and wildlife. Alaska Natives generally were viewed as extremely disadvantaged during the decade after statehood. The Yupiit of the Yukon-Kuskokwim delta region were among the most impoverished. Relative to other areas of rural Alaska, the availability of Western material goods was minimal, modern housing nonexistent, and educational levels low, and tuberculosis—as destructive as earlier influenza and smallpox epidemics—ran rampant.[9]

Events of the 1970s launched a new era in the Yukon-Kuskokwim delta and throughout rural Alaska. Passage of the Alaska Native Claims Settlement Act in 1971 transformed land tenure in Alaska. It extinguished aboriginal land claims statewide in exchange for title to forty-four million acres of land and nearly one billion dollars given to twelve regional and one nonresident for-profit corporations as well as to more than two hundred village corporations. These organizations administer the land and money received under ANCSA. The Calista Corporation (lit., "worker") was established to manage the corporate resources of the Yukon-Kuskokwim region.[10]

The 1960s and 1970s saw an explosion of social and political organizations in the region. The Association of Village Council Presidents (AVCP) took over responsibility as the regional nonprofit corporation. Since 1964 the Yukon-Kuskokwim Health Corporation (YKHC) has administered the majority of health-care services. In 1976 the "Molly Hootch decision" (*Hootch v. Alaska State Operated School System*, named for a Yup'ik student who sued the state for the right to be educated in her home village) mandated sweeping educational reform. Local high schools sprang up in communities that previously had sent their children to boarding schools in Bethel and St. Marys (Catholic Mission) or outside the region and often the state.

ANCSA paved the way for construction of the Trans-Alaska Pipeline, and the state's share of the oil profits gushed into public-works projects and social programs. Supported by ANCSA village corporation activity and state services, the villages of southwest Alaska experienced steady growth in both population and modern facilities during the 1970s and 1980s. Employment income and other cash transfers continued to support local harvesting activity.

The regional economy shifted radically as the delta's population coalesced at permanent sites. Along with the continued importance of subsistence harvesting activities, the most significant feature of village economy in southwest Alaska today is its dependence on government. Commercial fishing and trapping, craft sales, and local service industries provide only a small portion of the total local income. As much as 90 percent flows through the village economies from the public sector, including both wages and salaries and various state and federal transfer payments. In this mixed economy both transfer payments and wages are used to pay for the snowmobiles, skiffs, motors, ammunition, and fuel that contemporary harvesting activities require. Families need money to hunt and fish, and their success at harvesting animals is directly tied to their ability to harvest cash. At a time when the market economy of southwest Alaska continues to founder, hunting and fishing activities become increasingly difficult to afford.

Recent growing awareness of devastating social problems related to poverty and cultural dislocation has renewed among the people a sense of urgency to retain their uniqueness as Yupiit. Yup'ik efforts have concentrated on maintaining control of land, resources, and local affairs; improving residents' health and sense of well-being; and adhering to cultural and linguistic traditions. This renaissance reflects a certain nostalgia for the "old ways," which in political and economic terms refers to times after many technological improvements had been introduced but before the Yup'ik people had experienced subordination to federal and state control and related dependency.

Many present Yup'ik leaders, teachers, and community members seek to enhance their cultural identity as they simultaneously work to reduce dependency. As public monies declined in the 1990s, residents struggled to find solutions to village problems and ways to continue to live in their homeland. They sought increasing responsibility and authority for local health, education, and political and economic organization. Many Yupiit still see themselves as living in a highly structured relationship with the resources of their environment, not merely surviving off them. In a period of rapid change, they selectively appropriate Western means according to their own practical purposes.

Central Alaska Yup'ik continues as the primary language for most people living in coastal and tundra villages, though its use has declined in Yukon and Nushagak River communities. Public-housing programs have dictated the construction of single-family dwellings, yet elaborate patterns of interhousehold sharing of food and other goods, adoption of children, and hunting partnerships continue to nourish extended family relationships. People remain strongly committed to traditional harvesting activities and continue to share products of the hunt.

The last two decades have seen the revival of intra- and intervillage dance festivals, hosting of local and regional elders' conferences, and increased awareness of and concern for the preservation and use of Yup'ik traditional knowledge. Such activity reveals the desire of many Yup'ik people to take control of their land and their lives and to assert their pride in being Yup'ik. Many in the larger Yup'ik community, including parents, teachers, and students, are dedicated to communicating the Yup'ik view of the world, particularly to the younger generation. Their efforts, including work documenting traditional *qanruyutet*, reflect their desire to gain recognition for their unique past, parts of which they hope to carry into the future.

Qanruyutet Anirturyugngaatgen

QANRUYUTET CAN SAVE YOUR LIFE

A MONG THE MOST MEMORABLE ACCOUNTS OF THE COURAGE AND knowledge it took to draw sustenance from the Bering Sea is the tale of Atertayagaq. Like Homer's *Odyssey*, Atertayagaq's story is the epic tale of someone who faced many life-threatening situations as he journeyed to new and unknown places, eventually returning home. Unlike Odysseus, however, Atertayagaq was not a powerful man who used his wits and prowess to survive but a small boy who followed the *qanruyutet* (wise words or instructions) that he had been given and which enabled him to come home safely.

Atertayagaq | Someone Small Who Drifted Away

Atertayagaq's story is not a *quliraq*, or traditional tale. Rather it is a *qanemciq*, a historical account of events that took place within the memory of living elders. Many know parts of Atertayagaq's tale, but few today hear it in all its detail. In October 2006, John Phillip of Kongiganak, Roland Phillip of Kwigillingok, and George Billy of Napakiak gathered in Bethel with a handful of young people to tell Atertayagaq's story. All three had known Atertayagaq when they were young. John Phillip had been among the listeners when, as an old man, Atertayagaq stood in the Kwigillingok church and told his story.

Standing in front of the congregation, Atertayagaq first asked fourteen-year-old Peter Jimmie to stand beside him; he then told his listeners that this was how old he was when he drifted away. Atertayagaq had just acquired a kayak and was preparing for his first seal hunt. John Phillip noted: "One feels uncertain when one acquires a kayak and goes down to the ocean for the first time. But before we acquired kayaks, they always spoke to us about how we should travel on the ocean, about its currents and winds, and the waves and ice. Those [instructions] were all in his mind."

Atertayagaq and his family lived near Kuiggluk, just south of Kipnuk, at the time. To prepare him for the ocean his mother had made him fish-skin mittens and waterproof sealskin boots but had not yet finished sewing his seal-gut rain garment. His hunting partner told Atertayagaq to stay behind when the others left and come down the next day when he was ready, following their tracks. The next morning, even though his seal-gut rain parka was still not finished, Atertayagaq went to join his companions, towing his kayak behind him. Soon he came upon a small crack in the ice but continued on. Then he came to a new, wider crack where the ice had moved laterally. Since he did not know it was unsafe, he continued walking far out on the ocean. Soon a north wind started to blow and snow began to drift along the ice. When it was impossible to continue, Atertayagaq returned, following his tracks. Reaching the crack, he found that the ice had already drifted far from land.

Atertayagaq could see the shore ice, but the waves began to pick up. Having no seal-gut rain garment and thinking of the *qanruyun* (oral instruction) never to allow oneself to get wet, he was afraid to use his kayak to cross the open water. So he moved his kayak far up on the ice and sat down to wait. He had two orphaned nephews he was providing for, and remembering them he sobbed out loud. He said that was the only time he submitted to his desperation, not thinking of himself but of them.

When the weather cleared, he could see the two mountains below his village and headed toward them over the ice although they were far away. When he was hungry, he used one bullet to shoot a bearded seal, eating the meat raw. He conserved his matches as well, using them to light the slats of his kayak seat one by one and then adding seal oil to his fire to melt ice for water.

Atertayagaq had noted the position of the moon when he left and used that to estimate the time he was on the ocean. One month went by, and it was never windy. He lost sight of the mountains but continued to tow his kayak over the ice. As his boot soles wore out, he pulled the upper parts of his boots down and walked on them. He also killed a walrus and kept its intestines, thinking of the time the ice would disappear and he would need a water container.

During the second month he began seeing mountains that he had never seen before—probably on the Alaska Peninsula—and he headed toward them.[1]

Atertayagaq said that during this time he seemed to become related to the sea mammals. He said that when he saw them, he felt joyful, but when they were gone and he had not seen them for a while, he felt lonely and sad. Also he no longer had a human scent at that time but smelled like a sea mammal.

Once, while on the ice, a small walrus, a *puyurtaq* (one with small tusks), swam toward him from the north. When it got to him, it hung its small tusks on the edge of the ice right below him and faced him. It seemed to be telling him something, but he couldn't understand what it said. After it had been there for a while Atertayagaq spoke to it, and the walrus made noises in return. Then the small walrus swam away along the ice edge, continuing to make noise before it finally disappeared. Atertayagaq put his kayak in the water and followed the path it had taken.

Along the way, three kayaks appeared, as well as a number of people on the ice. Atertayagaq approached them, thinking they were hunters from Quinhagak, and as he got closer he could see their windbreaks held up by gaffs staked on the ice. Just as he was about to reach them, he briefly looked to one side, and when he looked back he saw only walrus sliding off the ice into the water. Atertayagaq quickly got on the ice, as he had been told that swimming walrus are dangerous. He regretted looking away, thinking that if he had continued looking straight ahead he would have reached them while they were in human form.

By the third month, Atertayagaq didn't mind being on the ice, although he felt sad when he didn't see sea mammals. The weather became warmer, and the ice became smooth. He began to suffer from thirst as the *elliqaun* (newly frozen ice) was salty and undrinkable. He put a ball of snow on the back of his kayak and, following the *qanruyun*, bit part of it and melted it in his mouth to quench his thirst. John Phillip observed: "Sometimes he would come upon those obstacles. He stayed down on the ocean, constantly recalling the *qanruyutet* and following them."

Finally land appeared, but Atertayagaq couldn't find a path to travel through the ice, as it was his first time down on the ocean. The sea mammals he ate began to taste like warm, cooked food. Then one day he woke to clouds and wind, which worried him. He saw something on the horizon that looked like a small dark cloud and thought, "Weather that will surely kill me is approaching." Instead, a sailing ship appeared. He knew what it was and that white people would be on board, but he was afraid of them. Considering all the times when he might have died, he thought, "I will give myself over to them."

Atertayagaq then waved his paddle in the air, and the ship—which was heading through the Aleutians on its way to Bristol Bay—lowered its sails and came alongside him. When they pulled him and his kayak aboard, Atertayagaq said that he felt as though he was surrendering himself to them. His experiences did not end there. The captain offered him his first drink of alcohol, which he thought was poison until the captain drank first. Later, when they met another ship, he

Atertayagaq as a lay pastor in Quinhagak, 1930s. *Moravian Archives, Bethlehem, PA*

and his kayak were lowered over the side to retrieve a box of cigars from the other vessel, and he remembered the apples and cookies that were thrown to him.

The ship continued north to Ugashek, just below Egegik, and the captain took him ashore and found him a place to stay with the local Russian Orthodox priest, who baptized him, giving him four names—Isaac, Ishaak, Qiatuaq, and Atertayagaq. It was at that time that Atertayagaq learned that there was a Christian God. Also, when a woman recognized the family designs on his parka, he found that although he was far from home, he had relatives in Ugashek.

When spring came, Atertayagaq accompanied others traveling north, gaining many experiences along the way. He finally arrived in Goodnews Bay in early summer and from there went on to Quinhagak at the mouth of the Kuskokwim and upriver to Nunapigmiut just as the king salmon started to run. Word of his return quickly passed upriver to Keggukar, where Atertayagaq's family was fishing. At first they thought the messenger had misunderstood, and they did not believe that Atertayagaq had survived until he actually arrived.

Later in life Atertayagaq became a lay pastor, working with the first Moravian missionaries in the area. The night before he left to go seal hunting he had dreamed about the path he would follow. He saw himself walking out on the ice and entering a building that looked like a store with two pictures on the walls. Years later, when he served in Quinhagak and entered their church for the first time, he recognized it as the building he had seen in his dream.

Ca Tamarmi Qanruyutengqertuq | Everything Has Qanruyutet

John Phillip began his retelling of Atertayagaq's story by encouraging the young people in the audience to listen: "We were like you young men, although we didn't live like you back then. We had a *qasgi*, and our homes were made of sod and not frame buildings. They spoke to us back then—you young men, listen now—before they gave us instructions in the *qasgi*. They said to us, 'Now live with discretion. Pay attention to things that you see, and listen to those giving instructions, although they aren't speaking to you directly. You will hear about things you will use.'" John compared his young listeners to Atertayagaq.

> The person in the story was probably younger than you when he first went down to the ocean. Someday when you are able, you will go down to the ocean. For that reason, we want you to pay attention to what we say.
>
> Before they spoke to us about something, they told us to watch a person's mouth while he spoke. They say we will forget if we look briefly away. That's a *qanruyun*. Before the story is told, I'm giving this instruction to you as a gift. I'm not the only one [who heard it]. Everyone heard that *qanruyun* in the past. They say a person who receives the *qanruyutet* can lead a good life by following them.

John Phillip noted that young men were admonished to listen to instructions they would use later in life: "They told us about everything that we would live through, and they would tell us that things that we will experience in life are in our future and we cannot avoid them, but will use that *qanruyun* when we experience it." Roland Phillip shared Atertayagaq's comparison between the provisions he carried inside his storage bag and the *qanruyutet* he carried inside his mind: "Atertayagaq said that while he was on the ocean, before he experienced the *qanruyutet* that he was given, his [*qanruyutet*] were inside that grass storage bag. He said that when he came upon a certain situation or obstacle, he would take that *qanruyun* and use it, the one in his mind."

When they were young, George Billy, John Phillip, and Roland Phillip were all taught to listen carefully to the *qanruyutet* and remember what they heard. According to Roland Phillip: "We, too, were instructed before we went down to the ocean and hunted. And since we were given instructions, we recognized what we heard about the ocean being a joyous place, how it is a dangerous place, and how currents vary, and what to do when a large amount of ice accumulates; we recognized those things as we experienced them."

John Phillip shared the admonition to follow the *qanruyutet* when in danger rather than one's own mind: "When we faced danger down on the ocean, sometimes our hearts pounded very hard. Even though we were afraid, they cautioned us not to show panic or act in haste. A person is told not to follow their own judgment down [on the ocean]. If he recalls his instructions when he encounters danger, one can bring the *qanruyutet* before himself, like Atertayagaq did, and attempt to survive using them when he comes upon bad weather. Those are some of the *qanruyutet* we were given." Roland Phillip added the admonition not to take risks: "They say if a young person, a boy or girl, is rebellious and breaks his instructions, he won't live long. But one who follows the rules will live an extra day [meaning a long life]."

No story captures so well the life-saving importance of the *qanruyutet* in the wilderness, guiding both practical and moral decision making. John Phillip noted: "Down on the ocean, Atertayagaq was guided by the instructions that he was given, that he had stored inside. He said that when he was drifting, he followed his *qanruyutet* at that time." Atertayagaq's work as a lay pastor and his retelling of his story in church later in life reflect his continued efforts not only to follow *qanruyutet* but to share his knowledge.

Cilkiaryaraq | Rigorous Training Process

Atertayagaq's survival on the ocean was also contingent on his behavior at home. Before they went hunting, young boys were expected to work hard doing chores in the *qasgi* and homes, preparing themselves both physically and mentally for their future lives. Roland Phillip noted that when he would dump the urine bucket of an elderly woman, she would tell him to pretend to shoot an arrow at what he had dumped, thinking of an animal that he wanted to catch. Roland Phillip also recalled his father's instruction concerning those who replaced floorboards over the *qasgi* fire pit after a fire bath: "When I was a boy, long before I started to hunt, my father told me to try to place the floorboards back on when the people who had just taken a fire bath were done. He said those sea mammals perched on the ice would be readily available to those who constantly replaced the floorboards. Sometimes I would get someone to help me, telling him, 'We could quickly replace the floorboards when they're done. The animals that we will catch will be on top of those when we hunt.'"

John Phillip described how boys would constantly clear doorways of snow during blizzards, thinking of the animals they would someday hunt:

> During winter, without sleeping at night, especially when there was a blizzard, they would clear the entranceway of the *qasgi* or the doorways of homes. Those who would be hunting were asked to carry out a chore with the anticipation of [catching animals].

Two men and a boy seated in a *qasgi* in the early 1930s. *Moravian Archives, Bethlehem, PA*

It is said they always cleared paths thinking of the path they would take in hopes of traveling safely. And they say that during snowstorms some people [shoveled] all night when [the snow] tended to fill [doorways], keeping an animal in mind and thinking of catching it without encountering danger.

Roland Phillip recalled the well-known story of Qanagaarniarun, the lazy son-in-law whose in-laws, in disgust, traded him for a roof board (*qanak*). Humiliated by their treatment, Qanagaarniarun resolved to mend his ways. At night he would constantly clear pathways and shovel entryways, thinking of the animals that he might catch: "While that person was shoveling, it is said his shovel would hit something down there and come to a halt, and he'd look and find a caribou antler poking out from the ground. He was then filled with antici-pation, for the animal that he had in mind would be in his path when he traveled to the wilderness. And sometimes while shoveling, when his shovel hit some-thing, he'd clear it a little to find a bearded seal snout. Since [his hope for success at hunting] was continually on his mind, the animals that he would eventually catch became visible to him." Roland was admonished not to carry out chores thoughtlessly. Rather, like Qanagaarniarun, one should constantly think of the animals one wished to catch: "They say the mind makes it possible for a person to catch an animal."

John emphasized that animals would come only to one who worked: "They told me that if I did not follow through on tasks, things that I would use wouldn't be readily available for me like they were for other people. They say things cannot be readily available to a lazy person. They say only a person who constantly works and carries out chores will have things that he needs readily available."

John Phillip recalled that when carrying out chores for elders and others, their thankfulness would push one to success:

> They said that the person who gave him instructions encouraged him to be successful. Indeed, sometimes when we brought a urine bucket outside [to dump it], they'd tell us, "So that you will catch a northern phalarope, so that you will catch a solitary sandpiper." When their words came true, when we'd catch something, we'd bring it over to our grandmother. They'd see their words come to fruition in that way also.
>
> And even though [the birds we caught] were extremely small, the elderly women received them with overwhelming gratitude. And they'd say, recalling the instruction they were given, "*Anirtaqulluk* [may it go that way (a phrase used as either a blessing or a curse)], it will be replaced by more," meaning that others [would be caught] in addition.
>
> They say that animals they were thinking about catching were readily available to those who carried out chores.

Boys were also encouraged to clear grass, charcoal, and other debris from the *qasgi* floor. At an earlier gathering, Roland Phillip (November 2005:280) told another well-known story of how the refuse a young man cleared in the *qasgi* provided a path for him in the future. One day, while hunting, the youth lost his kayak and drifted down on the ocean on an ice floe. The incoming tide brought him toward the shore-fast ice, but still too far to jump across. When he took a step in the water to check the depth, his foot touched something below, and he was able to cross to the shore ice. When he climbed up he looked back and saw the debris, including old grass mixed with charcoal, that he had cleaned from the *qasgi*: "They say that [debris] wasn't visible. But just as he looked back, those things that he walked on finally surfaced. They say he cleaned the debris in case he might encounter dangers, and not just for the purpose of catching animals." Roland concluded: "Since *qanruyutet* were part of their way of life, the instructions helped them, and they always had a purpose for following their instructions in different situations."

Qanruyutet | Wise Words and Instructions

When you are in need, the qanruyun makes you like an old man or woman, clothes you, gives you a cane, and brings you where you wouldn't think to go. The qanruyun holds your hand and takes you.

—Frank Andrew, Kwigillingok

Much has been written about the importance of observation and practice in learning the techniques necessary to thrive in the subarctic. Less well-known is the importance placed on verbal instruction. The narrative repertoire, including detailed rules for living as well as stories like that of Atertayagaq, was vast, and children could master it only through constant listening. In fact, among Yup'ik people, a stated ideal was that elders speak and young people listen. Elders who did not speak publicly about what they knew, but only shared it privately with their own children, were considered stingy and "jealous." Conversely, children who "failed to pay attention to the speaker's mouth" or left the room before the speaker was done were admonished that they would someday be found dead in the wilderness, "with their teeth gleaming at the end of a snowdrift."

The building blocks of Yup'ik moral and practical education were the *qanruyutet* (words of wisdom, teachings, or oral instructions, from *qaner-*, "to speak"). Elders explained every aspect of their lives with reference to these "rules for right living," which they referred to in a variety of ways, including *ayuqucirtuutet* (instructions and directions), *qaneryarat* (words of advice, sayings, lit., "those which are spoken"), *aarcirtuutet* (warnings), *inerquutet* (admonishments or prohibitions), *alerquutet* (laws or instructions, prescriptions), *elucirtuutet* (directions or instructions), *piciryarat* (manners, customs, traditions), *pisqumatet* (sayings, lit., "those they wanted one to follow"), and *eyagyarat* (traditional abstinence practices). These concepts are related but not interchangeable, and the term used depends on both the speaker and the topic addressed. Elders used *qanruyutet* most often when speaking of the instructions that they had learned as children and that continue to guide their lives today.

Many *qanruyutet* are stylized sayings, often including the phrase "they say," "they said," or "it is told." In Yup'ik this attribution to an ancestral voice is denoted with the enclitic *-gguq*, used to report what has been said by others if the speaker cannot claim complete authority for his statements.[2] For example, "*Ciutek-gguq iinguuk*" (Ears, they say, are eyes) is shorthand for the admonition that people follow the rules lest they be talked about in other villages and known by strangers for their misdeeds.

The basic tenets of Yup'ik moral instruction include multiple memorable phrases, both elegant and picturesque. This is no accident, as elders wanted them spoken about and remembered from an early age by everyone. These adages include short, graphic phrases such as "boys are like puppies," "women are death," and "a tongue hurts, even though small." All are considered *qanruyutet* or *qaneryarat* like their lengthier counterparts. Nick Andrew (November 2000:124) of Marshall noted that elders tried to keep admonishments short: "When a person hears an *inerquun* that is not long, it sticks inside the head." In her description of the adage "an idle person gets approval from sloth," Theresa Moses (June 1995:64) of Toksook Bay noted: "These were some of the things they told us to get our attention."

Elders explained the general features of *qanruyutet*. Although the instructions might be stated somewhat differently from region to region and speaker to speaker, elders contended that teachings were similar for all Yup'ik people, past and present. Hilary Kairaiuak (November 2000:71, 126) of Chefornak elaborated: "These *qanruyutet* and *alerquutet* [laws] that come from our ancestors do not change. Though spoken of in other places far apart, when they give their explanations, it is the same. . . . Because their instructions are the same, even though one travels to other villages, when one arrives and hears a person who is instructing, he recognizes that instruction that he heard in his hometown."

Reinforcing this view of an unchanging moral code, elders stated that one could not add to the *qanruyutet*. Peter Jacobs (November 2000:116) of Bethel remarked: "If I constantly spoke, adding information to the *qaneryarat*, those elders from long ago would say to me, 'You there, if you add to the *piciryarat* of our ancestors, these people will know what you are doing.' Our elders, those who spoke constantly, used to speak cautiously, thinking, 'They will know because they are knowledgeable.'" This truth claim is significant, emphasizing the importance of speaking from experience, of what one has both seen and heard.

Traditional *qanruyutet* functioned as moral guideposts and were often aptly paralleled with biblical teachings. Frank Andrew (September 2000:144) insisted: "The *qanruyun* that came from our ancestors is in our hearts, even though it is not written down. If we listen to it attentively, it is like the Bible. A person will not lose it while living. That is why it should not be hidden away from small children." Indeed, elders sometimes quote the Bible, chapter and verse, as proof of the veracity of particular *qanruyutet*. Contemporary elders continue to view *qanruyutet* as truths to be recovered and an essential moral code for the properly lived life.[3]

Elders maintain that every situation has an appropriate instruction. Frank Andrew (June 1995:70) insisted that "there are rules against every peril": "Teaching our ancestors' *qanruyutet* and living by them created tranquility among people. There were *inerquutet* [admonishments] for everything we did. There are many, many *inerquutet* we have to know for every hazardous situation. Though it seems to be a simple thing and fun to do, it's best not to mess with it if it has an admonition attached to it."

John Phillip (October 2003:63) noted how *qanruyutet* travel down the generations: "Just as we were told, we are serving them to you now, passing it down. They say the *qanruyun* travels down, continuing through the ensuing generations, and doesn't cease. We don't know our first ones who spoke of *qanruyutet*, but since [the instructions] have reached our generation, we heard them and used them."

After telling the story of Atertayagaq, John Phillip (October 2006:271) reminded younger listeners that later in life they would experience things that they heard about: "Those who came before us used to tell us, 'Listen closely to instructions you are given. If nothing happens to you, you will come upon it; it is in your future.' I recognize all the things that I've encountered now that I'm this age. That's what evidently occurs; they directly encounter the things [elders] spoke of. If nothing happens to you, you will come upon it."

Elders were both honored and cared for, in part because of the instructions they shared. John Phillip (October 2006:333) recalled: "They say that they loved those elderly people very much since they were their instructors. They say people tried to repay our elders for what they had done; they helped them and cared for them for the rest of their lives." Anna Agnus (June 2003:57) noted that it was as if elders in the past held a compass for them, guiding their lives.

John Phillip (October 2006:272) told the young people that someday they, too, would be elders sharing instructions:

> Indeed, what [Angutevialuk] said was true. Since [the elders] would not continue to give instructions, since they would one day pass on, he spoke to us so that we would receive our instructions. Later on, we, too, will be gone. You will surely serve the following [teachings] to people born later, your grandchildren and your great-grandchildren. That's why you should try to record the few *qanruyutet* that you know inside your heads.
>
> I speak to my grandchild just like I do to you, and I do not hesitate, as she will pass down what she learns to the next generation by putting it in books; you will read what we've said.

Michael John (March 2007:1229, 1240) noted that *qanruyutet* could help a person only if used: "My father and grandfather also gave me advice. They said that if I actually carried out the instruction that I was given with my body, if I used it from time to time, that I would come upon it one day. They said that I would be working toward something that would help me in the future. But they said it would not help me if I just listened to their instruction and didn't experience it firsthand. Instructions can only help if one uses his body to carry them out." Camilius Tulik (March 2007:598) declared: "Let us bring up the old ways and try to practice them."

Ca Tamarmi Ellangqertuq | Everything Has Awareness

Elders viewed teaching the *qanruyutet* as essential, and they "suffer" over the fact that many young people lack this knowledge today. To appreciate the depth of this conviction, one must understand the Yup'ik view of how the world works and the human condition within it. Contemporary elders' view of the world differs markedly from that of their *kass'aq* (non-Native) neighbors. They value different behaviors than do Euro-Americans, making use of a richly textured emotional vocabulary in which compassion, gratitude, and restraint play prominent roles.

A belief in the awareness of all things underlay the traditional Yup'ik view of the world. *Ella*, the word for "world," "outdoors," "weather," "universe," can also mean "awareness" or "sense." The point at which children gained consciousness of the world and began to remember their experiences was marked by the phrase "*Ellangellemni*" (When I first became aware). Everything in the Yup'ik universe was believed to possess some measure of awareness. Michael John (June 2008:96) stated it well:

> I heard that a blackfish was heading down [river]. When it came upon a large blackfish trap that it didn't want to enter, it would enter the one behind it. It would pass [other traps] and finally allow itself to be caught and enter one that it chose to stay in.
>
> They viewed the person who owned [that trap]. Every type of animal, fish, and seals that people hunt are aware of one who made efforts at following the instructions when that person was young. When a person used his body to carry out the instructions he was given, they evidently wanted to approach him.

Humans and animals inhabited a sentient world where even dirt and bones were aware of their actions. Frank Andrew (October 2001:162) described how a person would be impervious to sickness if he handled refuse and unclean things: "They say that everything has awareness. Because those poor *caarrluut* [pieces of refuse] cannot clean up after themselves and do things on their own, they feel gratitude toward the person who moves them to a place where people will not step on them. Their gratitude settles on the body of the person who does that task and gives him strength. That person will encounter good health, and sickness will not permeate his body." Moreover, animals viewed a person like that as extremely bright and not at all frightening, and they willingly approached that person. Conversely, people who were not living according to *qanruyutet* looked frightening, and animals shunned them.

Yuum-gguq Umyugaa Tukniuq |
They Say That a Person's Mind Is Powerful

Another essential aspect of the human condition is the possession of *umyuaq* (mind or mental activity), including a person's thoughts and intentions. Yupiit distinguish between a person's reason and feelings, thoughts and desires. They do not, however, equate the concept of mind with disembodied reason pitted against physical passions that the rational mind should control. Everyone has a mind. What distinguishes people—marking the difference between an untimely death and a properly lived life—is how they choose to use it.

In the Yup'ik view, a person's mind is inherently powerful, capable of pushing others toward both positive and negative outcomes. According to Theresa Moses (June 2000): "*Umyugaa-gguq tukniluni* [They say that one's mind is powerful]. They told us that a person's gratefulness is powerful, and their hurtful feelings are also powerful. If we hurt that poor person's feelings, they can shove us into negative circumstances. But if that person is grateful, it is like they are pushing us toward our own happiness. They make [all animals] more available to men every time they travel."

Umyugiuryaraq (the act of following one's own mind, both thoughts and desires) instead of the *qanruyutet* is socially reprehensible and guarded against by numerous social constraints. In their sentient world, Ellam Yua (the Person of the Universe) is always watching and will reprimand those who do as they please. According to Frank Andrew (September 2000:96): "They used that as a warning to us. If we keep breaking all of these rules that they wanted us to live by, Ellam Yua would make our lives difficult, allowing us to come to our senses."

Naklekun Umyuam-Ilu Atunritlerkaa | Compassion and Restraint

The Yupiit externalize and codify constraint instead of reason, relying on guidance from the ancestral *qanruyutet*. People decide for themselves and bear responsibility for their choices. By teaching *qanruyutet*, elders ensure that youngsters have what they need to make informed decisions.

It takes a lifetime to learn the multitude of *qanruyutet*, but they can be thought about in terms of a few basic "dos and don'ts" flowing from the concept of a powerful mind. The foremost admonition is to act with compassion, sharing with and helping those in need. Atertayagaq felt compassion for his nephews, for whom he hoped to provide. According to Jasper Louis (May 1993) of St. Marys: "When people who are not our own relatives get together, we should say things that could open their eyes. They may not understand at the time, but later in life they will realize that it was done out of compassion." People are also taught to control their own thoughts and feelings, avoiding private conflict and public confrontation. A well-known adage admonishes, "Braid your anger in your hair so that it will not come loose."

These two admonishments advocate neither selfless altruism nor passive acquiescence. Rather, both are responses to the Yup'ik understanding of the mind's positive and negative powers. To act with compassion elicits the gratitude of those whom one helps and brings the power of their minds to bear on one's future success. To act selfishly or in anger, on the other hand, injures the minds of one's fellows and produces dangerous negative effects. This immediate and tangible reciprocity is at the core of Yup'ik social and emotional life. Theresa Moses (June 1995:63) elaborated this relational morality and the power of a person's emotional response: "They say a person's gratitude is powerful. They say their feelings of humiliation are also powerful. They taught us how to use discretion among our fellows and try not to say things that might hurt their feelings."

The view of society that Theresa expressed is fundamentally relational rather than individualistic. A Yup'ik person comes into the world as part of a complex web of kinship, including living and dead humans and animals. How one treats others yields predictable consequences, for better or worse.

Eyagyarat | Abstinence Practices

Contemporary elders describe another class of rules, known as *eyagyarat*—the traditional abstinence practices following birth, death, illness, miscarriage, and first menstruation—that expand our understanding of restraint. Diverse *eyagyarat* practiced in a variety of situations have a common purpose: restraining and guiding behavior during life's transformations to avoid annoying or frightening human and nonhuman companions in a sentient universe. These practices are interrelated, allowing men and women in a vulnerable condition to coexist safely with their fellow humans and animals in a knowing and responsive world.

John Phillip (November 2001:152) emphasized how important it is for both a man whose wife has miscarried and a girl who has menstruated for the first time to avoid contact with the ocean: "If a woman miscarries, her husband is told not to go to the ocean. If he goes to the ocean, it will know." Another coastal dweller, David Martin (January 2002:131) of Kipnuk, described the personhood of the ocean, which not only knows about transgressions but reacts to them with anger and disgust: "The ocean gets mad and becomes angry just like a person. When those who do not follow abstinence practices upset it, the ocean boils up in anger, and the wind howls."

People's safety during transformative events rests on their invisibility in a sentient universe. Frank Andrew (September 2000:32) described how, following her first menstruation, a girl should climb a hill and throw moss or fine dust at *ella* (the universe or weather) to blind it. The ocean, he said, also has eyes that must be closed. In his explanation we see how the personhood of humans, animals, and the universe itself are interrelated in Yup'ik thought. Frank continued:

> They say that men practicing *eyagyarat* should not go down to the ocean until the red-necked grebes and ringed seals arrive in spring. When red-necked grebes start to defecate in the ocean and when the blood of ringed seals soaks the ocean down there, they say the *makuat* [eyes of the ocean] close and become blind. The grebes blind the ocean's eyes that it uses to look around. After the grebes arrive nothing will

happen to the ocean if they go down. The ice cannot be broken, and the weather is calm. The ocean down there knows everything that is going on.

Eyagyarat are an abiding expression of the personhood not only of humans and animals but of the land and sea. This understanding of the world as a knowing and responsive place, possessing both agency and will, finds expression in other ways. For example, when Frank Andrew (September 2000:174) was asked how to tell if the winter would be cold, he explained: "When grass grows long and in large quantities, they used to say that it was going to be cold all winter, that the land was insulating itself." Likewise, John Phillip (November 2001:175) described how the sun puts on a hooded parka and mittens in cold weather: "The ring around the sun is its parka ruff."

Just as people who ignored the *eyagyarat* could worsen the weather, good weather could result when a girl going through first menstruation motioned to the winds following the direction of the universe (clockwise, east to west). The world and weather respond in many contexts. John Phillip (November 2001:157) described a responsive world in which humans are held accountable for their actions: "When the weather is good and suddenly turns stormy, they say the weather is trying to kill a person. The weather is hunting for those who don't listen. When it takes the human breath away, the weather will quickly improve, like someone who doesn't feel pity for another. The weather suddenly calms down and gets good because it is thankful for taking a human life."

John Phillip concluded: "Some think the wise words are gone. They are still real today. Don't make the mistake of saying that we are becoming *kass'at* [white people] because we are now speaking English. The ocean and the land will not become like white people." In southwest Alaska, the personhood of the ocean and land, like the personhood of animals, remains—and they are Yup'ik, not *kass'aq*.

Tuvqakiyaraq | The Way of Sharing

Elders' understanding of the power of the human mind and the importance of compassion and restraint underlies what they consider appropriate responses in everyday life—how a person should think, feel, and act. One should not follow one's own mind, but rather the *qanruyutet* and *eyagyarat*. These rules and instructions teach compassion and restraint because one's actions have a powerful effect on the minds of others. In a sentient universe, actions always elicit reactions. Acts of compassion will result in feelings of gratitude and good fortune. Acts of willfulness and failure to follow the rules will result in hurt feelings and failure to catch animals. We make choices, and they have consequences.

One of the most fundamental Yup'ik admonitions—to share with and be compassionate toward elders and orphans—derives from this understanding of

Uqiquq (distribution of seal meat and blubber) in Tununak, 1980, given in honor of a man's first catch of the season.
James H. Barker

the power of the human mind. Frank Andrew (September 2000:94) described the power of gratitude to help a person to a better life:

> Now this *tuvqakiyaraq* [giving food to others], being compassionate. They say that those who are capable must help those less fortunate through sharing food and doing chores for them. We were admonished: "Even though an old woman wants to pay you, do not receive it."
>
> When an elderly woman or man is given something or helped, she is extremely grateful and thanks you with enthusiasm. And they give the person who helped them something beneficial, thinking of something in their minds that will aid him positively in his life.
>
> They say that the minds of orphans, elder women, and elder men become powerful with happiness. That is why they asked those who were capable to take special care of them, to give them necessities and food to eat. Those who help receive the ability to live a good life through their gratefulness. They say that an entity that cannot be seen compensates them, Ellam Yua.

Acts of compassion, then, directly affect one's relationship with animals. According to Jasper Louis (May 1993): "In those days, when we happened to catch game, did we ignore anyone? No, whenever we caught something, we gave someone part of it. After that, one tends to be fortunate in improving his catch." Paul John (June 2000) gave an even clearer articulation of this relational morality: "It is an instruction that if we always help orphans and we are capable of providing food for ourselves, there will be more of a chance that animals will make themselves available to us. It is like the help we gave them returns to us."

Many academic debates about sharing among northern peoples discuss the issue in terms of the circulation of objects, focusing on whether people engage in generalized reciprocity, balanced reciprocity, unbalanced reciprocity, or a "myth of reciprocity."[4] Yup'ik discussions of the ethics of sharing describe its consequences in terms of its nonmaterial return—the grateful thoughts it elicits—which are a necessary precondition to a hunter's continued ability to catch animals. The successful hunter shares, and recipients respond with good thoughts, which cause animals to once again present themselves. Compassionate human relations elicit the compassion of animals. Unrestrained, unregulated, selfish acts, on the other hand, bring about disastrous consequences, including food shortages that can affect the entire community.

In human social relations, the return for a compassionate act may take the form of powerful thought. In relations between human and nonhuman persons, the return gift is the approach of the animal as prey, also an act of compassion. Animals view compassionate and restrained humans as desirable, even pitiable, and approach them. Theresa Moses (June 2002) noted, "The animals viewed a young man's behavior through their windows that they were constantly looking through a long time ago. Even today, animals probably use water as a window to watch everything."

The hypersensitivity of animals—their ability to see and hear what occurs in the human world—must always be taken into account. Mary Mike (October 1994:34) of St. Marys described the importance of acting appropriately because the animals know:

My deceased grandmother used to admonish us about fish. If my husband catches something, I was not to think that it is of little value but to receive it with great respect. . . .

They said that those fish know what's going on around them. They know when a person welcomes them very nicely. Some people wouldn't welcome them so nicely. They would even step on them. They feel sad about the ones who step on them.

They had rules about all the things to be caught. All the things that we eat know who isn't taking care of them properly or doesn't welcome them with a grateful heart.

A man's relationships with women also affect his relations with animals. Paul John (February 1977) described how a good woman could make her husband a better hunter, whereas a scolding wife would cause him to be a poor hunter. Anna Agnus (June 2009:107) commented: "Some men are very good at catching things, but it won't multiply if someone doesn't work on their catch. Even if it's just one small thing, when she works on it without making any excuses, it's like she catches it with her hands as it approaches."

Yun'i-gguq Maliggluki Ella Waten Ayuqliriuq |
They Say the World Is in Its Current State Following Its People

Paul John and Anna Agnus are not alone in their belief that human social relations continue to affect relations between humans and animals. In December 2007 Paul John spoke to his contemporaries concerning recent changes in the moose population along the Kuskokwim and Yukon Rivers:

> At this time we hear that moose populations are low up north. They blame the people from Kalskag on up [the Kuskokwim River]. [Upriver people] tell the people of this [coastal] area that the land up north doesn't belong to them or that they want to have those animals themselves. Because they aren't grateful [that others are trying to subsist], [moose populations] aren't increasing. Since people who depend on those animals have disputed over them, they are becoming scarce.
>
> And about ten years ago, I constantly went with others to the Yukon to hunt moose. The people there only encouraged me and wanted me to catch one, and I wasn't the only one treated that way. At this time, their moose are starting to move down closer to the mouth [of the Yukon] and have increased in number. Because people aren't disputing over them and are encouraging those from other places to hunt them, the Unseen One has increased their numbers. That's why they tell people not to quarrel over animals that can be consumed.

Paul John also addressed the decline of the Bering Sea fishery due to wasteful practices, noting that when people throw unwanted fish back in the water, food resources will be harmed: "This adage, which tells that if something is abused it will dwindle, is certainly the truth. But if they begin to follow what we say, the numbers can begin to climb. Besides telling people not to waste, they also told them not to fight over it or their number would dwindle."

As is also the case for indigenous people elsewhere in the world, Yup'ik ontological and epistemological principles of interspecies responsibility regularly conflict with the human-centered, dualistic Western paradigm.[5] Many Yupiit still understand their environment as a world occupied by persons, where compassion and restraint ideally characterize all interpersonal interaction. Resource conflicts characterized by animosity by definition drive the animals away. Conversely, acts of compassion encourage their return. Paul John testified to the importance of cooperation and listening to one another to resolve resource management issues. The consummate politician, he emphasized the potentially positive role biologists might play in game management if they listened to what Yup'ik people were telling them: "I like the work of the biologists. They try to make us understand, not hiding things from us. . . . If we work together, it would be better."[6]

Many contemporary elders still view animals as active, ethical beings who respond to human thoughts and deeds.[7] They "suffer" over the fact that many young people no longer share either their understanding of how the world works or the importance of following the rules of compassionate and restrained behavior. The evidence of this inattention is everywhere apparent—people hoarding their catch, selling rather than sharing the berries they gather, ignoring traditional abstinence practices, acting violently under the influence of alcohol. Annie Blue (September 1997) of Togiak maintained: "We shouldn't be ignoring our ancestors' ways but help the younger folks understand them because they are at a loss. We talk about proper behavior out on the land and in our relations with animals. When a person goes out, he must always be mindful to do the appropriate thing. I don't think these ways are observed anymore. That is why fish which used to come in huge numbers have dwindled."

Elders emphasize the importance of communicating this view of the world to young people. Today, sharing knowledge is as critical as sharing food in both the transfer and transformation of Yup'ik moral standards. Admonitions to act with compassion and restraint remain foundational not only in Yup'ik interpersonal interaction but in their relations with their environment. Elders gather to tell stories like that of Atertayagaq to educate young people not only about the practical skills they need to survive but about the knowing and responsive world in which they live.

Though many elders remain committed to sharing *qanruyutet*, opportunities to do so diminish with each passing year. When John Eric, John Phillip, and others of their generation were young, they lived in the *qasgi*, where they listened respectfully to their own elders talking and teaching. Women listened in the homes as their mothers and grandmothers also spoke from experience. *Qasgit* continued to be built into the 1960s, but with the introduction of single-family dwellings and formal education, they were no longer used as homes and schools. Contemporary elders are the last to grow to adulthood in *qasgit*, where they received oral instructions that they continue to view as the moral foundation of a properly lived life. Theresa Moses (May 2003) noted: "The *qasgi* was the library holding all the information a person should know. A person went to the *qasgi* to listen to rules for living and placed what he learned in his pocket and went home." Paul Kiunya (May 2003) recalled: "When I was old enough to go to the *qasgi,* my mother would say, 'Okay, go over to the *qasgi* and try to steal something.' She apparently meant that when I listened to men teaching, though the instructor wasn't talking directly to me, if I learned something, it would be something that would help me for the rest of my life."

Roland Phillip (November 2005:326) shared the words of a Kwigillingok elder when he understood that the missionaries were constructing a school in their village where young people would learn to speak English: "When that elderly man learned what [the building] was for, he said, 'Behold! They are about to establish something that will cause people to become foolish and lose their common sense. They are starting something that will cause people to become irrational.'"

At the same time that contexts for learning decreased, opportunities for wage employment and involvement in the non-Native world were limited. Southwest Alaska continues to be the poorest region in the state, with 50 percent of families below federal poverty level, and the social ills that accompany poverty—alcoholism, accidental death, and domestic violence—plague communities throughout the region. The Yukon-Kuskokwim delta has the highest suicide rate in the state—five times the national average between 2003 and 2006—and the highest rate of sexual assault in the nation.[8] Joseph Jenkins (May 2003) of Nunapitchuk expressed what many feel: "Young people are so very painful to lose. When one sees them with their emotional problems, one cannot remain indifferent. Alcohol is especially destroying us."

Although the Yup'ik language remains strong in coastal and lower Kuskokwim communities (with over 14,000 first-language speakers), language loss has been severe in Yukon and middle Kuskokwim villages. Golga Effemka (January 2006:139) of Sleetmute sadly declared: "Upriver they don't comprehend in Yup'ik but only in English. It's because we elders don't teach them. When our parents raised us, they spoke to us in Yup'ik. They no longer speak [in Yup'ik] nowadays. And when speaking to them [in English], some get angry because they can't speak in Yup'ik." Commenting on the present situation, Paul John (September 2003:240) observed: "Our ancestors lived their lives and tried to make things better so we won't fall forward. We live like we don't have a good sense of awareness."

Roland Phillip (November 2005:350) complained that young people today are constantly noisy and rambunctious inside homes: "Because we no longer give oral instructions to our children and grandchildren, our homes are now their pockets. When coming indoors after playing, without saying anything they [get whatever they want] because they aren't given instructions. If we admonished them regularly, although they thought of wanting something they wouldn't be overindulgent, even with food, when they remembered their admonishments."

Elders also spoke of environmental changes they have observed over their lifetimes. Many commented on the warming temperature. Paul John (January 2007:11) recalled his own elders' observation: "They say they will no longer experience winter." David Jimmie (March 2007:224) of Chefornak remembered how when he was young he could hear the crunching noise of footsteps in the snow, a noise now rarely heard. And John Eric (March 2007:224) said that past travelers used fire-bath respirators (bundles of tightly packed wood shavings) to breath the cold air. Mary George (January 2007:41) of Newtok observed that salmonberries now ripen earlier than in the past. Lizzie Chimiugak (January 2007:303) of Toksook Bay warned that the sun's heat was becoming more intense. She said that she tells younger women that when tending herring hung to dry, one needs to constantly turn them to prevent the heat from "cooking" and ruining the fish.

The most noteworthy signs of warming temperatures are changes in sea and river ice. Many commented on what they have observed, including later freeze-up in fall and earlier breakup in spring; thinner, less reliable river ice; the disappearance of *cikullaq*, newly frozen ice that in the past formed along open water in cold weather; rougher sea ice along the coast due to freeze-up following fall storms; and fewer *evunret* (piled ice) as well as *evunret* appearing in places where they were not previously seen.

Paul John (January 2007:9) noted: "Because there is no longer any extreme cold weather, there is a lack of ice nowadays. And there are no longer any genuine *evunret* growing [along the ocean]. When ice formed a good distance out from shore, they caught many sea mammals and knew that they wouldn't be scarce. Now that condition rarely happens." Hunters note the retreating edge of the shore-fast ice. They also say that shore ice is sometimes patchy where in the past it was continuous. Joseph Patrick (January 2007:91) of Newtok said: "Back when I accompanied seal hunters in spring, the ice used to be solid and extend toward the north. Then after a number of years, I saw that the ice had started to form in patches. Because of the sandbars that have developed, the ice there can no longer stay in one piece."

The land is also changing. Nick Andrew (January 2007:16) observed that the land is sinking in some areas and "becoming thin," as it was in the distant past. Nelson Islanders were unanimous in observing that inland rivers are increasingly shallow, making traditional hunting areas less accessible by boat. Many attribute this to the increase in beavers, which dam rivers and streams. Another explanation may be a drop in the water table as permafrost melts away and as the snow in the uplands that feeds the island's streams decreases. Little baseline work has been done on Nelson Island, making it difficult to draw conclusions.

As the land and sea change, so do their inhabitants. Hunters noted that seals migrate earlier in the ice-free waters but that there are fewer of them than in the past. John Eric (January 2007:136) said that when he was young, he would throw empty rifle casings at spotted seals to scare them away when hunting for the preferred bearded and ringed seals. Today all seals are welcome. Elders observe not only fewer seabirds but fewer songbirds. Beluga whales, which regularly migrated along the coast, are also seen less frequently. The number and health of white-fish in inland rivers are said to have declined. Some species, most notably the indomitable beaver, have increased. And new species like the salmon shark are now observed.

Throughout our discussions, elders were eloquent not only in what they said but in how they said it. The most striking feature of their conversations was the integrated way in which information was shared. Elders did not distinguish between various human impacts on the environment, including the effects of commercial fishing or overhunting, and the "natural" effects of climate change. Instead, elders continually referred to the role played by human action in the world when describing changes in the environment or species availability.

At every gathering, at least one elder repeated the well-known *qanruyun* "The world is changing following its people." Peter Jacobs (January 2007:15) noted: "*Yun'i maliggluki-gguq ella waten ayuqliriuq*" (They say that *ella* has become like this following its people). Simeon Agnus (July 2007:581) agreed: "*Qanrutketukiit-am tua-i yun'i-gguq maligtaa nutaan ellam uum* [They say that the weather is following (the behavior of) its people]. Since its people are becoming bad, the weather is replicating their behavior." *Ella* has always been viewed as intensely social, responsive to human thought and deed. The Western separation between natural and social phenomena sharply contrasts with our Yup'ik conversations, which eloquently focused on their connection.[9]

The adage "The world is changing following its people" captures the Yup'ik view that environmental change is directly related not just to human action—overfishing, burning fossil fuels—but to human *interaction*. As Paul Tunuchuk (March 2007:216) said: "We are in this situation today because of not being dutiful to each other. It's as though we are sleepwalking." To solve the problems of global warming elders maintain that we need to do more than change our actions—reducing bycatch and carbon emissions. We need to correct our fellow humans. They encourage young people to pay attention to *qanruyutet*, believing that if their values improve, correct actions will follow.

The way Yup'ik elders work to correct their youth is to speak to them—as they did during village gatherings—sharing knowledge with kindness and compassion. As they say, "We talk to you because we love you." Stanley Anthony (January 2007) vehemently maintained the need to instruct the young, because, as he said, "The instructions aren't mine," implying they are not made up and should not be cast aside.

Nuna-gguq Mamkitellruuq

THEY SAY THE LAND WAS THIN

THE COAST OF SOUTHWEST ALASKA CONSISTS OF A BROAD marshy plain, the product of thousands of years of silting action by the Yukon and Kuskokwim Rivers. Geologists tell us that during the Pleistocene, 20,000 years ago, the Bering Sea coastline was far to the west, and the ancestral Yukon River flowed south of Nunivak Island, reaching the Bering Sea near the Pribilof Islands. Over time sea level rose and the Yukon migrated north and east to its present mouth on Norton Sound.[1] The lowland delta between the two rivers today is as much water as land. Thousands of thaw lakes dot the terrain, crisscrossed by interconnecting sloughs and streams. The sea is shallow, the land is flat, and fall storms can push the tide inland as much as thirty miles.

The Yukon-Kuskokwim delta is bordered to the south by the Ahklun, Kilbuck, and Kuskokwim Mountains and to the north by the Andreafski Hills. Between these ranges, the low volcanic vents and flows on Nelson and Nunivak Islands and outcrops of intrusive rock in the vicinity of Cape Romanzof and the Askinuk Mountains provide the only relief. Leaf and cone fossils of dawn redwood (metasequoia) and the bones and tusks of woolly mammoths sometimes visible in eroding banks on Nelson Island and elsewhere testify to a warmer environment 65 million years ago. Plants and animals lived and died, their remains covered by sediment. As the Pacific Plate moved under western North America, these sedimentary layers were folded and pushed upward, only

Rock outcrops at Cape Romanzof.
Jeff Foley

Looking west toward Up'nerkillermiut
along the north shore of Toksook Bay.
Ann Fienup-Riordan

to be eroded again by wind and rain. On Nelson and Nunivak Islands, volcanism between one and two million years ago cut these sedimentary beds and covered their surfaces with basalt flows, and the entire rock package was thrust up by faulting and tilted to the southeast.[2] The valuable mineral deposits of *uiteraq* (red ocher) and *qesuuraq* (vivianite), as well as *urasqaq* (white kaolin clay) found alongside coal seams on Nelson Island today, all testify to the rich environments of the past, transformed over thousands of years into materials Yup'ik people use to this day, both to decorate and to preserve their wooden tools and clothing.

Ciuliaqatuk | First Ancestor

Yup'ik elders tell their own stories of how the land was formed and altered through time. They share with geologists a deep appreciation for how much has changed from the days, as they say, *nuna mamkitellrani* (when the land was thin). Then the boundaries between the ordinary and extraordinary were more permeable, and people encountered unusual, sometimes frightening things. In these distant times there was no light or human life. Both were brought by Ciuliaqatuk (First Ancestor), Raven, the Creator. Nick Andrew (October 2005:311) declared: "Beginning from long ago, people have seen and thought of that Raven who has done much. They really cherish it."

Tulukaruk (Raven) is also said to have created many landforms, especially on coastal headlands where archaeologists tell us hunters first settled when they arrived in Alaska, before gradually moving inland and upriver. Linguist James Kari was struck by Yup'ik narrators' use of place-names in these Raven stories, noting that specific places are never cited in Athabascan creation stories.[3] Paul John (July 2007:270) told of how Raven created Qaluyaat (Nelson Island):

> They say that this place was once water before it turned into land. When I heard it as a story, the village of Kalskag inland is where the mountains start. The Kalskag hills are between the Yukon and Kuskokwim Rivers [where the rivers come together]; there are no mountains below it. I heard that Ciuliaqatuk tossed his small amount of ashes and said, "In the distant future, the descendants will live on it." They say that this area turned into land starting from the time that he threw his few ashes. That's the extent I know the story.

Michael John (July 2007:284) added detail:

> They said that this place here [Nelson Island] was once *evunret* [piled ice]. And they said Raven's daughter, when it was time, was fishing for tomcod along a small piece

of ice that had stuck to this [land]. They told [Raven] that the ice detached and his daughter floated away.

When they said the ice floated away while she was on it, and he saw her, he filled the bottom of his garment with land from the surrounding ground, and he splashed it along these *evunret*. And when he splashed it, this area here became land. That's how I heard the story.

Susie Angaiak (March 2007:1320) of Tununak agreed: "In the legends they mentioned that Nelson Island was once *evunret* that Raven turned into land. In the mountains up there [above Tununak] are rocks piled like ice." Rock formations known as Qairuat (lit., "pretend waves") along the north shore of Toksook Bay are said to mark where ocean waves hardened when Raven created Nelson Island.

Raven left marks throughout Nelson Island. Michael John (July 2007:285) noted tracks along the top of a boulder below Engelullugarmiut, said to have been made by Raven's kayak sled, and nearby are etchings made by Raven's daughter when she played with her story knife.[4] Simeon Agnus (July 2007:178) said that Raven stepped on all the points along the north coast of Toksook Bay, giving each a name, when bringing food to his father: "My, how large that Raven was as he made his strides. I wonder where he came from, but I never heard."

Nick Charles (December 1985), originally from Nelson Island, told the story of Raven shaping the land with his ice pick.

I also heard that Raven tried chopping the ice and digging somewhere above Tununak, at that place they call Cikuliullret [lit., "Ones Chopped with an Ice Pick"]. Perhaps he aimed to dig all the way through, and people lost in the ocean could return through that hole. He dug in and chopped, but his ice pick broke before it opened.

So when the tip broke off, he put it to the side and left it saying, "I'll just put it there." And he said that if one of the descendants found it, he would become wealthy. Up to now no one has found it.

They say that in the past Cikuliullret looked like it was actually chopped, and rocks appeared like they had been chopped with an ice pick.

Many also know the mountain above Tununak where Raven had his daughter sit in isolation when she had her first menstruation. Paul John (July 2007:259) recalled: "He said Qilengpak across there was the place where Ciuliaqatuk's daughter had her first menstrual period. Ciuliaqatuk had her go across to Qilengpak and had her sit there. And they say that the area that her blood soaked is *uiteraq* [red ocher, from *uita-*, "to stay, to remain," perhaps referring to Raven's daughter's confinement]." Michael John (July 2007:284) added: "They said that the upper side of Qilengpak had a doorway. But when *yuugarkaunrilnguut* [those who wouldn't live] went in and out of that place, they caused [the doorway] to shrink fast. That's what they used to say about it."

Nelson Island is one of the few places in southwest Alaska where *uiteraq* is found. Simeon Agnus (July 2007:258) noted: "Qilengpak is where people go and get *uiteraq*. They mention that *uiteraq* isn't visible for some people. It disappears. But they say it is displayed to some people, and the area surrounding them is very red. Although that one up there [Qilengpak] isn't a living being, it presents [red ocher] to some people and makes it absent for others. That's how that place up there is." Paul John (January 2006:204) added that people found different shades of *uiteraq*, some bright red and others pale. When they took it, they were to leave something behind, including a bra one Nunivak woman left as a gift: "Those who want to give an offering to that place where Ciuliaqatuk's daughter sat: it now has a bra."

Raven remains a giver of gifts. Joseph Jenkins (May 2003) recalled: "He was telling about the raven. That's what I experienced, too, when I started traveling to the wilderness. When I finally saw a raven, I yelled at him, 'If you have something on your back, remove it for me.' Because he probably heard me, the raven began to fall and started moving his wings around, and before he reached me, he flew away. Sometime, not long after Raven does that, you see a red fox or you see a mink. They tell us that if we believe in the *qanruyutet* and follow them, they will end up being true for us."

Raven also created mountains to the south of Nelson Island. Paul Kiunya (October 2005:314) explained: "Raven is not a bird when they speak of him. He is a person. It wasn't only here that he placed things on the land. There are mountains across from us that we call Ingriik [Two Mountains]. It is said Raven carried them on his back from Nelson Island and brought them over. He was supposed to place them on Qukaqliqutaat, but his backpack broke, and when the ones he carried fell, he put them where they are now."

Qairuat (lit., "pretend waves") etched in the cliffs on the north shore of Toksook Bay. *Ann Fienup-Riordan*

Some say that Raven created mountains to the north as well. According to Teddy Sundown (October 1975) of Scammon Bay: "The Askinuk Mountains were all ice once, *evunret*. The ice floes that are seen on the ocean, that's what these [mountains] once were. Now, one can see mountain rock edges that look like water dripping, but because it turned to rock it stopped. We can see the trails the water left when it ran down the mountain." Teddy Sundown also mentioned a hole in the rocks along Cape Romanzof that some say was made by Raven: "A very long time ago a person rowed wherever he went hunting, even out in the ocean past this place. Every time he came along that place down there, he would stop for a while and chop on it to make a hole. He tied his boat to the shore and had a bit to eat. They say that he's the one who made the hole in that place. Its name is Putulek [One That Is Pierced]. A person may go in there and come out the other side."

An'gaqtar | The Stone Lady

While Raven left his mark on coastal headlands, rock outcrops throughout the delta are said to represent people frozen in stone, often when they were experiencing a frightening situation. Many, like Qilengpak, are still animate and responsive to those who seek their help. Among the best known is the stone woman An'gaqtar, which can be seen on the coast near Togiak to this day. Frank Andrew (September 2005:45) explained:

> An'gaqtar was a person. The stone figure looks exactly like a person from a distance. It's standing at the base of a mountain, *ingrim ceturtellrani* [lit., "at the place where the mountain stretches out its legs"]. The area below the stone figure is all cliffs, and many boulders are along the shore below making it difficult to land.
>
> An'gaqtar lived in the Togiak area and turned into stone after losing her mind [when she could not find her husband]. The stone figure is visible on this side of Togiak, beyond Quluqaq, on the land that juts out from Kangiqutaq [Cove].
>
> It is a stone figure facing east with a baby on her back. She is wearing a squirrel parka. When you look at her from a distance, you can see her parka ruff very clearly on a perfectly shaped woman. It is said that the stone is huge and doesn't appear like a person when you look at it close-up.
>
> And those drawings on rocks, the etchings of humans and other life forms, they say that they are designs she created in the area across from Togiak after she lost her mind. They say she used her pointing finger to make figures of animals and people on the rock, which was soft like mud, [after walking on the water, which was hard like rock].
>
> After she went insane, she apparently left their village while everyone was asleep and disappeared. When they searched for her, they discovered that she had turned to stone.

When I went to Togiak for commercial fishing I saw the stone figure [from a distance]. They say there's a trail to get to it. And I've heard that even animals always stop to pay homage when they travel near it, circling it before continuing on. And when little bugs and insects come to it, they always go around it, following the direction of the universe, *ella maliggluku*. Even people are instructed to go around it when they come to it and not just walk by it and leave.

Evidently, the military tried to take it away one time, but whenever they tried to approach, she didn't let them find her by creating intense fog around her.

Traveling past Togiak in 1883, the Norwegian collector Johan Adrian Jacobsen encountered An'gaqtar and noted that the fifteen-foot figure was so lifelike it appeared as though human hands had made it. Jacobsen also noted that the figure was widely venerated and that no one passed without stopping to leave a gift.[5] Arctic peoples in Greenland and Canada were also known to make offerings to stones to further their hunting success.[6] More than a century later, John Phillip (January 2006:200) described the offerings he saw when he visited An'gaqtar while commercial fishing near Togiak:

People who want to give a small offering, keeping in mind that she is filled with gratitude, keeping in mind the commercial fishing that they take part in. I saw that one. There is a lot of money on top of it, the many gifts that she was given. It's amazing to see that [stone]. There are different types of candy bars, bullets, matches. People gave that one they called An'gaqtar many different types of offerings, all the way around it.

To this day, anyone who wants to can go to it. I saw it twice myself, and I followed that custom and circled it following the sun, *ella maliggluku*. And some people say that when they are hunting and pass by, they give her something, thinking about what they intend to catch, recalling her gratitude. And when they go to her in winter, sometimes when they look in her bowl, fish scales are inside. They say that occurs when there will be an abundance of fish [in summer].

Paul John (January 2006:203) noted that An'gaqtar's husband is at the upper part of the Togiak River, at the second Togiak lake: "He, too, has a bowl. When there are going to be many fish, it is obvious through the inside of his bowl [which will have many small fish]."

In an account translated by her grandson, Myron Blue, Annie Blue noted other stone figures related to the time of An'gaqtar. During her travels An'gaqtar came across some children wading in water and startled them as she approached. They began to cry and scrambled toward land, stumbling and turning to stone at Ivruaq, near Quluut, where they can be seen today. An'gaqtar also molded a humanlike figure called Tarunguaq in a place of the same name. Below she made a hollow rock where the sound echoes when slapped. Once two boys made fun of the hollow rock by urinating there, and they died soon after.

Teggalquurtellret | Those Who Turned to Stone

Annie Blue (September 1997) described other stone figures near Togiak, including a grandmother and her grandsons. The boys had been playing by the shore of the Ingricuaq River in a skin boat when the outgoing tide caught them and carried them away. Hearing their cries, their grandmother ran outside and yelled instructions to them. Then she wrapped her red-fox parka around her shoulders and started running along the shore, crying and singing to her grandchildren, telling them to let the wind blow them to shore. The grandmother followed the river until the boys disappeared behind the islands below their home. Then at the southernmost base of Ingricuaq Mountain, she sat down, with her fur piece still draped over her back. Annie Blue concluded:

> They say that rock was shaped exactly like a human. The upper back of the human-like rock figure was covered with red rock flowing like a shawl, and the figure had a lifelike head with a face. The rock figure was located at the point of Ungalaqliq. . . .
>
> The two grandsons apparently were pulled by the current up to Nunaalukaq Bay. I've heard that there was a rock that looked exactly like a skin boat with two human figures in it. One figure at the front appeared to be pushing away from the shore with an oar, and another figure at the back appeared to be pushing with a paddle.
>
> Then the same year as the Great Death [1918 influenza epidemic], two hunters heard noise coming from those two rock figures. The two rock figures were merrily laughing and whistling. And during the same year Unkuugiiq also lifted her chest up and turned around, as if to look at something, turning toward Tarunguaq. During the same time, Ugli was found bending over his bowl. All the stone figures changed positions just before the Great Death worldwide epidemic. I've heard that Unkuugiiq was taken by someone not too long ago, but apparently it broke through the bottom of the ship and sank.

Stone figures in other parts of southwest Alaska are likewise described as both changeable and responsive to human action and intention. Neva Rivers (April 2006) of Hooper Bay told the story of a man wearing a visor who climbed the Askinuk Mountains but, afraid to descend, died there and turned to stone, where he can be seen to this day. Teddy Sundown (October 1975) added detail:

> Elqialek [One with an *Elqiaq* (Visor)] is on this mountain. In summer long ago, just before a great coughing sickness from which many people would die, it seemed to turn its head, and the part of its head where the visor was shifted.

Upper: Granodiorite tors very close to Elqialek, located at the mouth of the Lithkealik River. *June McAtee*

Lower: Kokechik tor field. *Rob Retherford*

[Elqialek] is farther down the coast from Paimiut. It's the one close to sea level, and its children lead up the mountain. His wife is also above, breast-feeding a child.

Inland from Elqialek are other figures fighting over a walrus, trying to take the walrus from each other. The stone walrus looks just like a walrus. There are more on an island, called "those who fell on their backs." Once when they were playing tug-of-war they fell and lost.

All rock formations here have stories of how they came to be. Each has its own character.

Nunivak elders, including Andrew Kolerak, Andrew Noatak, and Peter Smith, also tell about people who turned into stone figures, now known by the names of those they once were.[7] Andrew Noatak shared a long account of Ayuguta'ar, who traveled from north of the Askinuk Mountains down the coast to Nelson Island and across to Nunivak Island. He initially stayed at Englullrarmiut, then traveled south to Cingiglaq (Cape Mendenhall), where four brothers, all skillful hunters, lived with their younger sister.[8] Ayuguta'ar married the sister and stayed with them through the winter, during which the brothers hunted daily and were always successful. In spring Ayuguta'ar went hunting as well, returning with a seal early in the day. His brothers-in-law finally returned, empty-handed. They were so distressed because he had outdone them that they scattered in all directions, each sibling transforming into a stone which henceforth bore his name. Their sister was so saddened that she took her baby and walked inland, where she also turned to stone with her puppy beside her, its tracks visible on the surface of nearby rocks.

Wassilie Evan (March 2004:126) of Akiak described a stone head located below the upper falls on the Kisaralik River.

> They say one of the people who was up in the mountains started to lose his mind. They went downriver, starting from Kisaralik Lake, and he was floating downriver in a small skin boat. When they got to the falls, they yelled at him, "Dock your boat! Dock your boat!" That one who was going crazy paddled, following the current, and then he suddenly went over the falls, along with his boat, and disappeared. Then he resurfaced right below the falls, and in the front of his boat was a stone mask. Then he brought it up [on land].
>
> They say that [after that incident] that stone mask would have white tears on it when something was going to happen to the people who came upon it. They say when they reached it, when they floated downriver, they would always go to it and wash its face, fixing it up before going on their way. They say when nothing was going to happen to those people, it wouldn't have tear marks on it.

Wassilie stated that this stone mask can no longer be found: "One day, someone from out of our area scavenged it and took it away. The one who took that pretend person died." Two stone heads, however, with equally dramatic origin stories are located lower on the Kisaralik River, just above Golden Gate Falls.[9] Some elders report that these "heads" were used as ballast for skin boats going through the falls and were placed below the falls after use. Both are small conglomerate boulders with white-chert pebble clasts for features. June McAtee observed that the cracks along the clasts would likely collect moisture, which could be interpreted as tears.[10] She also noted that changes elders observe in stone figures throughout the region make sense geologically. All these formations are

Stone head on the Kisaralik River, September 1988. Matt O'Leary, Pippa Coiley, and Dennis Griffin, US Bureau of Indian Affairs, ANCSA Office, Anchorage.

ephemeral, as the forces that created them—erosion, the freeze-thaw cycle, and massive wasting—are the same ones that are slowly destroying them.

James Guy (July 1988) of Kwethluk told the story of Tunumilek (One with Something on Its Back), a stone woman standing on a tributary of the Tuluksak River: "They say that Tunumilek [with her child on her back] was flown away by lice [that infested her body], and they landed with her there, and she became a stone when she died. They said that when she died, so did her child, and she became a stone." Fannie Nicori (June 1988) of Kwethluk told the story of a man infested with lice, which became so plentiful on his body that he flew out of the *qasgi* window, first landing on a tree and then on the mountainside, where he remains to this day. Fannie noted that the head has eyes, nose, and mouth of white rocks. When the wind is going to change direction, the stone man moves by itself and faces the wind. A long rock by his side—his kayak—also moves when "it's going to be lucky" and food will be plentiful.[11]

Golga Effemka (January 2006:194) told the story of a stone woman and child near Sleetmute, upriver from the mouth of the Holitna River.

They say that woman was starving and came from somewhere. She ran out of food and went down to the river. They were hungry. She lost her poor husband some-where on the mountain, and he sat down and is situated there. We don't see her hus-band, but his wife is holding a small child along the riverbank. When we travel, our elders instruct us to give her some of our provisions, anything, to place it right below her or, if we were having some tea right beyond her, [to give her some]. Sometimes she makes animals available to catch because the poor thing is grateful for what we have given her.

And they say that the logs downriver from her are the food left over from what she ate. They say the wood we gather is [animal] bones. Also, a small lake with a beaver in it is across from her. Although they hunt it, they cannot catch [that beaver], no matter what they do. They call it *qimaguyukaaq* [one who flees].

We also go inland from there, up the mountain to survey the surroundings from a high point. They refer to it as the place where she packs water. They say those who went there to look from a high point would pack water from there, but it's dry today.

And upriver from her is a large mountain along the river with little people. They used to hear them back when they went to spring camp, playing inside that mountain. It so happens that they were *ircenrraat* [other-than-human persons].

They say when she sat, that poor woman desired to turn into stone, as the poor thing was hungry, so the descendants who are going upriver give her an offering. You can see it from the river. They say even white people take out their wallets and throw money in the water there when told about it. How desirable, money!

Some of the white people actually comply, as the poor things want to catch animals. And when it's nearly impossible to catch anything, when giving an offering by going up to land in front of her and digging in the ground and placing it inside, wanting to catch an animal, she makes animals available. She made animals available to me a number of times, allowing me to catch before night. That stone is true. They say when she and her husband were starving and traveling from somewhere, the poor thing lost her husband. He became stone, but we don't see him.

U.S. Bureau of Indian Affairs (BIA) archaeologists recorded a similar account from John Avakumoff (1987) of Chuathbaluk, who discussed a group of stone people, one known as Arnassagaq (Old Woman), located on a ridge on the left bank of the Holokuk River upstream from its confluence with the Kuskokwim. This site is within Ingalik territory but close to the present boundary with Yup'ik-speaking people about twenty-five miles downstream. According to John Avakumoff, a man, woman, and child were turned to stone as they traveled east from the Aniak River drainage over the Buckstock Mountains into the Holokuk River drainage during a time of famine. When the family crested the mountains and viewed the food-rich Holokuk River below, the woman was so excited that she ran down the mountain toward the river, carrying her child. As her husband watched, his wife and child turned to stone as they approached the river. Soon after, the husband was also turned to stone. John Andrew (July 1988) of Kwethluk told a slightly different version: "[Joshua Phillip] told me the story about the lady with a child up in Ulukaq [Holokuk]. The couple was coming home from their camp, from the hills, the man was rowing down in a skin boat, and his wife and child were going on the ground right through the hills here. And he lost her in the hills somewhere, couldn't find her. After he got back to where they were going, he went back to look for her. And they found her on the side of the mountain, but she had turned into a stone, too. If you see it to this day from a distance, it will be the shape of a woman with a child on the back." John

Avakumoff added that this event was followed by a "great sickness," which—as in the story of An'gaqtar—may refer to the 1900 or 1918 influenza epidemic or the smallpox epidemic of 1838–39. John Avakumoff also noted that present-day travelers leave small offerings of food on the riverbank below the stones. John Andrew (July 1988) cautioned: "If you go hunting in that area, if you don't pay homage or leave something behind for the lady and child, you'll never get what you go after."[12]

Arnassagaq is one among many stone figures documented by BIA archaeologists in the central Kuskokwim region in the 1980s. Jack Egnatty (1987) and Nick Mellick Jr. (1987) of Sleetmute and Pete Bobby (June 1987) of Lime Village all reported a stone family on a ridge near Cotton Village on the Hoholitna River. Pete Bobby (June 1987) also identified a stone family on a hilltop above Sts'a on Tundra Lake. When a Sts'a man's grandchild was murdered by Kuskokwim people with whom the boy had shared food, his family retaliated by digging a hole through the mountain and killing many Kuskokwim people in a blood feud. After the slaughter, when the grandparents were nearing death, the grandfather instructed his family to burn their bodies (indicating that they would continue to participate in the lives of the living) and told them that they would be there to help them in the future. Their bodies were burned, as the grandfather had asked, and the next morning one of the family walked to the top of Sts'a and saw the grandparents there as stone figures. Pete Bobby (June 1987) noted that to this day the stone people help those in trouble: "If you gonna starve, if you need help, you go to the rocks and tell 'em, and they gonna help you."

In conversations with elders from the middle Yukon, Nick Mellick documented half a dozen stone people.[13] A figure of a woman holding a baby stood on a hillside three or four miles downriver from Paimiut until it fell down in the 1940s, and another figure stands halfway between Paimiut and Holy Cross. Nick Andrew (March 2004:526) also recalled the story of people turned to stone when lice flew away with them: "Upriver beyond Russian Mission and downriver from Paimiut, there are rocks with smaller ones in front. They say those are the ones that the lice flew off with. They turned into stone on that mountain and are clearly visible up there. It is as though they are right in front of one another, looking for and picking lice from one another's heads."

Elsie Tommy (March 2009:333) of Newtok described a similar group of stone people located on a high point just above Iqallugtuli on Nelson Island: "They call them Tengutellret [from tengute-, "to fly away with"], the ones the lice flew off with. When we were girls, they looked like real people, and some were carrying babies on their backs. Their belts were rocks. They even had light-colored rock designs that shimmered. The rocks [on their backs] resembled small rivers; they say those were their hair. And some had sparkly rock designs on their heads. They say those were nits."

Nick Andrew (January 2006:207) also described the rock nest located on a cliff top visible from the Yukon River near Marshall, said to have been the home of *yaqulegpak*, a giant bird. Accounts of such a bird were reported by Edward Nelson over a hundred years ago, and sightings of enormous birds occur to this day in southwest Alaska.[14] As anthropologist Ken Pratt notes, this is an old and venerable tradition.[15] According to Nick Andrew, a large village was below the bird's nest, but the bird never bothered the people. Then one day, a woman went to pack water, wearing her tundra-hare parka with the fur on the outside. Mistaking her for prey, the large bird swooped down and carried her up to its nest to feed its young. The woman's husband climbed to the nest with his bow and arrow: "They say that when it jolted from the impact, it broke off the lower section of the nest, and today there is a chunk gone from its side. They said those bird fledglings resembled elderly men [seated wearing parkas], and they had started eating [the woman]. After it had pushed [a section of the nest] with its feet, it flew away, and they found that [bird] dead up north. Then they killed those [fledglings] inside [the nest]."

Frank Andrew (June 2005:145) also knew the story and added detail:

They say very large boulders surrounded that enormous nest and different types of bones—walrus bones, caribou bones, beluga bones, the bones of different types of sea mammals—were inside and outside the nest.

They say when it grabbed [the woman], only her boot soles showed amongst the fur of her parka as it brought her up [to the mountain].

While [the bird] was away, her husband went to [the nest], bringing his bow. When he got to them, he found a good rock and tied a skin line around it for a harness to hold him in place. He then crouched down and kept watch. When it wanted to, it came holding on to a whole caribou.

When it landed and dropped [that caribou], the one with fledglings went up along the edge of the nest, on top of a boulder. When it opened its tail feathers, [the man] shot an arrow at the lower part of its torso. When it suddenly flew, he shot another arrow at it before its wind reached him. A very large boulder flew when it suddenly moved back, and the man flew around in the air when it suddenly flapped its wings, but he'd fall when his harness tightened and held him. When [the bird] left, it went toward the east. Both [birds] never came to that place again. That's how they tell that story.

They talk about those two large birds though, one who laid eggs every year. They called that place Unglutalegmiut [Place with an *Ungluq* (Nest)]. And it is said those two never bothered people. They nested every year. They call those large birds *meterviit* [eagles] here.

Nelson Islanders tell a similar story about a huge stone nest at Nialruq, home of Tengmiarrluk, a huge bird. According to Paul John (January 2006:206):

> They say that large bird that resided there never hunted people. But the wife of one of the great hunters—they say she had a nice parka made of a small caribou—while she was picking berries, it grabbed her and took her away. When her husband suspected what had happened to her, since he was a strong man, he went to it and threw a *nuusaarpak* [three-pronged bird spear] at it, intending to hit its head. Since those bird spears have three prongs, they say that *nuusaarpak* broke off its beak when it landed.
>
> They say that four fledglings that hadn't grown real feathers yet were in its nest. It is said that while they were inside the nest covered with down feathers, they were as large as elderly men sitting with their arms out of their sleeves and tucked into their parkas. I never heard that [the fledglings] survived. They probably starved to death.

Yuguat | Pretend People

The rock nest on Nelson Island has since been destroyed, the stones piled to form *yuguat* (pretend people). Some say these *yuguat* were originally intended to frighten approaching warriors. Others describe *yuguat* as place markers used by travelers to orient themselves. The work of human hands, *yuguat* are not considered sacred and are formed and reformed to this day. Peter Jacobs (January 2006:216) described stone piles, known as *caputnguat* (imitation weirs), near Cuukvagtuli:

> There are many stones called *iingarnat* [volcanic rock with eyelike holes, from *ii*, "eye"] in my village. Downriver and upriver from [Cuukvagtuli], those appear as though people piled them; they have always been there. They called them *caputnguat*. They don't do anything, and they don't mess with people.
>
> They also call those [stones] located downriver between the three hills *caputnguat*. They say when those stones see a person eating next to them who isn't going to live, they sing a *yuarulluk* [shaman song] to him, and here they're merely stones. I heard once that they sang to a person, and he died before a year passed.

In many parts of the delta, surface rocks were scarce. Where they were available, they were sometimes used to mark graves. Annie Blue (October 2006:61) noted: "There were many wars fought on the Togiak River. On the rough terrain there are many piles of rocks. That's where warriors killed [their enemies], and they'd bury them and cover them with rocks. They called them *teggalqirat* [rock cairns, from *teggalquq*, "stone, rock"]." Similar cairns can be seen on the hill behind Goodnews Bay, also said to mark graves dating from the period of bow-and-arrow warfare.[16] Cairn burials, as well as burials marked by a single large rock, were common on Nunivak Island.[17] Peter Smith (September 1988) of Nunivak mentioned two large rock piles on the island to which passersby add a rock to make them larger.[18]

A large stone head between 100 and 450 years old was found on the islet Qikertallgar just south of Nunivak Island. Unlike other stone people found on Nunivak and elsewhere, the head's surface was shaped by human hands, with lips, nose, and eyes pecked into the surface.[19] A story related by Andrew Noatak in 1986 associates the origin of the stone head with the legendary giant Mell'arpag. While juggling with four stone balls, one collided with another in the air, causing the balls to disperse over the island, one landing on Qikertallgar.[20] In their detailed discussion of the stone head, Ken Pratt and Robert Shaw conclude that the head was likely created and used in association with Nunivak burial practices, in which the placement of human skulls on a high point played an important part.[21]

Running through these many accounts are recurring descriptions of stones as persons, both figuratively and literally. Many are viewed as capable of movement, hunger, gratitude for offerings, and foreknowledge of events, especially sickness and death. They are depicted as both sentient and responsive to human requests. From the majestic An'gaqtar to the lowly *uiteraq*, these landscape features can alternately hide or reveal themselves to human travelers. Together they animate a world in which even a lowly stone merits respect.

Yuguat (lit., "pretend people") at Nialruq on Nelson Island. *Janet Klein*

Ella Alerquutengqertuq

THE WORLD AND ITS WEATHER
HAVE TEACHINGS

A MONG THE RICHEST, MOST EVOCATIVE WORDS IN THE YUP'IK language is *ella*, translated as "weather," "world," "universe," and "awareness," depending on context. Contemporary Yupiit may use *ella* to denote "atmosphere," "environment," and "climate." Clearly, the Western concept of ecosystem as an integrated system of natural and cultural phenomena is not new to Yup'ik people.[1] Paul Tunuchuk (March 2007:216) exclaimed: "Everything inside *ella* has customary teachings and instructions attached to it—air, land, and water. And they mention that we must treat it with care and respect. What will become of us if we don't treat it with care?"

Ellam Yua, the Person of the Universe, watched the world. According to Herman Neck (June 1988) of Nunapitchuk: "I understand Ellam Yua to be one who watched all people at that time." Ellam Yua not only watched with boundless sight but observed transgressions and meted out punishment. Here it is not simply that the universe is aware but that universe and awareness are synonymous. This view of the world motivated many personal ritual acts in daily life, for example, the proscription against young girls taking their dolls outside during winter lest *ella* see them and turn the weather stormy. To produce good weather, a person might turn a worm inside out, impale it on a stick, and place it on the open tundra.

Ella would see it and bring out the sun to dry the worm. Following first menstruation and during mourning, women went about belted and hooded to hide themselves from *ellam iinga* (the eye of *ella*) and bind themselves off from a knowing and responsive universe. Ellam Yua possessed all the human senses—sight, taste, smell, hearing, and touch—and people alternately curtailed and accentuated their actions to influence it. Inappropriate human action always was directly tied to cosmological upset and subsequent disaster, while appropriate action and self-control were prerequisite to a successful harvest and long life.

Recent descriptions of Ellam Yua have a Christian ring, reminiscent of a heavenly father and almighty, all-seeing creator. Although *ella* is probably an ancient concept, the idea of a "person of the universe" may be more recent. Still, Ellam Yua differs in significant respects from its Christian counterpart. Rather than a deity in human form, Ellam Yua is a genderless, sentient force. *Ella* is a key concept in Yup'ik cosmology and epitomizes a transformational and interconnected view of the world that many non-Yupiit are only recently beginning to appreciate.

Ellam Cumikellra | Observing the Weather: "Those Up There Do Not Lie"

John Phillip (October 2005:319) noted: "They say that *ella* does not try to surprise its people. *Ella* usually tells its people ahead of time what it will do. But since we aren't aware of it, we say that the weather has suddenly become inclement when it changes abruptly. They say that if we were observant, we would know. We would use our eyes and recognize that process."

Men and women alike recognized the truth of John's words. A person's first act each morning was to go outside and check the world and its weather. Many had strong memories of this ubiquitous but essential task. John Alirkar (January 2007:420) of Toksook Bay recalled: "Since men had instructions to follow, when they'd go outdoors from the *qasgi* in the morning, they'd look at the sky, forecasting what the day's weather would be like. They eventually learned to read and predict the weather." Roland Phillip (November 2005:116) remembered, "When I went outside in the morning, I'd become wide awake when I shivered from the cold." Anna Agnus (July 2007:510) concurred: "From the time we were small, when a person who went out early in the morning came indoors, they asked, 'What does the sky look like?' They wanted us to scan the sky."

Nick Andrew (October 2003:88) pointed out another reason children were woken early and told to go out: "The ones who continue to sleep are lazy people, following their own minds. They will lack things and envy their peers who always wake up in the morning. They say sleep deprives them of things. The ones who follow our ancestors' instructions don't lack things because they know that they have to try. Sometimes I woke up without shoes on top of the snow." Nick pointed out that knowing the weather also enabled a person to prepare properly for traveling: "They wanted us to know what clothes to wear when we traveled. I pity some of those who don't do that when I see them. They leave in good weather, and when the weather begins to rain they come home soaking wet. But those who go out in the morning to look at the weather are always careful when they go." Paul John (October 2003:88) recalled his training: "Our ancestors always told young people, especially young boys, to watch the weather. They really urged them to be aware as they lived. I had a paternal uncle who asked me every morning, 'How do you think the weather will be today?' Because of that I always tried to go out every morning and observe and predict the weather. One day after he asked me, he said, 'Now, I think you will know about the weather.' He never asked me about the weather again."

Paul John (October 2003:89) elaborated on the critical role observation played in learning to understand the world and its weather.

> Sometimes as one was about to go outside someone would instruct, "Don't forget to observe the sky out there." And remembering what they said, although he wasn't quite sure what he needed to observe—the clouds, the horizon, and the sun's emergence—he would observe those things. And if he noticed something in the morning and watched all day, he would begin to understand what the weather would do before the day ended.
>
> Also as the sun gradually went down, if he saw some sign beneath the sun toward the south, he would begin to understand the changes it would make before daybreak. So by using his eyes, not his ears, a person can begin to recognize the behavior of *ella*. Even though he hears about it, he can't truly identify it.

George Billy (February 2006:511) recalled: "Ellam Yua is very grateful to whoever is the first to go out and check on the weather. I believe in that, even at the present time." Although no longer a routine part of life, checking the weather in the morning is not lost. George Billy continued:

> Nowadays thinking about that sometimes I test my grandchildren by waking them up and saying, "Go ahead and check the weather!"
>
> And sometimes, in bare feet, he would go outside. When he came in, [I would ask him], "How is the weather?"
>
> He would answer in English, "Foggy!" [*laughter*]

Akerta | Sun

Ellam Yua is not alone in being referred to as personlike. According to oral tradition, the sun is a transformed woman who fled to the skyland while being chased by her brother, who became the moon and continues to pursue her.[2] Variations of this tale are told all across the Arctic, and to this day people metaphorically refer to the sun and moon as living beings. They describe an eclipse as *akerta nalauq* (the sun dies or withers) or *iraluq nalauq* (the moon dies or withers), using the verb base *nala-*, a term for the death of plants or animals. People say *akerta pit'uq* (the sun rises [or catches game]) as the sun begins to come up. *Akerta aqumgauq* (the sun is sitting down) designates the winter solstice, and *akertem ayarua* (the sun's walking stick) is a sun column caused by ice crystals in the air.

Observing the sun is essential to understanding *ella*. Elders noted that, like weather generally, the sun signals what is to come: "They say the sun gives you signs when it will be bad twenty-four hours before it happens. The sun declares that *ella* is going to be a certain way" (John Phillip, October 2005). Paul John (June 2009:74) said: "They say dawn shows what the weather will be like during the entire day."

George Billy (February 2006:513) compared the rising sun to an eager groom: "The sun in the morning is so glad that it is like one who is running toward its bride. The sun helps its people." As the short, dark days of winter gave way to spring, the sun's return began to be felt. John Phillip (October 2005:323) explained:

> In March, when it's becoming spring, the sun appears up there, and where it hits, it will melt a little. When that happens, those people of the lower coast say that the sun is now good to face. They said that we could travel to the ocean if we wanted although it was cold.
>
> During fall, the sun is lower and daylight is short. It doesn't take long until it sets. But beginning from the middle of winter, [the sun] starts to get higher and daylight gets longer. Then it gets warmer, and [its heat] starts to melt [snow and ice] until it's all gone. The sun works on the land and melts the snow. That is the sun's work.
>
> Then the *qaninerra'ar* [new-fallen snow], if it stays, if the sun is really shining, the top gets damp and melts. And even though snow is thick, the sun works on it and it becomes soft and melts. When there is no more sun, it gets cold and freezes. Then when it becomes daylight, when the sun appears, it melts. That's also the sun's work.

Matthew Beans (December 1985:130) of Mountain Village recalled: "As daylight lengthens in spring, they say when the setting sun is slipping slower in the west, there will be lots of fish the next summer. If there will be few fish, it goes faster as it goes north. But when it sets slower, these people like that sign."

Both morning and evening, the sun's condition was carefully observed as a weather predictor. Paul Kiunya (October 2005:127) stated: "When I woke in the morning and looked out, I'd see that the horizon would be really red. Then the weather would get bad before night." John Phillip (October 2005:206) also examined the horizon at daybreak: "I'd go outdoors and see that the horizon was bright, but the area underneath was very red. They tell us that when the sun rises [lit., "catches"] causing the clouds to turn red, even though the weather is good, it will get bad before day's end. But when the horizon is bright in the morning, the weather might be good all day. Our ancestors used those signs as indicators, and that's what they said about them."

Frank Andrew (September 2000:203) said: "When it is extremely red during the morning, *erenret-gguq malngugtai* [they say that the daybreak is affected by warming weather that brings wet conditions]. Even though it is cold during winter, some days it is no longer cold. When the weather is going to turn out that way, dawn is extremely red." Nick Andrew (October 2005:204) noted: "In the morning when it's cloudy, if the horizon appears as though it is split or cracked and is purple at dawn, the weather will get very bad before the day is over. [The clouds] are split when something is going to happen to the weather. But that purple color means that [the weather] will be very wet. At first daybreak when you look at the horizon, you will see a split cloud. It's a warning."

Roland Phillip (November 2005:104) noted that the sun is a weather predictor year-round: "If the horizon is red in spring and fall before the sun becomes visible, that indicates the same poor weather conditions. But if it is white and not reddish, it indicates that good weather is approaching." Paul John (March 2008:595) elaborated:

> When the dawn is red and the cloud above the sunrise is clearly visible and extended out a great distance, bad weather will hit before night.
>
> No wonder those who observed [the dawn] closely would admonish their young people who were preparing to leave, "*Ella nangenrunrituq; ayanrilngerpet uita assirikan ayagniartuten* [This weather is not the last; stay and don't leave, so you can go when the weather gets better]." [They did that] because they understood what the weather would be like through the sunrise and clouds in the morning.

Paul Kiunya (October 2005:127) also observed the setting sun: "When it's clear outside, just before the sun sets, [the horizon] becomes red. They say *kingyarluni akerta* [the sun is looking back] because it will be good the next day." John Phillip (October 2005:206) added: "When it is going to ask that it be good, it is said it looks back to the place where it originated."

Roland Phillip (November 2005:80, 106) noted: "When the sun rises and there are rays pouring down underneath it with a reddish area visible, it is also written in the Bible that the weather will be bad." Indeed, the adage "red sky in morning, sailors take warning" is of ancient origin. In the Bible (Matthew 16:2–3), Jesus is quoted as saying, "When it is evening, ye say, fair weather for the heaven is red. And in the morning, foul weather today for the heaven is red and lowering." The cogency of this saying is partly explained by the fact that weather systems generally travel from west to east in midlatitudes—though in Alaska more often

southwest or west moving north along the coast. When the sun rises in the east ,
it illuminates the mid- and high-level clouds of an approaching weather system
to create a red sky in the morning. Alternately, if the sun sets as a weather system
exits and high pressure is building, then it illuminates the departing clouds, creating a red sky at night with fair weather to follow.[3]

Elders say that the sun puts on clothing to keep warm during winter. According to Theresa Abraham (March 2008:157) of Chefornak: "They say our sun up
there puts on warm clothing, putting on gloves and a hat. When the area around
[the sun] is bright, they say *maqarqelluni* [it is putting on warm clothing]." Nick
Andrew (October 2005:382) recalled: "They said when the sun has a ring around
it, it is wearing its ruff. When the inner area looks dark, the weather is going to
get warmer, but when it looks pale, it will get colder."

John Phillip (November 2001:162) referred to the ring as a "pretend parka ruff"
with mittens alongside when the weather was going to be cold: "There would be
two red things alongside it, and they said that those were its mittens. When those
mittens were visible, they'd say it was going to be cold. They say that the sun is
putting on its mittens." Paul Andrew (November 2005:98) of Tuntutuliak agreed:
"Although the weather is good, if the sun has a ring around it that is dark inside and
a mitten on the south side, the weather is about to get bad."

Yup'ik people were not alone in using these atmospheric optical phenomena
as empirical means of weather forecasting. The sun's mitten (sun dog or parhe-
lion, "beside the sun") is a bright circular spot on a solar halo (known as a nim-
bus or icebow). Both are primarily associated with the reflection or refraction of
sunlight by small ice crystals making up cirrus or cirrostratus clouds located in
the upper troposphere, but ice fog and ice crystals close to the ground can also
produce them. The dark inner ring (called a halo) is caused by ice crystals in cirrus clouds. When cirrus clouds are in advance of a storm, they thicken and lower
into altostratus clouds, forming a pale ring (corona), indicating an approaching
rainstorm or snowstorm with warmer, windier weather. Halos and sun dogs also
occur with cirrus clouds that are well to the side of a storm when warming is not
occurring. Near the Bering Sea, they are often associated with the jet stream and
so indicate an approaching storm and warming. Cirrus clouds and halos can also
mean there are no low clouds to insulate the ground, so temperatures are very
cold and any changes would have to be warmer.[4]

Agyat | Stars

*At night when the stars were very visible and bright up at Kayalivik, when we'd look up
and talk about the stars, we'd try to figure out what they were. They would instruct us
to try to learn their names.*

—Elsie Tommy, Newtok

Stars are only seasonally visible in the northern sky, disappearing under both summer's midnight sun and muted by reflected moonlight on snow during winter. During fall and spring, however, the sky is full of stars.[5] Although thousands are visible, only some are named. Paul Kiunya (October 2005:325) remarked: "Our co-inhabitants [white people] who observe the stars name them. But those I heard about from our people don't have many names." This may be because in the past stars were viewed as neither objects nor spirits but as holes in the sky world, through which spirits might look down.[6]

The few names given, however, animate the night sky with stories. The Milky Way is known as Tanglurallret (Showshoe Tracks) or Tulukaruum Tanglurarallri (Raven's Snowshoe Tracks). Frank Andrew (June 1995:23) recalled: "They said that the long cluster of stars on a white band was the path Raven traveled on snowshoes. They say when Cormorant ran away with [Raven's] wife, the white band was the way Raven traveled back home on snowshoes after pursuing them." According to John Phillip (October 2005:310):

> They look like my footprints. When it's cloudless, they have color, and its path is bright. The stars aren't large.
>
> They say that Raven was always messing around like us. He would kick the sea anemones as he traveled. His tracks up there go in the same direction as our land, going north.

John Phillip (October 2005:319) also described how Raven's Tracks bend in the direction the wind will blow from:

> Sometimes Raven's Tracks don't appear straight. Then they say that the wind has inflated Raven's Tracks, causing them to bend.
>
> Even though the wind was from the north, someone would say, "Those snowshoe tracks up there are beginning to be inflated from the south." They'd also say that they were inflated from the north.
>
> They say that is the wind's path, that the wind bends them. But sometimes they are straight when the wind doesn't affect them. Raven doesn't always stay still.

Roland Phillip (November 2005:116) agreed, adding that Raven's Tracks could be used to predict the weather: "They evidently aren't always straight; they curve when the south wind or the north wind bends them. When I started to observe them, I began to use [Raven's Tracks] to tell coming weather conditions, and what they forecast would actually occur."

Twinkling stars also indicate wind. Joseph Patrick (March 2007:1198) recalled: "They say if the stars are flickering at night, that indicates coming wind. And although it's already windy out, if the stars aren't twinkling, the weather will become calm and windless." Bob Aloysius (October 2005:316) of Kalskag added: "Sometimes at night when you go out and look at the stars, they have small tails in the direction the wind is blowing to, even though it's calm. When it's going to be windy from the north, the tail will go toward the south." In fact, stars will shimmer when there are strong winds aloft, such as a jet stream. This is often the precursor of an approaching storm and winds at the surface.[7]

The three stars in Orion's belt were known as Sagquralriit (Those That Spread Out, Scatter) and were visible just before sunrise in November and December. Paul Kiunya (October 2005:305) described how people used Sagquralriit to tell time: "When those stars appeared, it wouldn't be long until daybreak. In winter, when it doesn't become light right away, when those three stars reach a certain point, since our ancestors got up early, they would say after going out [and checking the stars] that the Sagquralriit had gone over there and that it was time to get up. When daybreak was nearing, some people traveled in the dark, returning when it was just becoming daylight." Thomas Jumbo (March 2007:661) of Nightmute added that travelers used the three stars as a compass when traveling at night.

Martina Wasili (March 2008:378) of Chefornak recalled two stars lifting up the sunrise, possibly referring to the stars Altair and Tarazed, which rise above the horizon in the northeast in the predawn hours from mid-December through March.[8]

> After drinking tea, my father would get ready and leave. Then my mother would tell me to go check to see if the sun was coming up. I'd go outside and see. Two small stars situated along the horizon were used as indicators. One was larger, and the area underneath would start to turn red. They said that those two small stars were lifting the sunrise.
>
> After watching for a while, when I'd go inside and tell my mother that the sun was just starting to rise, she'd be happy.

Paul Kiunya (October 2005:303) described the two stars whose appearance announced lengthening daylight in spring: "Back then, they talked about the two stars they used for guidance. After the two stars are gone all winter, they are the first to appear before the sun. They used those two stars to measure the increase in daylight. When those two stars appear in spring, daylight will get longer."

The Big Dipper and North Star were also visible in the northern sky. Nick Andrew (October 2005:303) explained: "It seems like there is one that looks like a dog up there that has a snout, ears, and tracks. In my village we call that Tunturyuk [from *tunt-*, "caribou"; i.e., Big Dipper, Ursa Major; also Qaluurin (Dipper)].[9] When it goes toward daybreak after midnight, it begins to bend forward. Long ago, they said that they used that for telling time." John Phillip (November 2001:162) added: "Tunturyuk is also a tool used by travelers, and it faces north. It begins to bend forward around midnight, when it is going toward daylight."

Nick Andrew (October 2005:303) then described Agyarrlak (the North Star, Polaris), also known as Erenret Agyaat (Day's Star) and *agyaq pekcuilnguq* (the star that doesn't move), used for guidance when traveling at night: "There is a star up there above the [Dipper's] handle. People who travel at night use that to navigate. It doesn't move and is always visible, but the others move." Theresa Moses (November 2001:160) noted that Agyarrlak had in fact moved once, long ago: "They call that big star that is red Agyarrlak. Because he lacked food, Agyarrlak moved down there. That star was red and visible there. Then it really moved here."

Many recalled using Agyarrlak as a guide. "When some are down on the ocean and it gets dark, they use that star. Since that star faces land, they use that and go up to land, even though it's dark" (John Phillip, October 2005:307). Phillip Moses (January 2007:440) of Toksook Bay described how he was instructed to use Agyarrlak as his guide when returning to Nelson Island from Kasigluk:

> When Ayagina'ar and I traveled by snowmobile in the direction of the ocean, there was a large star that was low in the sky, larger than all the rest.
>
> Just when it started to get dark, my son-in-law Ayagina'ar wanted to leave [Kasigluk], and we had no compass. That elderly man told us, "Now, when you leave, use Agyarrlak over there as a guide. Face the area beyond it, just to the north when traveling."
>
> Using his instructions, we arrived [at Nelson Island] without any problems. We came from afar, and we never veered from the path. He told us that we would head straight for our village if we traveled using his directions. We arrived and didn't lose our sense of direction.

Elders named other constellations, including Ilulirpiit (Large Fish Trap Funnels), Kaviaret (Ursa Minor, Pleiades, lit., "Red Foxes"), Tunturyucuar (Little Caribou), Qerrun Ayemnera (Broken Arrow), Tengqulluuk (Parka-Hood Tip), a curved line of stars called Qagtellriit (Those with a Northern-Style, Curved Parka Bottom), and a straight line of stars representing the bottom of a Kuskokwim-style parka.

Some stars were viewed as responsive beings. Elena Charles (November 2001:160) of Bethel remembered calling out to the star Kaviaruaq (Sirius, lit., "Pretend *Kaviaq* [Red Fox]"), asking it to change its parka. In fact, Sirius is the brightest star in the night sky and does indeed change colors as it is viewed through earth's atmosphere low on the horizon. Some use it as a weather forecaster. A bright red color indicates approaching cold weather. When it does not twinkle, mild weather is said to be on its way, but when a storm is coming, it flickers dramatically.

Bob Aloysius (October 2005:316) identified Iralum Qimugtii ("the Moon's Dog"; possibly the planet Venus), which moved closer to the moon in cold weather: "In winter the moon has a dog. The moon's dog is a star. When it's going to be warm for a while, it stays far from it. When it's going to be cold, it stays close to the moon. Back then, we also slept by dogs when we used to sleep outside." Paul Kiunya (October 2005:303) recalled the star that always performs slow, old-style Ingula dances: "When that star appears up behind our village, it always twinkles. They call that Ingularturayuli [One Who Always Does Ingula Dances]. They have a song for that star: 'The star back there, Ingularturayuli, *Aarranga*.'"

Stars not only change their clothes and dance but also defecate. *Agyat anait* (star dung) are seen as shooting stars or *agyulit* (comets, lit., "ones that go") in the sky, and puffball mushrooms are their earthly remains. This makes sense because, as John Phillip (October 2005:317) quipped, "Feces are all different." John explained: "Sometime while you're out, you will see a star defecating. It will show a little storm [tail of a shooting star] and diminish. They defecate just like us. [*laughs*] Some of them are very bright when it happens." Moreover, the direction in which the feces fly can foretell a weather change. John continued:

> If a strong wind was coming from the north at night, when someone came in he would say, "A star just defecated, and its feces flew from the south. The wind direction is going to change to the south."
>
> It did that, even though there was a strong north wind way up there. It would fly from the south, traveling against the wind. They say that would be the next wind direction. That was the sign of warm weather. They'd say the south wind was about to blow because the star feces drifted from the south.

John Phillip (November 2001:160) advised: "Keep track of the stars sometime when they excrete feces. When it's dark, the star will excrete one that looks exactly like feces."

Kiuryat | Northern Lights

The night sky is also enlivened by *kiuryat* (aurora or northern lights), appearing in colorful, constantly moving waves in the upper atmosphere when energetic particles from the sun collide with the earth's magnetic field. Nick Andrew (October 2005:185) noted how bright northern lights appear in cold weather. This makes sense, as clear nights suitable for viewing the aurora are usually cold. Phillip Moses (January 2007:618) agreed: "When it's going to be extremely cold, those northern lights appear. They make the cold visible. When they clash with warm air, they make noises by acting erratically. Though they appear once in a while nowadays, they don't appear like they used to." Michael John (June 2008:214) recalled: "Back when it was cold, the northern lights lit up the sky down to the ground. They would say they were forming a ruff."

Lucy Beaver (November 2001:9) of Bethel recalled: "They said when [northern lights] were very red, they were cold. Just like when people get cold, they turn red. At the beginning of winter, when they come and are very red, it's a sign that it's going to be a cold winter." Roland Phillip (November 2005:110) reported that cold weather made the northern lights bleed: "Sometimes when that occurred, my father would point them out to me, and I would see them emitting a red color. When they are like that, they say that the cold weather is causing them to bleed. And sometimes when there are northern lights, the bottom is very dark with the upper part white. People of the past would say that the weather was going to warm up. They said that the cold weather wasn't affecting the northern lights when the bottom section was dark." Joshua Phillip (December 1985:108) of Tuluksak noted: "When it is black below the northern lights, that is the sign of warm weather. And sometimes they get very red, looking like waves in the sky. They would say the weather was about to stay warm for a long time. But if they stayed down close to the horizon, they'd say the coldness was pressing them down. The weather would stay cold if they were low."

Nick Andrew (October 2005:187) understood their red color as a sign of impending bloodshed: "Back home [on the Yukon], they said that they would be red when people were going to war and a lot of blood would be shed. They said they were bleeding." Neva Rivers (November 2001:9) heard the same interpretation when she was young: "Long ago when we saw red northern lights, they said that a lot of people would die; there would be lots of bleeding."

Many coastal people interpret red in northern lights as blood that will be spilled when animals are hunted the following spring. Raised in Chevak, Lucy Sparck (October 2005:187) of Anchorage recalled:

> Let me tell you of what Nacuk'aq's dad, Ulruan, once said about those northern lights that were getting red at that time. They didn't know where it would happen, but they knew that blood would be shed.
>
> When spring came, they went to their camps. He said a lot of walrus came, and the men hunted them. Since there were lots, blood was down on the ice, and there was a lot of blood on the ocean. He said they saw that [red color] because it was in their future. They finally understood why they saw redness in the sky.

Northern lights are remarkable not only for their color but for their movement and, like wind, the whistling sounds they make. "They are never without sound," recalled Lizzie Chimiugak (January 2007:620). Dick Anthony (January 1996:93) of Nightmute had also heard them when he was young: "When there are northern lights in winter, it gets really cold. Sometimes when they seem to be close, that kind of sound [*Dick whistles*] is audible from them. One night when we went dipping for needlefish at Urrsukvaaq, it was really cold. Half the sky was covered with northern lights, and their whistling sound was loud when they were igniting like flames."

Children were told not to respond to these whistles or the northern lights would take them away. Paul Kiunya (October 2005:186) remembered: "Back when I was a boy, the northern lights made sounds. They warned us not to watch them because they said that they took people. They would try to scare us young boys. Nowadays, there aren't as many northern lights because it no longer gets cold. The cold makes them appear powerfully. They make *ugg'* sounds, and their colors change up in the sky." Theresa Moses (November 2001:146) noted: "I heard that they take those who watch them." Peter Jacobs concurred: "If they were too close, we were told not to go outside." Scientists do not know what gives people the sensation of hearing sound during auroral displays, as the upper atmosphere where they occur is too thin to carry sound waves. One hypothesis is that electrical charges traveling to the earth's surface produce the sound. Searching for an answer to that question, we may learn as much about the brain and how sensory perception works as about the aurora.[10]

Northern lights were also said to respond to wind. Frank Andrew (September 2000:203) noted: "When the northern lights are bent toward the east, the east wind that is going to hit puffs them up. But if they are bent in the other direction, the approaching north wind causes them to look like that." In fact, northern lights occur independently of weather on the ground. That said, the sky has to be clear to see them, and if coastal locations require winds to clear the clouds, then one would need winds to see the aurora.[11]

During discussions of northern lights, elders also mentioned observing what they described as *ella qupluni* (the sky splitting) and the dead looking down from above when people are about to experience hardship. Martina John (June 1995:23) of Toksook Bay explained: "It happened in winter. One of the women at my home saw it split at one o'clock in the morning. They were taking a steam bath when the dogs began to bark. They went out and saw a rainbow in the night sky. And while they were looking, it divided in half. On the edge of the divide they saw what looked like saw-blade teeth. Then from that edge they saw something jumping and swishing out. They could hear the sound as it did that." Paul John followed with an interpretation: "I heard that when it did that, the people who had starved to death would be looking out to the earth." Theresa Moses added: "They say that the faces of people who had starved would be looking out when there was going to be a food shortage."

Neva Rivers (May 2004:262) had also seen the sky split.

One night my mother and I were making baskets together. So then we went out to relieve ourselves when we were about to go to sleep. While we were doing so—it was cold, calm—we started to hear something crackling.

When we looked up, the sky had divided. The edges were like saw blades. Its middle was clear. It had divided from west to east. Between the crack it was clear, and there were many small stars up there. And then it closed again.

My mother told me that the sky had split. She said that those who are around them are those who have died by starvation and cold and that they have looked down. That's what she said to me.

Joe Lomack (May 2003:332) of Akiachak mentioned a related phenomena: "Last fall the sky above the ocean had this phenomenon, *erneruaryaraq* ["pretend dawn," possibly the sun appearing before it rises due to a strong inversion], indicating coming famine. It is said the heavens open and those who experienced famine look down from above when famine is about to occur. Before famine, food becomes readily available to people, but they throw it away, deviating from their customary behavior."

Elders also described Itqiirpak, the legendary Big Hand with mouths on each fingertip, said to rise from the ocean as a huge, red light. John Phillip (November 2001:162) recalled: "You know when the sun sets it's red? That's how Itqiirpak looks. They say it has that color." When she was young, Neva Rivers (May 2004:262) saw a flame on the horizon that she identified as Itqiirpak.

When there is a death on the ocean or if one is adrift, the wife of that Itqiirpak does indeed tell people. They said that if a person from Hooper Bay was going to perish, it would surface in front of the village as a fire. And if it appears farther north [of Hooper Bay], they will hear that a Scammon Bay resident had perished.

So I have seen that Itqiirpak as a fiery thing. It only lets itself be seen by a certain person. Not many people see it.

I saw one while we were living at Qauyagmiut, the spring camp, down at the source of Nuvuq. That fire down there was burning up like a big candle. Pretty soon its base seemed to be resting just like the candle. It was pointing toward land, but it was a fire.

Soon it went down, and then there was a beam of light going up. But it was a dim flame like a flashlight in fog. It was touching the ocean and going straight up. And then it slowly went out.

Then the second one [appeared], and I told my companion and he said that it was an Itqiirpak emerging. Though I never saw them before, I was scared and went back home when we were playing at night. My companions did not believe me.

Iraluq | Moon: "They Say the Moon Held the Weather"

Moonlight reflecting on snow was a significant source of illumination. The full moon always appears opposite the sun, so generally in winter, when the sun is low, the full moon will rise high during the night, providing abundant light. People also used the moon to predict coming weather conditions. According to John Alirkar (January 2007:420): "[By looking at the moon], they knew if the weather was going to be recurrently bad or good. When the moon progresses toward a full moon, if it is sitting on its back, they say that it is filled with coming bad weather. But if [the moon] is tipped and bent forward, the weather would mainly be good." Frank Andrew (June 1995:23) held another view: "Down at my home, when the new moon appeared on its back, people usually liked it. They said it appeared like that when it was holding things and was a sign of a productive month to come. And when it was standing straight up, they said it wasn't holding anything."

Paul John (June 1995:23) explained that the moon's position bore different interpretations: "People in different areas have different ways. Once I heard someone from the Kuskokwim talking about the moon's behavior. He said that when the new moon was on its back, it was a sign of bad weather ahead during the time when the moon cycle was ending. He said the new moon was holding upcoming bad weather. Down on the coast they believed it was holding upcoming good hunting and fishing. It was the sign of a plentiful food supply in the future." Wassilie Evan (December 1985) noted the same mixed message: "Sometimes when it was upright, they would say it was spilling out bad weather during that month. But if it was on its back, it was keeping the bad weather inside, and weather would be good. But some would say that when it was upright, it was spilling out animals; there would be plenty of game and fish."

Theresa Abraham (July 2007:509) described another lunar condition used as a weather predictor: "My mother asks me during cold weather if the moon has developed a ring around it. And when I told her sometimes during winter that it had a small ring around it, she said that it was about to get warm. But she said a very large ring indicated bad weather conditions." Theresa's mother is correct, as moon rings, just like sun dogs and halos, are caused by ice crystals in cirrus clouds and often indicate approaching warmer, windier weather.

Along with predicting weather, the moon was an important timekeeper. In fact, *iraluq* means both "moon" and "month." John Phillip (October 2005:329, 367) explained:

> This moon was our ancestor's *iralissuun* [calendar]. Because of that, they also call it *iraluq* [month] in Yup'ik. When the moon is full, they say *iralvagtuq* [there is a big moon], and it is bright at night. . . .

It starts this way and then becomes a half moon. It continues to progress, and when it turns toward us, they refer to that as *kingyarluni* [looking back], half of it is left until it's all gone.

Frank Andrew (June 1995:23) explained how people named the lunar cycle:

When the moon was just beginning to appear, they had a name for it, and when it was half full they called it *qupnginarluni* [its halfway point] and *muiqatarluni* [it's about to fill up]. And when it was completely full, they called it *tenguqmikarluni* [perhaps from *tenguqe-*, "to strain," as when defecating or lifting something heavy]. . . .

Just when it began to recede, they referred to it as *qessaqaurrluni* [from *qessa-*, "to be lazy"]. *Nulegluni* [from *nuleg-*, "to crack or split"] was what they called it. After they said that it started to become lazy, they'd say it had turned around to begin going back. When the moon travel cycle ended and the new moon appeared, they'd call it another name.

Months were named for different seasonal characteristics and activities—for example, Kepnerciq (cutting time), Iralull'er (bad month), Kaugun (hitting [of fish]), Ingun (molting [of birds]), and Cauyarvik (time for drumming). Frank Andrew (September 2000:181) gave a clear explanation for the names of the months in the Canineq (lower Kuskokwim coastal) area:

They said that Kanruyauciq [January, from *kaneq*, "frost"] was a time of extreme cold, and it stopped being windy. And the wild rhubarb and wild celery would become as tall as people because of the frost. They call it Kanruyauciq in my village for that reason. And among Akulmiut [tundra people], because it was harder for them to catch food during that time, they called it Iralull'er [bad month].

And during Kepnerciq [February] in my village, the weather was no longer cold. Snow melted, and lakes got deeper. They started to hunt from the ocean, and they began moving to spring camps. And back when they had tunnel entrances into the houses, they bailed out the water. When they were unable to bail anymore, they opened up the *kepneq* [house corner, from *kepe-*, "to cut off"] and went in and out through that. That is why they called February Kepnerciq in our village.

And March, Tengmiirvik [time that birds come], is a time for the birds to start flying, and they fly down in the States; they start to migrate this way.

And they arrive here in Alaska in April, and they start to be seen and fly around here in this area in Tengevqapiaq [actual time that birds fly]. And they get ready to nest.

When May arrives, they all lay their eggs. That is why it is called Kayangut Anu-tiit [coming of eggs].

June is the time when king salmon arrive. When they get caught in nets, they hit them on their noses and kill them. They call it Kaugun [hitting (of fish)] for that reason.

In Kangnit Ingutiit [time when one-year-old geese molt, July] birds that are getting old and those that no longer nest are getting ready to molt.

In August, when their babies are ready, the ones that nest molt and fly along with them. August is Aanat Ingutiit [time of molting mothers].

The Caninermiut who were cutting fish on the Kuskokwim River return home by rounding the bend [at the river mouth]. September is called Uivik [time of going around].

And the thin skin on caribou antlers quits growing and sheds during October [Amiraayaaq, "little shedding"].

When November came, the skins of their antlers started coming off; their horns shed their skin. November is called Amirairun [shedding of velvet].

When December came, they started their ceremonies with drumming, and they call it Cauyarvik [time for drumming]. Those are the meanings behind the names in my village out there. Some of the names are different in other villages.

Just as coastal, Kuskokwim, and Yukon residents experienced different weather patterns and seasonal activities, people from different areas did not always use the same name for the same month. In fact, different people frequently used the same name to designate different months. For example, in the Kuskokwim area, Iralull'er is November, Kanruyauciq is January, and Kepnerciq is February, whereas in the Bristol Bay region Iralull'er is January, Kanruyauciq is February, and Kepnerciq is March. Elders noted that sometimes these differences led to quarreling. Frank Andrew (June 1995:23) gave a particularly lively account:

People back in those days said that months didn't always come in the same order. Sometimes one of the men's months appeared earlier than the usual time. And one of the months came late.

When Kangirnaarmiut was inhabited, people used to tell a story. They said once at summer fish camp two old men suddenly began quarreling about their months.

The two old men were surrounded by rose hip plants. As the people watched, the two old men argued more and more. Then suddenly one of them poked the other with his walking stick, yelling that his month should come first. After the old man who was hit placed his walking stick aside, he stood up and grabbed the other man. They began to fight, and whenever they both sprang up from the ground, they'd clutch each other.

Then one landed on the other, grasping him. After the old man tried to free himself, he told the man on top, "Even though you have insisted on appearing first, I've ended up lying on rose hips [implying bravery]." After they fought, they reconciled.

They say sometimes months don't come in the usual order. Sometimes some of the moons appeared later than usual.

Frank Andrew (October 2001:296) later explained the argument's origin: "If there were five [months], they would be stated in this way. However, the other person's [months] would be stated earlier in the year. And this one would use a name for one that was before it. When they would squabble, they *iraluirautetull-ruut* [would argue over which person's name for a month was correct], saying that his peer's was not the right one. That happened frequently. They both wanted their knowledge to be the truth. They also had those types of people among them."

The moon was said to respond to human action. Paul Kiunya (October 2005:324) explained: "When the moon has an eclipse, they say that it is because the people below are going to go through famine or a catastrophe. They say that there is an eclipse because of the people."

In the past, shamans were said to travel to the moon.[12] Again, Frank Andrew (June 1995:23) gave an example: "They say that the old shaman Cunguqsur traveled to the moon periodically. He said that the designs on the moon, visible to the naked eye, were the shadows of mountains. My goodness! The shamans of that time were so speedy. They've said that the moon was very far away from earth, but shamans traveled to it in just minutes." Paul John followed with a story of how the irascible Bethel elder Chief Eddie Hoffman set non-Native officials straight on this point.

> The Standard Oil Company official used to come to Bethel occasionally. One time several of them arrived. Then Eddie Hoffman drove down to Standard Oil to fill up his oil truck. As the visiting officials were praising the final arrival of men to the moon, Eddie walked in the door. The white people continued to talk about the amazing men who had traveled to the moon, and they mentioned how many days they had traveled to reach it. As soon as they finished talking, Eddie spat out a cuss word. Then he yelled at the men and told them that for centuries Yup'ik people had already traveled to the moon, reaching it in five minutes. [*laughter*]
>
> Then the white men suddenly stopped talking and began looking at each other. And, of course, [Eddie] was yelling and talking loudly.

Ellam Cimillra Allrakumi | Seasons

Intertwined in discussions of the lunar cycle, elders shared abundant detail on the cycle of seasons—*uksuq* (winter), *up'nerkaq* (spring), *kiak* (summer), and *uksuaq* (fall, lit., "pretend winter")—and their movement from cold to warm to cold again. Paul Kiunya (October 2005:352) began: "Iralull'er [January] is the month that gets cold, and its coldness has a grip on all the land. Snow drifts, making the coldness worse. That month is bad. Kanruyauciq [February] isn't as cold as the one before it. And Kepnerciq [March] is the time that it gets warm within the year. It is so warm, it gets slushy and we walk in water." Bob Aloysius (October 2005:368) noted the upriver etymology for Kepnerciq: "In March we from upriver say *uksuq kep'uq* [winter is cut]. Kepnerciq is our name for March. Winter keeps getting cut, and soon there is no more winter."

Winter conditions did not always end right away. Paul John (March 2008:601) explained: "When cold winter conditions continue past their usual time in spring, preventing people from subsisting, they refer to it as *ellam agyaraa* [the weather is extended, from *age-*, "to go from one place to another"]. Although it is time for the weather to get warm, it cannot get warm. People were unable to obtain food when that occurred, and they couldn't go down to the ocean. When the weather stayed cold beyond its usual time, those poor people evidently experienced famine." At such times, they say, a compassionate person is revealed. Paul John (December 2007:353) explained: "After initially being referred to as *qunutungarliq* [a stingy person] during good times, finally during times of food shortage, the fact that they are capable of sharing with others becomes known."

John Phillip (October 2005:363) noted how activities changed with lengthening daylight: "People would begin to get ready in the *qasgi* during spring, preparing their kayaks before it began to melt. They prepared beginning in February and tried to be ready by March to go down to the ocean when the sun got warmer to face." John Eric (March 2008:200) also noted the warming power of the spring sun: "During [March], they said the following about the sun back when they hunted seals with kayaks along the ice edge. They said *uqrutat ilukeggiut* [the leeside of windbreaks have become comfortable]. Although a light wind was blowing from the west or northwest, some people would use canvas windbreaks. The windward side of the windbreak was bad, but the inner side, where the sun was shining, became warm and comfortable. One gets a tan and then goes up to shore." John Phillip (October 2005:112) also noted the effect of reflected sunlight on the human face: "When one is outside in spring the sun up there weathers [one's face]. When the sun's heat comes and changes our weather, the snow out there darkens a person."

John Phillip (October 2005:377) spoke of June breakup along the coast:

In June, people went to fish camps by moving from the coastal area when the path became clear. But the Qaluyaarmiut [people of Nelson Island], who lived outside our area, since they have earlier breakup, I would see them in spring when the ocean was ice free.

Before we got ready, a boat with a sail would pass by. My grandfather would say, "See those people down there! Those Nelson Islanders try to go upriver first so they can cut fish." They were the first to arrive and would welcome those who came after them since the shore-fast ice in their area goes away before ours does.

Roland Phillip (November 2005:99) spoke about the relationship between spring and fall, and how good weather in spring presaged bad fall weather and vice versa:

> I always tell others that the weather won't be good during fall if the weather was good during spring and not windy and rainy. The weather would be constantly good [during the end of spring], and conditions ideal for those who were cutting and drying fish, and the fish don't have a lot of moisture. When August approaches, it becomes rainy and the wind blows. From then on, since our village is on low ground, floods tend to come from the south. The wind mainly blows from the south all fall.
>
> If spring weather is good, fall weather will be bad. If spring weather is bad, fall weather will be good. Those two constantly switch.

Elders also described fall freeze-up. According to Paul Kiunya (October 2005:356): "In my experience, it freezes in November. During the end of October it gets cold, but *angitaqluni* [freezes and thaws, lit., "it loosens"]. It also gets cold and warm in November. It only freezes and thaws when there are no more mice." Nick Andrew (October 2005:358) explained: "They say when the mice dig [dens] early, the weather will freeze and thaw. When there are lots of mice running around, it means that they haven't finished gathering their food for winter, so it won't snow soon. As soon as they disappear, it will get colder and freeze."

John Phillip (October 2005:369, 114) spoke of the early onset of winter in October:

> Sometimes when it tries to become winter right away, it freezes within the first week of October.
>
> Then sometimes it becomes summer again and continues to get warm and then cold, back and forth. It would get very watery in the wilderness when I traveled home after hunting, and sometimes my dogs would walk in water. . . .
>
> When it's going to freeze and thaw, there are many voles. They used those [voles] to indicate that sporadic weather was coming.

Bob Aloysius (November 2005:113) characterized male and female winters: "When the weather got cold [in winter], they presumed that it didn't have a female. So they called it male. But when weather stays warm and doesn't get too cold, when it *kiagyugagami* [lit., "tends to want to become summer"] and is slow to become winter, they seem to call it a female [winter]." Frank Andrew (September 2000:203) noted: "People used plants as indicators. When these plants grew thick in large quantities before winter, they said that the land was insulating itself and that it was going to be cold all winter long." John Phillip (October 2005:114) added: "When lots of grass grows, *maqarqelluni-gguq* [they say (the land) is putting on warm clothing]."

Amirlut Qilak-Ilu | Clouds and Sky

Clouds were watched as closely as the sun and stars. Sophie Agimuk (January 2007:472) of Toksook Bay recalled: "From time immemorial those Yupiit paid attention to the weather, and they knew by the clouds when it was indeed going to change." Sometimes clouds indicated moisture rising from the land. According to John Eric (March 2008:185):

> This [advection fog] — when the sun is just appearing along the horizon, the land is still visible but appears as though it is covered by something. People in the past mentioned that it is moisture from the land. It also indicates coming snowfall.
>
> And when the land's moisture rises higher in the sky, they say it is gradually lifting. That's what my grandfather Amaqigciq and those younger than him said about it.

John Eric (March 2008:193) also recalled the adage that fish breath increased the sky's moisture: "You know when fish open their mouths constantly or flounder gather in shallow places when they are abundant. They say when there is an abundance of those, their breath adds to the moisture and becomes a cloud. But people talked about these clouds and said that when [fish] relentlessly breathe from underwater, the sun has a difficult time [getting rid of the moisture], even though it's sunny."

Amirluqtaat (scattered clouds, stratocumulus) are said to indicate *quuneq* (calm weather) throughout the day. John Eric (March 2008:186) said: "When [clouds] appear loose or scattered and are clearly visible, they say the weather will stay calm all day. My wife's deceased grandfather also told me that if I paid close attention to this type of clouds in summer, although I had no one to inform me about the weather, I would see what the weather would be like the next day and the day after." Edward Hooper (March 2007:1377) added: "In the past, when the weather became warm, calm weather approached from the direction of the sunrise. Clouds that were extremely scattered would arrive here, and then it would get calm."

Cagnilriit (cirrostratus clouds, from *cagni-*, "to be taut"), on the other hand, indicate coming wind. Lizzie Chimiugak (January 2007:473) noted: "When clouds look like they've been smeared with a pencil, they say that they are taut." John Eric (March 2008:187) explained: "People in the past said about those [clouds], 'My, how taut they are.' The wind is causing them to look like they are stretched like rope pulled taut." Nick Andrew (October 2005:278) contrasted high, stretched-out clouds preceding wind with high cumulus clouds indicating calm weather: "Although the weather is good and windless, the clouds begin to stretch out high in the sky. After a while, it starts to get windy. Then sometime later, even though it is windy and the weather is bad, the [clouds] begin to roll. That means that the weather will improve and won't be windy. Sometimes in spring when it's calm there are very white, pure clouds that roll and are high. That's how they predict that there won't be wind."

Cloud color is also significant. Looking at the photograph of stratocumulus clouds transitioning to cumulus (see color insert), John Eric (March 2008:187) commented: "When Nagyuk's father came from Kipnuk, he said, 'How warm the weather is.' The clouds had this type of blue hue, but here it was winter and there was a lot of snow around." The appearance of dark clouds known as *tenguguat* (lit., "pretend livers") indicated coming wind. These are possibly lenticular (lens-shaped) clouds, which form over mountains as winds start up and can stretch downwind like a series of waves.[13] Roland Phillip (November 2005:106) noted: "I look at the sky. If the clouds are dense, our elders say *ellam ilua tenguguartuq* [the sky above has *tenguguat*]. When the clouds are not gathered together, they say it won't be calm weather. Those *tenguguat* that aren't very large indicate wind. Although they don't indicate strong winds, it isn't calm." Theresa Abraham (March 2008:196) added: "When it's about to get windy, those *tenguguat* appear above those large mountains [on Nelson Island]." When Paul Kiunya (October 2005:327) asked Kipnuk pilot Andy Fox about *tenguguat*, Andy replied that they are not good as a path for planes: "If he goes through them, his plane will dance [experience turbulence]."

Joe Lomack (May 2003) recalled another dark cloud and its associated adage: "We were told to take good care of our wives. Never to be like an *iqalungaq*, a dark cloud that is terrible to see." Dark clouds over the ocean can indicate approaching wind, while smooth clouds presage calm weather. Paul Kiunya (October 2005:124) explained:

> Long, dark clouds that appear from the direction of the ocean, high in the sky, are signs of [approaching] wind. Sometimes it passes above us, and sometimes it hits the next day.
>
> Sometimes *uyunguryak* [haze, heat waves] appear down on the ocean. They say when they form and are smooth, they are signs of calm [weather]. Even if it was windy, our elders said it would become calm.
>
> And sometimes when it was very calm while we were down on the ocean during spring, after a while, the ends [of the sky] would turn red. That apparently happened when it was about to get windy the next day.

John Eric (March 2008:191) noted that during summer clouds with a dark bottom and a light top indicate approaching rain. When the cloud bank darkened as the front moved in, John recalled: "People used to say, 'The [sky] up there is getting heavy and smooth. Since it is getting smooth, it will start raining.'" Roland Phillip (November 2005:106) also associated dark clouds with wind and rain: "Wind is especially predicted through clouds. A dark cloud that is extended along the edge of the northern sky [indicates wind]. The north especially indicates coming weather conditions, or even the area above the ocean." Sophie Agimuk (January 2007:472) noted the converse: "Even though it's windy, when the cloud's underside is not disturbed, the weather will get calm. And when the cloud's underside gets suddenly dark, the weather will become inclement."

John Phillip (November 2001:162) described how open water reflected on the underside of clouds indicates coming bad weather:

> Then *qiugaar* [reflection of open water, blue hue] shows a picture of the *kangit* [bays]. It's very blue. Then slowly the *qiugaar* gets cloudy. After a while conditions down toward the ocean get bad, and the water shows it. That is a sign that the weather will get bad. Bad weather continues to come closer.
>
> You will wake in the morning and find it windy. Soon clouds approach us. Then they say the wind is being fed. It is feeding it to be stormy.

Roland Phillip (November 2005:109) described how clouds moving inland from the ocean indicate approaching bad weather: "Sometimes in spring I go out in the early morning when it starts to get bright, and it's very calm. I see clouds moving either from the ocean [west] or from the north. Those [moving clouds] indicate that bad weather is coming before day's end and that the wind will pick up. Sometimes I don't travel to the ocean, since I know what it means when clouds are moving fast. Then when [the clouds] lower and disappear, [they indicate], 'You there, go ahead and go [hunting].'"

Edward Hooper (March 2007:1374) mentioned that small clouds can indicate weather change on Nelson Island:

> When a small, dark cloud along the sheltered side of the mountain appears and disappears, that indicates really bad wind. In half an hour [wind] will arrive.
>
> The small clouds that indicate weather change are still used today. A small cloud will form, and then we'll see that it has disappeared right away. Then it will form again. That indicates wind and that a person shouldn't stay where he is [but head home], but at the time the weather is calm [and doesn't appear as though it will change].
>
> Clouds tend to form quickly in these mountains and above our village. When seal hunting, they are visible from down on the ocean. One shouldn't stay [but head to shore].

John Phillip (October 2003:65) also described how quickly moving clouds predicted wind on the lower coast: "Sometimes when you watch the clouds, when it's going to be windy, they travel very fast. When that happens, they say *takner-aarqatarniluku* [(it's going to be windy) for a long time]. Soon they travel slowly, even though it's windy. And they say that it won't be windy for long, when the clouds slow down."

Wassilie Evan (December 1985:108) described clouds predicting wind on the middle Kuskokwim River:

> The places where some people live are surrounded by mountains. Even if it's very clear, a cloud would begin growing on top of that mountain. And the cloud would meet the wind, and as the cloud grows it would begin stretching. And sometimes the cloud would be black. They say that cloud indicates that strong wind will come from that direction.
>
> In summer when the cloud comes down, they'd say that the weather would be bad for several days. But in winter when the cloud was black, they'd say it was going to snow. Some days when that cloud grows, that mountain would appear smaller, a condition they call *uyungerrsirluni* [from *uyunge-*, "to squat down"]. It is saying that a strong wind is coming. Sometimes that mountain would become visible and clear. They said good weather was coming.

Nick Andrew (October 2005:199) noted that a cloud bank along the coast indicates approaching cold weather:

> Downriver around [Marshall] on some evenings, the area toward the coast will have large distinct clouds, straight from the lower coast toward the north. It is thick and looks like fog, but it's not. It seems like a sharp cut and looks like a wall.
>
> Those of us from between the Yukon and Kuskokwim Rivers who go to spring camp in the tundra area use those to predict [coming weather conditions]. They say that when the *caniun* [cloud bank toward shore] is thick, it will be cold for a long time. When the *caniun* is not thick, they say that it's going to melt and become summer.

John Phillip (October 2005:387, 388) described how clouds in the south indicated approaching bad weather:

> On the lower Kuskokwim coast it will be windy from the north. Then we will see a cloud down there in front of [the wind], and although it's windy, [the cloud] will stand and counter the wind. Then the wind will get stronger. When that happens, they say that [the cloud] is trying to counter the wind. Sometimes the cloud will overcome it, and when it eventually arrives and covers the sky, the weather gets bad. . . .

When the cloud appears [from the direction of the ocean] and covers the sky, heading north, they say it supplies the north with the weather that it will provide. Even though [the wind] goes against it, [the cloud] goes up by itself, and bad weather comes.

Traveling on the lower coast in 1897, Moravian missionary John Kilbuck experienced the rapid onset of bad weather that John Phillip described: "[My guide] directed my attention to a hazy streak along the horizon, and said that I should watch, for it was rising towards the zenith, and that soon we would see it turn black. He then told me to look at the sun. I did so, and I could see far up in the air bright particles flying almost in a straight line. I looked at the horizon again, and there, sure enough, was a dark cloud rising higher and higher rapidly. Without delay we turned back. . . . For almost two days the storm raged terrifically."[14]

The color of *qilak* (sky) also indicates changes in the weather: "Keep an eye on the sky. In winter it will look pale when it's going to be cold. Even though it's cold, when the sky is dark blue, the coldness will be less intense" (Nick Andrew, October 2005:386). Peter Matthew (March 2008:160) of Chefornak noted: "If the sky along the horizon starts to get light after being dark blue in winter, they said that it would start snowing. But these days it doesn't really snow real snowflakes but only *nungurrluk* [ice fog]." Paul John (September 2009:351) added an explanation of *ingna*: "We Yupiit call the area between the east and south *ingna* [lit., "one over there"], the place where bad weather grows if it is going to be inclement. When it turns from blue to white, that means that bad weather is coming. That's why one can tell at dawn that bad weather will arrive by afternoon."

Phillip Moses (January 2007:471) gave an example of a dark sky indicating approaching wind: "One thing that never changes is the mouth of Akuluraq [Etolin Strait]. When the sky gets dark in spring, it will get windy and dangerous. Even though they are experiencing good weather, those who are hunting would go toward land." John Phillip (September 2009:350) explained the significance of a rough edge at the horizon: "When weather will be bad, the blue sky in the direction of the ocean is no longer straight. Approaching rain becomes evident."

Anuqa | Wid

Many elders describe the calm, windless weather they enjoyed in their youth, especially in spring. Frank Andrew (September 2000:173) recalled: "As I observed the world, there wasn't wind during spring in our village. During spring the ocean would be like the inside of a house. Some days, there was no wind from any direction. It got a little windy here and there. And when the emperor geese arrived, it would be hot and windless." Peter Jacobs (October 2003:119) recalled: "When the wind was always calm in winter, there was a saying. They said the wild rhubarb [covered with frost] would get as tall as a person because there was no wind."

Winds have names corresponding to the directions from which they come.[15] North or north wind is *negeq, negeqvaq,* or *piakneq* (from up there); south or south wind is *ungalaq* or *qacaqneq;* east or east wind is *calaraq, keluvaq, kiugkenak,* or (on Nelson Island) *ungalaq;* and west or west wind is *kanaknak* or *keggakneq.* On the Yukon, a southeast wind is *kiugkenak,* and a southwest wind is *uaqnaq* or *ungalaqliqneq.* John Phillip (October 2005:279) noted: "When the wind is coming from the area between [north and west], we say, '*akuliignek anuqlirtuq, negeqvankuk akuliignek*' [it is windy between those two, between the north wind (and the west)]."

Winds of different strengths and speeds also have names. John Phillip (October 2005:280) said: "Wind speeds are different. They would say *anuqsa'artuq* [it's breezy] or *anuqlirtuq* [it's windy]. Sometimes, when it has been foggy, the wind blows away the fog." Winds of different speeds make different sounds. David Martin (January 2002:130) noted: "They call the whistling of the wind *culu'uggluni,* and it can be heard inside the house. The sounds are different, some loud and some quiet." Winds even had different shapes. John Phillip (October 2005:389) recalled: "They call the wind that does this circular motion *ull'uyaq* [whirlwind]. It goes around and blows things upward."

John Phillip (October 2005:280) noted that along the coast, prevailing winds are from the north and the south:

> *Negeqvaq* [north wind] and *ungalaq* [south wind] are the only strong winds. When it's windy from the south, there is a strong snowstorm. They say that it floods because the strong wind pushes [the water] and fills the Kuskokwim River. Places downcoast from us flood because the wind is strong.
>
> In my village, they built the airport running north to south. Airports are like that along the coast; they face north and south, the most powerful wind directions.

Roland Phillip (November 2005:109) explained how strong north winds blow for a number of days:

> They warned us that the wind continues blowing when it comes from the north. But they say it doesn't usually extend beyond three days, although the wind sometimes blows from the north for a long time.
>
> The south wind doesn't usually blow for a long time. And in summer, the south wind isn't dangerous at all. I'm not afraid of it because if my motor breaks down, the south wind will drift me ashore. But the north wind can drift me out to sea.

Edward Hooper (March 2007:1377) cautioned: "They said that one mustn't think the wind will get calm when it comes from the north; they say that stronger wind is approaching." Frank Andrew (September 2000:203) noted: "They said when the south wind blew, it was gone in less than two days, but the north wind goes on longer."

A south wind could be a real gift during fishing season. Paul John (March 2008:457) explained: "When herring first arrived in Tununak [in spring], they say that there used to be a south wind that [Edward] Hooper's grandmother bequeathed [the people]. They say that before she died, she mentioned that although she wanted to bequeath something to her grandchildren, she was worried that she had nothing to pass down to them. She told them that she wanted to give them a wind that they would have, so that they could fish during good weather." Joseph Jenkins (May 2003) added: "*Ella-gguq ag'aqami keggaknermek anuqsaaralartuq* [They say when the weather extends beyond its usual time, a breeze will blow from the south]. The south wind is a good wind. At the ice-fishing places at Cuukvagtuli, when the wind is blowing from the south, ice particles continually appear from underneath, filling the water."

Conversely, Paul John (March 2008:594) shared a story of the north wind's power: "One time, many of us were traveling by sled from Chefornak [to Toksook Bay], including Canaar, who was carrying Nick Canaar and his deceased younger sister Qimagaq in his sled. He was traveling behind us the entire time, and we came upon wind. After a while, the one behind us yelled out, 'You up there, the wind has blown the two that I was carrying away!' When I looked back, I saw Qimagaq and her older brother being blown across the lake by the wind. [*laughter*]"

The north and south winds leave different marks on the land. Whereas the north wind creates snowdrifts called *iqalluguat* which can be used as landmarks when traveling, the south wind smoothes the land. John Eric (March 2008:199) explained: "Since the south wind is moist, when the snow is constantly blowing along the ground, it will fill these [gaps in the snowdrifts]. Since the south wind has moisture and is warm, it smoothes the rough snow, and even hilltops are very smooth."

Although potentially good, the south wind, they say, is a liar. Nick Andrew (October 2005:387) recalled: "While it is windy from the south and improves they would tell us not to leave. They say that the south wind is a liar. After pretending to get better, the weather gets bad again."

East and west winds also had predictable effects. John Phillip (October 2003:119, 2005:111) explained: "They say that the east wind in spring *kiagyuuguq* [brings *kiak* (summer)]. In spring the west wind *uksuryuuguq* [brings *uksuq* (winter)], and it will be cold and it doesn't melt. . . . They say that in fall the west wind *kiagyuuguq* and the east wind *uksuryuuguq*, helping to go toward winter. They switch like that."

Wind in spring was also said to bring animals. Frank Andrew (October 2003:119) recalled: "When there was mainly east wind, the elderly men would say, 'Maybe there will be many animals this spring.' The [sea mammals] like to face the wind in spring." John Phillip (October 2005:297; 2003:57) commented: "Also in spring, even in winter, sometimes when wind blows from the north, they say that it is bringing animals. . . . It is true that baby sea mammals go up near shore, facing the wind. The ones that reach shore ice for the first time stay for a while." John Phillip (October 2005:284) also mentioned that hunters in search of sea mammals would likewise face the wind.

> If the [wind] isn't following the shore and comes toward your face, they say *qacartaa* [it blows against it]. No matter what direction it's coming from, if it's blowing against that and facing it, *qacarrniluku* [they say that it is blowing against it].
> Sometimes when [sea mammals] surface in a place that the wind is facing, [hunters] stay at those places, even though the wind blows against it. Although it's uncomfortable, they stay there trying to catch them. Sometimes animals like to surface in places that the wind faces and is blowing against.

John Phillip (October 2003:57) talked about the relative merits of different winds in opening or closing the sea mammals' trails: "Sometimes in spring when the wind blows from the east, they say that it keeps the area toward Bristol Bay open [ice free], making the animals more abundant. They are eager for the opening of the ice because it is their trail. These are conditions which allowed them to be abundant. But they don't like it when the wind constantly blows against [the shore ice]. When the ice piles, the animals stay farther away and are unable to go up to shore."

At times the wind follows the shore: "If a person wants to hunt, he will ask, 'How is the place where you are staying?' When the wind follows the ocean shore, the edge of the ice, they say *cenirnirluni* [(the wind) is following the shore]. Even though they don't mention the [wind direction], they say that the wind is following the shore" (John Phillip, October 2005:283).

Wind was also said to blow fish into rivers in spring and summer. Nick Andrew (October 2003:119) explained: "In spring on the Yukon River before the fish arrive, if the wind continues to blow from the north, they always say that the Yukon would get fish from the north. When fish arrive, the king salmon are small and their backs look blue. They say that those fish are from the north. But if the wind had blown from the south, they say that there will be lots of fish, that it will bring fish in." Matthew Beans (December 1985:109) noted: "There are two sets of fish on the Yukon River. When there's lots of south wind, the salmon are usually big. They say that they are the ones which come into the river on the south side. When there is hardly any south wind, they would not be big. They call those *'luqegglirviutaat*." Lucy Sparck (October 2005:296) recalled what Chevak elders shared: "In my village, they said that the winds pretend to catch fish. Then when winter ended, they said the wind would bring logs."

Peter Jacobs (October 2003:119) noted the wind's effect on the Kuskokwim:

> When I was young, there were two people originally from Bethel. Those two would say that when the wind was blowing from the south, the king salmon were swimming, and that when the wind suddenly changed to the west, it would bring in the king salmon. The wind shakes the fish. They say that wind would blow the fish away, too.
>
> This spring there wasn't much water. It seemed that when the wind suddenly changed to the west, the water began to enter. Then my wife said to me, "The fish are pushing the water." When they arrived, there were lots of fish.

John Phillip (October 2005:285, 297) described how wind brings fish into rivers in the Canineq area:

> For us who live on the lower coast, when the wind is coming from the south and blowing against us during spring and it floods, it brings our fish, the tomcod, far upriver in great numbers. . . .
>
> When it was windy like that in spring, they say that it is bringing the king salmon. The fish at the mouth of the Kuskokwim River, the king salmon, happen to be around in winter before they go upriver. They speak of *cikut aciirturtaitnek* [those that went under the ice] or *aciirturtet* [first run of king salmon under the ice]. The fish are always ready at the mouth of the Kuskokwim River, even in winter.

Winds in fall have always been associated with flooding on the lowland tundra of the Bering Sea coast. Sophie Agimuk (January 2007:529) explained: "Long ago, that tundra used to flood. When it started to blow from only one direction, these people said, 'If it starts to blow from the south, it will be dangerous and a flood will be inevitable.' Truly, if it has been blowing continually from the north, then the water is ready to come up, and it will flood when it switches to the south wind." John Eric (March 2008:251) noted how these floods cleaned both land and sea: "When all the birds [have left] this coastal area, when the wind starts to blow from the southeast, they say that it is cleaning the ocean. And they say that it cleans the bird feces too. After flooding, when going down to the ocean, it's clean and the bird feces are gone. That occurs in late August."

Occasionally, flooding occurs after freeze-up, and in the past these dangerous storm surges could inundate low-lying homes. John Eric continued: "Once in a while it flooded extensively after freeze-up. It would move the river ice down along the coast, and ice inside lakes was moved to other areas. Sometimes [ice chunks] the size of tables were on the tundra. When it floods, it covers the coastal area."

John's stepmother, Maria Eric (March 2008:251), said that she had experienced two big floods in her life, one while she was living in the old village of Cevv'arneq:

> When we woke, I went outside and saw that the ice along the Cevv'arneq River was situated across from our village, lined up and down the river. The ice had been brought up to land along low areas. Water was spewing, making a loud noise, and it looked as though someone was spilling it.
>
> We experienced [flooding] at Arayakcaarmiut also. Although the place we were staying was higher than other areas, we went inland with sleds when the water started to rise.

The water's condition indicated approaching wind. Peter Jacobs (October 2003:56) noted: "They say when the water and lakes look dark, it will get windy." Peter John (March 2007:1200) of Newtok recalled the converse: "When [my uncle] was about to take me to the ocean to hunt, sometimes when it was windy my heart would pound, not wanting to go. After looking around, he'd tell me that the water was turning light and that we should go because the weather would get calm and windless while we were there. And just as he said, when we'd get down to the ice, after a while the weather would start to get calm."

Ruffles on the water's surface were also "the wind's work." John Phillip (October 2005:388) explained: "Sometimes even though it's calm, there will be a small wake on a pond that looks like a fish. They say that those small wakes belong to the wind. The small wakes move, even though there's no wind." Frank Andrew (September 2000:203) said: "Those streaks on the water's surface that look like small fish quickly taking off are also indicators that wind will suddenly hit. First it gets a little windy, then suddenly becomes calm, and then a strong wind arrives."

Nelson Island elders explained the behavior of wind around their mountains. Lizzie Chimiugak (January 2007:477) noted how variable the winds are on different parts of the island: "They say that some areas might be windy but some parts would be calm." Phillip Moses followed with some examples: "Sometimes the people of Tununak say that they are windy, but here [at Toksook Bay] it is really calm. Their winds are different. And if [Toksook Bay] continues to be calm, and Nightmute is windy, some at Nightmute doubt [Toksook is calm] when they keep having violent winds day after day."

Nelson Islanders paid attention to their mountains to predict coming weather conditions. Clouds gathering in a valley to the northeast of Toksook are a well-known sign of approaching wind. Lizzie Chimiugak (January 2007:482) explained: "Sometimes clouds gather in that source of Al'auciq, even though the weather is calm, and then this place gets windy. But even when it's windy, when [Al'auciq] clears, the weather will get calm. It is a real weather predictor." Paul

John (March 2008:594) said that clouds gathering at Al'aquciq could also indicate coming snow.

Paul John (March 2008:597) noted that clouds forming on the mountain-tops predict wind: "Down the coast, when clouds start to hang over the mountains, it means that it can get windy. They refer to that as *ciknengluteng ingrit* [something closing in or descending over the *ingrit* (mountains)]. That was their sign not to stay on the ocean." Simeon Agnus (July 2007:505) described how mountains were used to predict wind on Nelson Island: "A small cloud will grow along the sheltered side of our mountains down there, the Nuuget, and it will immediately disappear although it's calm and not windy. That means that it's about to get windy." Simeon (March 2007:529) also noted: "They observe Ugcirraq [Cape Vancouver]. When they see a cloud descending from [and covering] its top, they know it will get windy. The [Nightmute point] does that also. When our point here is seen from Umkumiut, it stretches out when the weather is fine. And when it's windy, it shortens." Chefornak residents also commented that clouds over Nelson Island indicate coming wind. Frank Andrew (October 2003:134) described the same phenomena in the Canineq area: "The top of those hills out near the ocean is a good indicator of wind. When the wind is going to start blowing from the direction of the ocean, the hill that is visible down on the coast gets lower and its slopes disappear."

Clouds are not the only indicators of coming wind. Simeon Agnus (July 2007:505) described using a jet contrail: "The trails that the jet took up in the sky are dense or tight. They can use [the jet contrail] to predict coming weather conditions. When the jet trail disappears, the sky is windy up there. But when it's calm, they are visible for a long time." Simeon (December 2007:57) also noted: "When butchering an adult bearded seal, when it is split down the belly and it suddenly sprays [blood], it means that it is going to get windy for sure."

Another indicator of windy weather in spring and fall was *inivkaq*, a superior mirage—an object appearing higher than it actually is—which is the result of temperature inversion, when light is bent, or refracted, as it travels through layers of air with differing densities. *Inivkaq* forms when warm air moves in above very cold air on the ground. Paul Kiunya (October 2005:167) noted: "When there's *inivkaq*, all the bays become visible in the sky; they rise. Cliffs that are a great distance away are also clearly visible, and mountains and hills appear elevated." Roland Phillip (November 2005:102) said that both mirages and "the pouring of the sun" presage windy weather:

> *Inivkaq* occurs during spring and fall. Hills or steep things on our land get higher, and even [distant] villages are visible.
>
> They say *inivkaq* is a sign of bad weather or wind during fall. And another sign for bad weather is *akertem qurrlurarallra* ["the pouring of the sun," indicated by

visible rays coming down from the sun]. They say rays coming down from the sun during fall indicate coming bad or windy weather. But they say during spring's end, *inivkaq* is a sign of good weather. And *akertem qurrlurarallra* is also a good sign during spring. When *inivkaq* occurred in late spring, I was so eager to go out to the ocean, as it would be calm all day.

These are things that my father spoke of long before I started to observe them. When I became observant, when the weather was the way he predicted, I'd say, "That is what my father said would occur."

John Eric (March 2008:156) noted that on the coast south of Nelson Island *inivkaq* presaged wind from the south: "After the weather had been cold for a long time, when *inivkaq* occurred they said that the weather was about to change and that south-wind [warm] weather conditions would ensue." Although people still see *inivkaq*, it is no longer considered an accurate indicator. According to John Eric (March 2008:154): "These days, although that occurs after a period of cold weather, it is no longer used to predict weather change. And although Nelson Island up there [to the north] rises because of the mirage effect, the weather out there never changes."

Elders noted that when the mirage quivered, it indicated windy weather. According to Nick Andrew (October 2005:169): "That *inivkaq* moves sometimes, like heat going up, when it's going to get windy. But if it's not going to be windy, it only quivers a little bit. Back home they call it *uyunguryirluni* [heat rising, mirage], like when you look at a fire and see heat rising." John Phillip added: "They also call that *uyunguryilria* on the coast. We see *uyunguryak* [heat waves] during summer and spring."

People also paid close attention to changes in wind direction. Paul John (December 1985:103) explained: "They call this condition *angialuni* [from *angi-*, "to become loose, unravel"] when it goes back and forth. They said that it was a sign of bad weather. If the wind is from the east and turns toward north [i.e., begins to come from the north], they say the wind is fetching bad weather. *Ella-gguq tua-i pairrluku* [They say it is meeting *ella*]. But if it goes toward the west, they are glad. The weather would be good. They say *ella* will not go to it." David Jimmie (March 2008:161) also described the wind turning: "When a wet south wind turns toward the ocean [i.e., begins to blow from the west], they say that the wind has turned in a good direction and [the weather] will improve. But when it turns from behind our village [i.e., it begins to blow from the east], [the weather] won't improve right away. They say that the wind has turned in the wrong direction."

John Phillip (October 2005:384) described how people liked the weather and wind that came from the west:

Then sometimes when it was cloudy, the west got brighter first. When that happened, they said that good weather came from that area. They were thankful when clouds cleared from the west. They said that when the center cleared up first, it removed its hood. And sometimes the cloud, the bad weather, heads down, and the weather clears and becomes good. They said that when it was from the west, it was the place where *picirkaq* [whatever would be] was coming from, whatever the weather would bring. They liked that when it happened.

Wind from the west isn't very common. And the weather is hardly bad when it's windy from there. The wind isn't very strong and doesn't blow for a long time. Whenever I build a home, I face the door west.

Ivsuk wall'u Ellalluk | Rain

//

They say when it rains it gives the land water to drink. —John Phillip, Kongiganak

The Bering Sea coast experiences rain as well as wind, especially during late summer and early fall. Rain is known variously as *ivsuk*, *ellalluk*, and *cellalluk*. *Minuk*, *ivsirrluar*, and *kanevvluk* designate light rain or drizzle, while *anngiinaq* (lit., "one that keeps going out") is a light rain that doesn't quit. A hard rain is *tup'agtuq* or *ivegpak*, while *tagevkara'arluni* is rain that gets bad and then better. And morning dampness or dew is *yukutaq*. After rain, *agluryat* (rainbows) are frequently seen in the summer sky. John Eric (March 2008:192) noted: "If there is moisture, one of these [rainbows] will form."

Endings on *ellalluk* and *ivsuk* can also distinguish between hard and light rain. John Phillip (October 2005:128) explained: "When it's really raining, we say *ellarvagtuq*. The water pours hard, and really makes you wet. They say *ellallirtuq* [it is raining]. And when it rains a small amount, they say *ellalliksuartuq* [it is raining a little]. When there's low visibility and it's wet out, we say *minirpagtuq* [thick drizzle]."

Edward Hooper (March 2007:1378) described spring rains in the past: "They also mention that the first rain that occurs following winter comes with calm weather. If there is heavy rain, warm weather will follow. It no longer does that these days."

Elders described how wet, rainy weather prevented traveling. Peter Jacobs (October 2003:61) reflected: "When it really rains in winter or spring, it gets windy and wet. They told us that bad weather would follow the heavy rain, and they said not to travel until afterward. They said that when it was bad again, it wouldn't last as long as the first one."

John Phillip (October 2005:292) described both steady rain and hard rain: "When it's going to rain for a while, the clouds look smooth, and the sun doesn't show. And sometimes the sky shows rain in the distance. A dark cloud arrives and it starts to rain hard, getting everything wet. When rain comes from the south, they call it *tagevkaq* [to head up (to shore)]." In summer, hard rain can be accompanied by *kalluuk* (thunder and lightning). John Phillip (October 2005:294) shared a colorful adage: "Sometimes you can tell when there will be thunder. The sun gets very hot. Then we will see large clouds, and some will be dark. Then it will get darker outside, and it pounds with thunder and begins to rain hard. When rain begins to pour, they say that the thunders are urinating."

Thunder is said to "drip" lightning, and the "thunder's drops" can cause tundra fires. John Phillip (October 2005:295) continued: "Some time ago, over in my village, there was heavy thunder in summer when people were at camp. After a while, when it was getting dark and lightning hit down there, that thunder dripped [lightning hit]. It hit the water inside the Ilkivik River, right below the place where those people were camping. It began to boil where it hit and boiled for quite a while. They call it *kucirluni* [dripping]. Some of the lightning fell there." Lightning can also kill. Lizzie Chimiugak (January 2007:623) recalled: "They suspect that the late mother of Mariq was killed by lightning. She went to pick berries before a thunderstorm and failed to come home. So they found her at that place, and she was dead. This happened long ago."

Elders said that thunder in the past was stronger. Sophie Agimuk (January 2007:621) recalled: "It used to be really powerful. When it thundered, the land did not stay still. It has gotten weaker at the present time." Phillip Moses added: "It did that when we were living down at Umkumiut. It really used to roar. It was scary. Nowadays that does not occur as much." When non-Natives harnessed electricity (also called *kalluuk*), some said that they caused the thunder and lightning to diminish. Lizzie Chimiugak (January 2007:623) concluded: "They claim that white people have taken *kalluuk*. They suspected that was why it became weak. Because they keep taking it, it probably is getting scarce."

Nick Andrew (October 2005:295) recalled: "I know of only one time when thunder happened in winter. There was a shaman named Anguqtaryaraq who died recently. When he was getting sick, he said that when he died he would be a guide to the thunder. Then when he died in winter it thundered. Because it was probably easy for that shaman, he wanted to drive [the thunder] when he died."

Taituk | Fog

When I was young we would take our shotguns and go hunting when it was foggy because the birds would fly so close to us.

—Nick Andrew, Marshall

Taituk (fog) was also a common condition. Paul John (October 2003:115) likened fog to the breath of bearded seals: "During hunting season in spring, they said that the breath of the bearded seal that someone caught had come to life." Peter Jacobs agreed: "That's what I heard, too. People on land would say when it suddenly became foggy, 'The people down on the ocean caught a bearded seal!' They held that responsible for [the fog]."

John Phillip (October 2003:95) described how the sun can dissipate fog: "Sometimes we would get up with fog rising from lakes. When it is foggy I see the sun barely showing. And the sun begins to show as it begins to heat up, and [the fog begins to disappear]. The sun dissolves it." John Eric (March 2008:194) commented: "The sun's heat burns [the fog]. When there is fog along the land in summer, when the land's surface starts to heat up, the land also helps the sun [make the fog disappear]. As in spring when the sun starts to heat up, there is *uyunguryak* [haze, heat waves]." Breezes can also disperse fog. Peter Jacobs (November 2001:162) recalled: "I would get up in the morning and it would be calm and foggy. Soon a light breeze would come. The old men would say, 'Look, it will clear the fog away and push away our bad weather.'"

Traveling in fog can be hazardous. John Phillip (October 2003:95) explained: "Traveling in fog is like traveling in a snowstorm. Sometimes a person dies when lost in the wilderness in fog." Paul John (October 2003:94) compared traveling in fog to traveling in high winds: "When it is so foggy that you can't see anything, they say fog is like high wind because you can't tell which direction to go. Truly, even if it isn't windy, when you leave the shore-fast ice, one who doesn't pay attention will get lost."

John Phillip (October 2003:94) noted that fog is vapor and one gets damp traveling through it. Paul John (October 2005:201) explained that fog can originate from either the ocean or the land: "Sometimes, fog grows from the ocean, and sometimes it grows from the land up there. It is noticeable when the [horizon] suddenly turns white, and then that white area approaches and arrives. You could tell when fog is building; fog is noticeable although it's far away." They say fog that arrives from the ocean won't easily dissipate. John Phillip (October 2003:94) explained: "That is what they say on the lower coast. Even in winter when it grows from the ocean it is foggy for quite a while. Sometimes when it is foggy and becomes windy, it blows the fog away."

Hunters were advised to return to shore if fog came from the ocean. John Phillip (October 2005:202) observed:

> Although it's foggy these days, our young men aren't cautioned not to go; they just use those GPS [Global Positioning System], the manufactured shamans of the white people, and travel.
>
> But back when we didn't have compasses, that [fog] was dangerous; sometimes it was foggy all day when it came and covered the area from the ocean and there was a light wind. Back when we'd travel with kayaks, we'd go to shore because it wouldn't

pass right away when it came from the ocean. But when we were on the shore-fast ice, there was no worry, and we stayed there although it was foggy.

Radiation fog—which forms over the land's surface at night as air cools to the dew point—was less threatening and would quickly dissipate. John Phillip continued: "When fog forms from the land, when it appears and goes upward, that's different. When the sun appears not long after, it is said the sun will bring it down." Roland Phillip (November 2005:107) explained that fog from the land was not to be feared:

> While we're down on the ocean, we'll see fog rising from the land. They tell us not to be afraid of that fog. They say that it won't be there all day and will be gone before the day is done.
> But they told us to fear fog rising from the ocean's surface. When I'm out on the ocean hunting in spring, I'm always looking out for fog. If it seems like it's going to form, following my father's advice, I try to get to a safe place before I encounter a hazardous situation, not considering it more important to catch an animal.

Fog could also indicate wind. Roland Phillip (November 2005:114) noted: "Those in Tununak and Nunivak Island especially were afraid of fog hovering above Etolin Strait. When that occurred, they said the weather would not be calm. That was a warning sign." Simeon Agnus (March 2007:564) commented: "They call our village Nightmute, the place of constant strong wind. Those Qukaqliqutaat [Mountains] up there [above our village], as long as there is fog between them, the weather will not become calm. It becomes calm only when the fog is gone." Laughing, Simeon exclaimed, "What in the world are those two mountains? Are they *angalkuk* [two shamans]?"

Ellaliutulit | Those Who Predict the Weather: "Ella Did Not Surprise People"

Contemporary elders agree that people of the past knew the weather. Simeon Agnus (July 2007:582) compared them to meteorologists: "After searching the sky, they'd say, 'The weather is going to be calm all today and tomorrow.' What they said would transpire. Now that we've started to hear white people who are weather forecasters, [our ancestors] weren't any different in their predictions." This was no small feat, as coastal weather conditions can change rapidly and, as on Nelson Island, are closely related to specific areas. Mark Tom (March 2007:1198) remarked: "Those before us talked about those things without looking at books. They were powerful." Nick Andrew (October 2005:278) quipped: "If the weather is going to be bad, they knew by the clouds. Today we are all becoming white people. We only listen to the weather station."

Close observation was the foundation of their knowledge. Sophie Agimuk (June 2009:81) declared: "Those before us used only *ella* to predict conditions." Paul Kiunya (October 2005:327) explained: "After a long winter, as it became spring, our ancestors knew how it would be by observing the trees and plants and by observing the clouds. Today the white people would call those people scientists. Those who observed the weather would predict how it would be, and they knew when it was going to get windy or calm."

Such foreknowledge was important to keep people safe: "They let us hear those things so that we would not travel in bad weather" (Frank Andrew, September 2000:203). Paul John (October 2003) elaborated: "Those people in earlier times apparently watched *ella* very closely as a safeguard when they used kayaks. They observed those signs and knew what it would do. When they knew it would be good, they would leave [to hunt], even though it was windy. And even if it was calm, they would take shelter when they saw it would be bad. They say it is not wise to go along with *ella*."

Frank Andrew (September 2000:203) described how people observed designs in the sky: "Those weather indicators look a certain way, the *qaralit* [designs] of the world up there, the clouds and the sun's rays. Only bad weather, wet weather, or the wind causes them to change. Good weather doesn't change them but lets them appear calm." Many noted that these designs do not lie. Sophie Agimuk (January 2007:473) remarked: "By observing the setting sun, they knew how the weather would change. Truly, when one takes time to observe, it is obvious, as those [signs] do not lie. If one pays close attention to it, the way the weather is going to change is obvious." Paul Jenkins (March 2007:349) of Nunapitchuk added: "When I observe the moon, it doesn't lie and has been that way since long ago. The edges of the mountains also don't lie. And in my observation, the river mouths don't lie."

Bob Aloysius (October 2005:393) used trees as a weather indicator: "In winter, when it's going to get very cold, we'd be in bed and after a while there would be a loud crackling noise. Trees begin to break when the weather is going to get cold. . . . And the trees around the mountains will look pale when it's cold. Then these [trees] start to get darker and easy to see. When it's going to get warm, they said that the trees get really dark, not green but black, especially these areas where there are spruce." John Phillip (November 2001:162) also used trees as indicators: "I always observe those mountains that we see from our village. When it's going to get warm outside, the trees are visible from our village and look clear and blue. That really is accurate."

Weather observation and prediction were both routine and essential parts of life. Every part of the natural world was carefully observed and communicated. John Phillip (November 2001:162) said simply: "Anything is a tool; even mountains serve as tools to predict the weather." Men needed to understand coming weather before venturing from home. Roland Phillip (November 2005:79, 115) explained: "Before they went down to the ocean to hunt, our ancestors knew about the weather by the appearance of the sun and by the edge of the sky. . . . They say that predicting coming weather conditions is not just limited to the daytime; they also say that the evening can indicate what will come during the [following] day."

People in different areas paid attention to different conditions. According to Peter Jacobs (November 2001:162): "Truly each village has different signs. When we were leaving Nunapitchuk in a very bad storm, the men of the *qasgi* would say to us, 'Just keep going, it will clear up farther away. The mountains from the Yukon let us have the wind.' When we traveled, we would go out to very beautiful weather when our home was very stormy. Each village has instruments to predict. They use what they see." Paul Kiunya (October 2005:126) noted that coastal and inland areas differed: "Our ancestors [on the lower coast] had ways of determining [weather conditions]. Elders in upriver villages also knew when bad weather was going to improve. After being windy, they had observers who knew when it was going to be calm through their own determinants." Phillip Moses (January 2007:419) mentioned those who noted weather conditions after the fact: "It's also true that after not having mentioned [what the weather would be like], when it started to rain, a particular person would come inside the *qasgi* and shake [the rain] off his skin boots. While he was brushing his boots, after not having predicted the coming rain, he'd say, 'Yesterday, since there were signs [of coming rain], it has occurred.' When the conditions occurred, he'd mention it. [*laughter*]"

Everyone was taught to observe the weather. John Phillip (November 2001:162) explained: "They taught young people about the weather right from the beginning so they would all know. Not just one person was taught. Everyone who listens and observes learns. They wanted us all to know." Particular people, however, excelled as weather predictors: "That person who would constantly observe the weather is referred to as *ellaliutuli* [one who reads and predicts weather conditions], one who knows the weather" (Simeon Agnus, July 2007:507). A few elders told of shamans using their special powers to predict the weather, but that was not common. Frank Andrew (September 2000:223) recalled:

> They say Alaqteryaq used to speak about those two small stars [in the east] that were extremely dependent on the weather. I used to see that short man. That person really knew *ella* because he was competent using those indicators. . . .
>
> That's why he was their leader when they went hunting. They did not travel down to the ocean before he announced what the weather would be like all day. When he finally announced it, they would go down, even when it was windy, when he said it would become brighter.

And when they woke in the morning, they would see that the weather was good with no wind. After observing *ella*, he would tell them to paddle without traveling too far out, that it would get windy before night. They said he was always correct in his predictions. He learned those signs.

Mark John (November 2001:162, March 2007:564) recalled advice he was given:

My grandmother's late younger sister Cakuucin used to tell me about the weather. Once she came up to me and said, "You didn't leave." I told her that after waking that morning, since things hanging to dry outdoors were blowing from the wind, I went back to sleep and had just woken. Then she told me, "If I was you, I'd leave. The weather will improve during the day and be good." She also told me that Kangirrluar [Toksook Bay] would burp out calm weather during the day.

[My brother and I] quickly prepared and left. When we reached the bay, it was dark because there was a slight wind. We traveled along the edge, and just when we reached the tip, the area ahead of us started to shine. We went down away from the shore and reached the ice, and the ice was all in one piece and looked safe. We caught animals. After us, other people started traveling. . . .

Then one time I was about to check my net, and the weather was calm when I went outside. Just as I finished preparing to leave, she hurried toward me and said, "You mustn't leave." She said that during the day it might start to get blizzardy. Although I didn't want to, I unloaded my things and stayed. I would think as I was just sitting, "It would have been good if I had left." During the day, I saw that it started to get a little windy. Then it got extremely blizzardy. She knew the weather.

Paul Kiunya (October 2005:124) spoke about his grandfather: "Since some of our ancestors were scientists, they were knowledgeable about *ella*. I had a grandfather who was like that. Once I asked him, 'Grandfather, how do you know about the weather?' He made an *eng* sound at me: '*Eng*, why don't you wake up early in the morning and observe the whole periphery of the sky, so you will learn about it in that way?'" Simeon Agnus (July 2007:507) was advised by his mother: "Since my father died when I was small, my mother would look up at the sky before I traveled and say that it won't get windy all day. Her predictions always came true. Although she was a woman, our parents and their parents knew the weather."

Contemporary elders continue to share what they were taught. Nick Andrew (October 2005:312) said, "I still see [those thick clouds] from my village of Marshall down there. When it's thick, I tell them that it's going to be cold for a while. Some smile and ask me, 'How do you know?' I tell them that I didn't learn them myself, but I know from those who used to speak and from their *qanruyutet* [instructions]."

A person can also use his body to predict changing weather conditions: "Our ancestors used their feelings as well as their eyes to predict the weather. In the morning my grandmother would tell me not to use only my sight but to keep track of how it felt with my body" (Mark John, November 2001:162). John Eric (March 2008:195) gave a humorous example: "When the weather was about to get cold and bad weather was coming, this [finger] cut off by a Skil saw had sharp pain. You should cut off [your finger] so that you will know the weather. But when the weather was about to improve, it would feel nice and itchy, but here there aren't any lice around. [*laughter*]"

John Phillip (October 2005:208) described smelling the weather: "Sometimes when I go outside, my nose smells what appears to be cold weather. That seems to happen when it's going to be cold and form frost on trees in the night. That also indicates when the weather is going to get bad, when it's going to get cold. Sometimes I feel it through the frostbitten spot on my nose."

Not all forecasts came true. Paul Kiunya (October 2005:327) noted: "In spring after it looks like it's going to do that, it doesn't happen. They call that *pivaguarluni* [forecasting changes that don't happen as predicted]." Frank Andrew (September 2000:223) recalled: "These are the things that our *ellaliurtet* [weather predictors, meteorologists] spoke of. I came to realize that even though the *ellaliurtet* among the *kass'at* say that the weather is going to turn out a certain way, sometimes it doesn't become a reality. Those people of the past were like that also."

Ellaliuryaraq | Changing the Weather

As weather played such an important role in people's lives, it is no wonder some sought to influence it. As we have seen, bad weather was often viewed as a social response by *ella* to human actions in the world. Conversely, careful and attentive human actions could set it right. People relied on their *angalkut* (shamans) to check the weather during spiritual journeys. Paul John (May 2003:84) jokingly recalled: "That *angalkuq* from [the lower coast] they called Kelissayagaq—one of his grandson's brought him tea and begged him for good weather when he went mink trapping. At that time, the weather became stormy. When his grandson came home, he went to [his grandfather] and showed his indignation. He answered his grandson, '[It was that way] because the great big *ella* wasn't about to be bossed by me.'"

Angalkut directed masked dances and other ceremonial activities aimed at promoting good weather during the coming harvest season. Paul John (May 2000:37) explained: "Masks were used as prayers to Ellam Yua to request an abundance of food and animals in the coming season. They also honored the winds and asked for good hunting weather." Like the all-seeing *ellam iinga* (eye of *ella*), these elaborate masks, many surrounded by hoops designated *ellanguat*

Kuskokwim mask representing "Negakfok" (*negaqvaq*, "north wind"), the "coldweather spirit that likes cold and stormy weather, used in dances in March and with a sad expression on account of the approach of spring." Along with four "Negakfok" masks, Kuskokwim trader A. H. Twitchell collected and described masks representing the south and east winds, probably from a single event held in Napaskiak in the early 1900s. 9/3430 National Museum of the American Indian, Smithsonian Institution.
Barry McWayne

(models of the universe), functioned as eyes into a world beyond the mundane, making the unseen seen.[16]

An *angalkuq* might construct and present a *ellanguaq* (model or pretend universe) to influence the weather. When he was young, John Phillip (September 2000:20) observed the presentation of a *ellanguaq* made by the *angalkuq* Murak in Anuurarmiut:

> A person from one village usually requested something from a certain individual in the other village [before a dance festival]. One man [named Murak] was poor and didn't have anything to give. He had two sons. This happened in the village where I was born, and I remember it clearly.
>
> His cross-cousin [from Kipnuk] knew that poor man was a shaman. When they got ready for the dance festival, his cousin asked the poor man to grant him a seal in the coming season. He directed the request to the poor man's son. The second request was directed to the younger son, and it was for good weather during the coming seal-hunting season.
>
> Since that old man had tools, he constructed round wooden hoops in the *qasgi*. I saw him making them from straight-grained wood. Then he carved a wooden model of a bearded seal and hung it in the middle. Then they hung the *ellanguaq* [in

the *qasgi*] with a string attached to it, and when someone pulled the string up and down, the wooden hoops came down and went up again. That was the *ellanguaq* I saw personally at a dance festival. People from Kipnuk [were invited]. . . .

That *ellanguaq* had feathers attached to it. It was as large as the *qasgi*, and the hoops got smaller in size as they went up [toward the ceiling]. And [the highest hoop] up above was situated along the center of the skylight.

[His cousin] requested those things from him, since that was the practice. Some requested things that could be acquired before the event took place. Since [his cousin] knew that poor man was a shaman—perhaps he was testing his skills. When the *ellanguaq* came up and down with the drumbeat, it looked very beautiful decorated with swan down.

That [*angalkuq*] presented the *ellanguaq* at the dance festival. After those items were mentioned, they sang the song, and the first verse talked about the brother bringing many bearded seals close to shore down in the bay. The second verse talked about the younger son bringing calm weather in spring. When the song ended— since he was his *iluraq* [cross-cousin, with whom it is usual to have a joking relationship], he added a ridicule song. He sang, "May their poor father push those [two sons] in the sled. Even though he claims to be a shaman, we are not sure about that."

When spring came many, many bearded seals arrived, and the weather got calm. Bearded seals were abundant, and they stayed through summer and into fall. Hunting was so good during fall that it felt like spring.

Frank Andrew (May 2003:175, 219; June 2003:182) also witnessed the presentation of a *ellanguaq*, this one decorated with duck down and the tail feathers of long-tailed ducks.

I watched them when they used a *ellanguaq*. They tied wooden strips together into many hoops and let them descend, using a rope. When they started a song, the puller kept moving [the *ellanguaq*]. When they hit the drums, when that *ellanguaq* was pulled, it seemed like the *qasgi* fell flat! Following the drumbeat, it went up.

They only did that sometimes during the Messenger Feast, using the *ellanguaq* as a request for good weather. The inside of the *qasgi* was so attractive when they had a *ellanguaq*. . . .

When the drummers hit their drums, the [hoops] up there fell down. The women would scream when they were afraid it would hit them. When they struck the drum the second time, when it got faster, it would slowly go up toward the ceiling. When it got to the ceiling, it stopped rising but made a jerking motion.

It didn't touch the floor. It would suddenly stop right on top of people and never touched them.

I think I saw them use a *ellanguaq* in the *qasgi* twice in this village. I also saw them do that at Qinaq. They danced in December, not just at any time of year.

Pilot Station artist Milo Minock's drawing of a potlatch on the Yukon River, showing an *ellanguaq* being raised and lowered during the dance. Minock wrote: "The bird up at top is a dry'd owl, when the song is near at the end the bird is lowered and its wings move up and down—there are three men making the bird move when it's time to. This may be repeated 2 times in one night—after every one dances." Ink on bleached reindeer hide (length: 28 cm), *J. and P. Kline Collection, D8c-3, Alaska State Museum. Cameron Byrnes*

Growing up in Pastuliq at the mouth of the Yukon, Willie Kamkoff (April 1997:102) of Kotlik saw the presentation of a *ellanguaq* painted with red ocher and decorated with models of the moon, fish, small animals, a rabbit with a hawk above it, even feather depictions of mosquitoes and flies:

> When they did that, come summer, there was an abundance of food, and the Yukon River was inundated with fish. When they sang one song verse, that hawk would suddenly come down on that rabbit, and sometimes fur would fly. They danced vigorously, and the inside of the *qasgi* vibrated when they beat the drum. They performed all night until daybreak.
>
> That *ellanguaq* was an ultimate petition for everything. When they had that in Pastuliq, they included images of all living creatures, even all the flying creatures.

Elena Charles (September 2000:20) also saw the presentation of a *ellanguaq* in the tundra village of Nunacuaq: "The one I saw had hoops made of twine. Candles were placed on them and were lit. That's why the *ellanguaq* was beautiful to

Dancing in the Kipnuk *qasgi* in 1933. Note the line decorated with owl feathers hanging above the drums, possibly a *ellanguaq*. *Augustus Martin, Martin Family Collection, Anchorage Museum B07.5.1.A23*

look at. In the middle hung a model of a man with waterproof boots, seal-gut garment, and fish-skin mittens. In time with the drummers, the person sitting against the wall pulled a string to let the *ellanguaq* go up and down again. It was absolutely joyous to behold."

When he was young, John Phillip (January 2011:86) watched a shaman use a paddle to change the weather: "Over at Kwigillingok, they asked someone to change the weather when they were going to have a Messenger Feast, when the weather was continually bad. While he stood inside the *qasgi*, someone was also outside watching. The one inside the *qasgi* took a paddle and used it to turn the wind around. The person outdoors would periodically say through the window which direction the wind had reached and told him to stop when it had reached the west."

People also employed private ritual acts to influence or work on the weather, *ellaliurluteng*. Paul Andrew (November 2005:117) recalled his grandmother asking him to face north and *ella allguruarluku* (pretend to tear the weather). Frank Andrew (June 2005:221) explained: "Sometimes when the wind hadn't changed direction, they'd tell those *aglenraraat* [girls going through first menstruation],

'Go outdoors and put saliva on your pinky fingers and [pick up] the wind and move it a little.' She would bring it down to the ground. She did it to all the winds [moving her pinky fingers from the sky down to the ground]. When she was done, she replaced her hood and went indoors. Then the weather would get calm. It was nice weather."

People might also circle something *ella maliggluku* (clockwise, east to west, following the direction of the universe) to affect the weather. Rita Angaiak (July 2007:594) of Tununak recalled: "One time, back when I was unaware of things that were shameful, they had me remove my clothes, and I'm not sure how big I was at the time. When I went outside, they told me to go around [the home] *ella maliggluku*. I ran around the house and then went inside. They said they were trying to make me change the weather during that time."

In some areas people threw charcoal toward bad weather to clear it. Edward Kinegak (July 2007:387) of Chefornak explained: "I know of a person in Akiachak who had some charcoal, as they always take charcoal with them from the stove. When they lose their trail while traveling and they throw that toward the bad weather, the weather suddenly clears and opens for them." John Roy John (July 2007:390) of Newtok recalled: "Once I arrived at a place that I wasn't thinking of going to. I was actually heading to Tununak, but I arrived behind our village following the trail that I took. Since I didn't understand what had happened to me, I told my grandmother about it. She said that I must have encountered something that altered my consciousness. Then she told me that if that happened again, I should cut the trail behind me and I would arrive at my destination at that time. That's how I understood *ellangperun* [an encounter that alters one's consciousness]."

Albertina Dull (June 2009:110) of Nightmute described *ellam equtaqluku* (the world shrinking on people) during encounters with *ircenrraat* (other-than-human persons): "A person whose world shrunk was walking, looking for a village. A slough had some sourdock growing alongside. When he looked closely at the ones he thought were sourdock, they were actually small trees." Lizzie Chimiugak (June 2009:30) recalled her personal understanding of *ella* when she was young: "When our world came to mind, it seemed that I was standing in the very center of something rolling. Then I thought, 'When I leave and we arrive at Talarutmiut, they have their own world. And they also have their own world at Nightmute. And when we travel to Umkumiut, there is also another world.' Since they covered us with a tarp inside a sled when we traveled, and we never looked around, I thought they were different [spheres]."

Michael John (July 2007:590) described his father's method of changing *ella*:

This is what I used to witness. When it was constantly windy, [my father] did what he called *angyarrlugluni* [process of using an *angyarrluk* (raft) to change the weather]. He'd make three dolls out of grass. Then he placed them inside something, even inside a dog bowl, inside the bottom cut piece of a five-gallon bucket, and filled it with grass. In spring, he'd get dead grass from the previous year, since they easily ignite.

Then he'd also soak a small piece of [seal] intestine in water and tie one end. He placed the end [of the intestine] on one of those [dolls] and had the wind inflate it.

When [the intestine] inflated fully, he'd take some twine and bind it so that it wouldn't deflate. Then he'd place [the intestine] in the middle of those three dolls, among the grass down there.

After doing that, he placed some oil on it and ignited them with a match. When they suddenly caught fire, he'd have the wind on a lake or river blow them away. And that [inflated intestine] made an obvious popping noise when it would suddenly ignite. He'd have the wind blow the bottom cut piece of a five-gallon bucket to the other side of the lake. Then the next day, we'd wake to find the weather calmer. And that's what he'd say, that it had blown up the wind.

Then he'd go and get that [five-gallon bucket] and after emptying it, he wouldn't do that again. Then he'd leave, paddling his kayak.

Rita Angaiak (July 2007:594) had a similar experience: "I didn't know what they were doing when they brought me along because of the weather, [but later I learned that] they were trying to make the weather improve. After constructing a small *angyarrluk*, they filled it with grass. When they were done, they lit it with a match and had it float a good distance away. Then the people with me started to make me cry, and I cried. They were probably trying to make me feel sad to part with that thing that they had placed in the water to float away. That's what they did to me."

Michael John (July 2007:591) also described how his father used a ptarmigan head to make the weather colder and therefore better for hunting: "My father wasn't the only one who did this, but I saw others do this also. During spring when the weather was warmer, after the ptarmigan arrived, they'd cut the ptarmigan head along its neck and didn't pluck it. When it got windy, when it used to get cold from the north wind, they'd stake it toward the north and then burn its feathers. They called it *nenglengcarrluku* [trying to make the weather colder]. Sometimes it would turn out exactly as they anticipated. In the morning we'd wake to find that the snow had formed a very hard crust, and conditions were such that one's feet didn't sink in."

Ellam-gguq Caqatallra Nallunaitetuuq |
They Say Ella Tells a Person What It Will Do

Bad weather was an ever-present possibility when traveling on the Bering Sea coast. To avoid its consequences, people closely observed it and learned to read its signs.

A person might perform certain actions, like tearing the sky, to cause the weather to change. Most important, people were admonished to follow their *qanruyutet* regarding both action in the world and interactions with one another. All the lessons described above could help a person stay safe. Nonetheless, if caught in a desperate situation, one must act with confidence, reassuring those who had less experience. Roland Phillip (November 2005:113) shared his experience when caught in an ocean storm: "It was obvious that [my grandson] was somewhat desperate. Each time we stopped, he would say, 'Grandfather, what are we going to do?' I'd cheerfully say to him, 'Nothing will happen to us; we won't get lost.' I constantly comforted him because I know that we have to reassure those who aren't as confident."

Many describe their experiences in *ellarrluk* (impassable weather). One of the most memorable, retold by many to this day, recounts how a number of seal hunters resting on piled ice in spring at the mouth of the Kuskokwim were washed away as their comrades helplessly watched when a strong wind suddenly came up. These men did, in fact, have warning, as other, more knowledgeable hunters paddled past, advising them to move to safer, sheltered ice. The men, who some say were sitting by their kayaks playing cards, did not listen and died as a result. The lesson is not only to know the weather but to pay attention to the actions of others more knowledgeable than oneself. They say *ella* has teachings and tells what it will do beforehand, and one must listen to *ella* to survive. As important, one must listen to one's elders and peers, whose experiences and admonishments will help one interpret what *ella* is saying.

Ella | The World and Weather

///

AKERTA, IRALUQ, AGYAT-LLU | SUN, MOON, AND STARS

agsaq/agsat. Star/s. See also *agyaq.*

agyam anaa/agyat anait. Meteor/s, puffball/s (lit., "star's feces").

agyaq/agyat. Star/s. See also *agsaq.*

Agyarpak, Agesqurpak. Venus.

Agyarrlak. North Star, Polaris. See also Erenret Agyaat.

agyuli/agyulit. Comet/s (lit., "ones that go").

akerta. Sun. See also *macaq, puqlaneq.*

akertem ayarua. Sun column (lit., "sun's walking stick").

akertem qurrlurarallra. Pouring of the sun, indicated by rays coming down
     from the sun.

ciqineq/ciqinret. Ray/s of the sun, sun's glare.

Erenret Agyaat. North Star (Day's Star). See also Agyarrlak.

Ilulirpiit. Large Fish Trap Funnels (constellation).

Ingularturayuli. One Who Always Does *Ingula* Dances (star).

Iralum Qimugtii. Moon's Dog.

iraluq. Moon, month. See also *tanqik*, *unugcuun*.

Kaviaret. Ursa Minor, Pleiades (Red Foxes).

Kaviaruaq. Sirius (Pretend Red Fox).

kiuryaq/kiuryat. Northern lights, aurora. See also *qiuryaq*.

macaq. Sun. See also *akerta*, *puqlaneq*.

puqlaneq. Sun. See also *akerta*, *macaq*.

Qagtellriit. Those with a Northern-Style, Curved Parka Bottom (constellation).

Qaluurin. Big Dipper (lit., "Dipper"). See also Tunturyuk.

Qerrun Ayemnera. Broken Arrow (constellation).

qiuryaq/qiuryat. Northern lights, aurora. See also *kiuryaq*.

Sagquralriit. Those That Spread Out, Scatter; three stars in Orion's belt, seen just
    before sunrise in November and December, used for telling time.

Tanglurallret. Snowshoe Tracks (from *tanglut*, "snowshoes"), Milky Way. See also
    Tulukaruum Tanglurarallri.

tanqik. Moon. See also *iraluq*, *unugcuun*.

Tengqulluuk. Parka-Hood Tip (constellation).

Tulukaruum Tanglurarallri. Raven's Snowshoe Tracks, Milky Way. See also
    Tanglurallret.

Tunturyucuar/Tunturayaaret. Little Caribou (constellation).

Tunturyuk. Big Dipper, Ursa Major (from *tuntu-*, "caribou"). See also Qaluurin.

unugcuun. Moon. See also *iraluq*, *tanqik*.

ELLAM CIMILLRA ALLRAKUMI | SEASONS

ellanglluk. Poor weather in which outdoor activity is still possible.

ellarrluk. Season of famine, weather that is so bad that outdoor activity is virtually
    impossible (lit., "bad *ella* [weather]").

ellaliurtet/ellaliutulit. Weather predictors, meteorologists.

kiak. Summer.

uksuaq. Fall (lit., "pretend *uksuq*").

uksuq. Winter.

up'nerkaq. Spring.

AMIRLUT | CLOUDS

amirluq/amirlut. Cloud/s. See also *qilaggluk*.

amirluqtaat. Scattered clouds, stratocumulus clouds.

cagnilriit. Cirrostratus clouds (from *cagni-*, "to be taut").

caniun. Cloud bank toward shore.

iqalungaq. Terrible, dark cloud.

qilaggluk/qilaggluut. Cloud/s. See also *amirluq.*

qilak. Sky, heaven.

qiugaar/qiugaat. Reflection/s of open water, seen in the sky as a dark-blue line (from *qiu-,* "to become blue").

tenguguaq/tenguguat. Dark cloud/s, possibly lenticular clouds (lit., "pretend liver").

uyunguryak/uyunguryilria/uyunguryiit. Haze, heat wave/s.

ANUQA | WIND

anuqessuun. Wind sock.

anuqlir-. To be windy.

aternir-. To be windy from shore.

calaraq. East or east wind. See also *keluvaq, kiugkenak, ungalaq.*

canirtaq. Windbreak.

cenirnir-. To be windy along shore.

culu'uggluni. Whistling of the wind.

inivkaq/inivkat. Superior mirage/s caused by temperature inversion.

kanaknak. West or west wind. See also *keggakneq.*

keggakneq. West or west wind. See also *kanaknak.*

keluvaq/keluvaraq. East or east wind. See also *calaraq, kiugkenak.*

kiugkenak. East or east wind. See also *calaraq, keluvaq, ungalaq.*

negeq/negeqvaq. North or north wind. See also *piakneq.*

piakneq. North or north wind (lit. "from up there"). See also *negeq.*

qacaqneq. South or south wind. See also *ungalaq.*

qamaneq. Place lacking wind or water current (from *qama-,* "to be windless").

quuneq. Calm weather.

quunir-. To be calm, windless.

uaqnaq/ungalaqliqneq. Southwest.

ull'uyaq. Whirlwind.

ungalaq. South or south wind, east (on Nelson Island). See also *qacaqneq.*

uqenqar-. To be windy at the back.

IVSUK | RAIN

agluryaq/agluryat. Rainbow/s.

anngiinaq. Light rain that does not stop (lit., "one that keeps going out").

cellalluk. Rain. See also *ellalluk, ivsuk.*

ellalluk. Rain (from *ellallir-,* "to rain"). See also *cellalluk, ivsuk.*

ivegpak. Hard rain (from *ivsir-,* "to rain"). See also *tup'agtuq.*

ivsirrluar. Light rain, drizzle (from *ivsir-,* "to rain"). See also *kanevvluk, minuk.*

ivsuk. Rain. See also *cellalluk, ellalluk.*

kalluk/kalluut. Thunder and lightning (from *kallir-*, "to thunder").

kanevvluk. Light rain, drizzle. See also *ivsirrluar, minuk.*

minuk. Light rain, drizzle. See also *kanevvluk, ivsirrluar.*

nungu. Fog. See also *taituk, umta.*

nungurrluk. Ice fog.

tagevkara'arluni. Rain that gets bad and then better.

taituk. Fog (from *taicir-*, "to be foggy"). See also *nungu, umta.*

tup'agtuq. Hard rain. See also *ivegpak.*

umta. Fog. See also *nungu, taituk.*

yukutaq. Dew, morning dampness, moisture.

Nunavut

OUR LAND

ELDERS RECALL THE TIME WHEN, AS THEY SAY, THE LAND WAS thin. Then the boundaries between the ordinary and extraordinary were more permeable, and people encountered unusual, sometimes frightening things. Tommy Hooper (March 2007:1330) told about the time that his mother heard an *amikuk* (legendary creature) calling on the tundra and escaped it only through prayer: "They say when those [*amikuut*] surround them, the area around that person turns into water, and they cannot do anything and die. Those first people, our parents, those before us, encountered strange things in the past sometimes. Some encountered terrifying situations."

Today the land is thicker. Simeon Agnus (July 2007:501) explained: "Every now and then someone encounters a ghost today, but it isn't as prevalent as it was in the past. Some people mention that it's because the land has gotten thicker. They said when the land was thin in the past, many extraordinary things were around."

Nuna Nallutaituq | The Land Knows

Although the land may be thicker now, many still view it as sentient and knowing, capable of responding to human actions in the world. John Phillip (January 2006:273) clearly explained the need to respect the land: "Since this is our land, our ancestors used it with honor. They were careful with it, as it was their source of sustenance. And those *aglenraraat* [ones who had their first menstruation] or those whose family members had died, they told them to honor our land also. They were instructed through those situations as well."

Elders emphasized the importance of following *eyagyarat* (abstinence practices) to ensure safety in a knowing world. As described above, these rules circumscribed activity for those undergoing difficult or transformative circumstances, keeping people covered and restricting their contact with the world around them. Frank Andrew (September 2005:284) explained the prohibition against stepping on islands: "The islands over there are alive. They are very sensitive and knowing. If a *kenegnarqenrilnguq* [one who has had a miscarriage or suffered the death of a family member] goes near them, they will immediately sense it, even in winter. That is why people were warned not to step on them with bare feet. They could walk on them with boots on but not directly with their bodies, and they were also warned not to go in salt water. And girls menstruating for the first time were warned the same way."

Movement over the land was also restricted. According to John Phillip (January 2006:275, 287):

> When one of our family members died, they told us to try to stay in a place [and not travel] until the *kanarat* [four or five days following the death of a male or female relative, respectively] were up. When they were in the wilderness and their family member died, they left [the deceased] there and didn't bring him along. That's why we see small crosses out in the wilderness sometimes while we're traveling. They honored that since *ella* was aware of that person's circumstances. That was another *qanruyun* back then. . . . And when a dead body was brought somewhere, even in the sky [with an airplane], the weather got bad.

Those who ignored *eyagyarat* might injure themselves. They also injured the land. Joseph Jenkins (May 2003) explained:

> When we were young, we were told to treat our land with respect. Apparently, it was by way of *eyagyarat*. The dos and don'ts of these practices are many.
>
> Our ancestors took good care of our land, and there were no adverse changes to the Yukon and Kuskokwim Rivers. When we first started arriving in Bethel, the river down there had a small creek coming in and a very fine shore. By not taking care of it well, they have made it like it is now.
>
> Foremost among the admonishments is that a young woman undergoing her first menstruation was cautioned not to go near the shore. If she is on the riverbank or a beach, a *quugaarpak* [legendary creature said to live underground] will surface, and the shore will be churned up and folded over.
>
> We have seen the village of Tuluksak. Their land does not change. Yet it became

a village long ago. Yet for the people downriver from there, because they disobeyed the traditional abstinence rules, places along the banks of the Kuskokwim River have eroded badly. These things were pointed out to us so that we would know about them.

Changes in their landscape, many believe, resulted from human action and attention or their lack. In contrast to their Western counterparts' fervor to subdue the natural world for economic gain, elders emphasize the importance of self-control and knowledge of the world and its inhabitants, who both know and respond to the choices each person makes in life.

Nuna Enem'ini Uitatuuq | The Land Stays in Its Place

Not only does the land know, but—unlike the ocean—it is knowable. Frank Andrew (June 2001:37) explained:

> They told us that the ocean cannot be learned; that it is not like the land. They said that the top of the land doesn't move, and it cannot be floated away by water or melted. We were knowledgeable about those sandbars and islands that do not move around. But the currents where the ice breaks have no way of being learned, and there is no set day for it.
>
> But they said the land stays in its place, and the steep parts stay in their places and do not change. And the plants do not change. They stay in their places every year. Some parts of the land are starting to sink and disappear, but these lakeshores that have a lot of red-barked willow do not disappear. Being knowledgeable about them is not confusing, because they stay in their places, the land. That is what they used to tell us.

The land, they say, has numerous instructions. Joe Maklak (March 2008:131) of Chefornak explained: "There are many oral instructions for the land and the water down there. If a person who knows the land and ocean speaks to you, he will talk all day."

Nuna Qantaqaput | The Land Is Our Bowl

Many recalled the adage that compared the land to a bowl. Joe Maklak continued: "If an elder talks about the land, he would say that the land is our bowl. We eat a variety of food from it. The ocean is also like that. When fish arrive during summer, they go and lay their eggs up all the rivers, including the Kuskokwim River. Animals, seals, and birds come and give birth here in our land, no matter what they are."

Bowls also had rules surrounding their use. In comparing the land to a bowl, elders imply not only that the land contains subsistence foods but that care of the land promotes their availability. Alex Bird (December 2002:63) of Emmonak recalled: "Back in those days they taught us things about the bowl and how to use it. The bowl was directly connected to hunting. When we used the bowl, we thought about the food that would be placed inside it in the future. We'd ask that it be filled with food that we wanted to eat; we'd continue to fill it and feed others." Frank Andrew (December 2002:63) added: "Following my instructions, I always licked my bowl after I ate. They told us that our bowl was the pathway for all foods that come to us. We were told not to leave it without licking it."

As with a bowl, the most important instruction was to keep the land clean. According to John Eric (January 2007:21): "Since the people of the past lived by treating the land with respect, they didn't contaminate it. They didn't discard things [on the land] because the land is our source of food. We should try not to litter it with food refuse or anything that will make it unclean."

Scraps and bones should not be left on the ground for people to step on. Peter John (March 2007:1213) recalled: "There are many admonishments about the land. They tell us not to leave food scraps along the ground but to dig a hole and place them under the tundra." Some said that out of respect for *pissurviim yua* (the person of the hunting place) people should not litter their camps. Nick Andrew (January 2006:275) explained: "They were told never to step on these foods and animal bones and to keep them in a safe place if we weren't going to eat them. They gave the eatable things to the dogs. They would gather burbot bones all winter. And when the days became long, when preparing food for dogs, they cooked them, boiling them for a long time. They never made a mess out of animal bones. And they never kept them out in the open but always gathered them and brought them out to the wilderness [to discard them]."

Bones themselves were aware of their treatment. According to Paul John (January 2006:278):

They said that foods were the responsibility of women. It is said that if a man brings the animal he caught to the village, it is as if he abandons it, as his wife is the one who must receive it. His wife was especially responsible for watching over the bones. They would try not to make a mess out of them and to keep them together in one place.

We men were also instructed that if we ate out in the wilderness, we were not to discard the bones from our food or leave them in the surrounding area. If it was summer, we would dig in the ground and cover them up after placing them inside. They say that food is easily filled with gratitude. They say that food, even a bone, knows a person who takes good care of it and is grateful to that person. And although we think that these fish have died for good when they arrive, since they return through their souls, they will come again when it is time for them.

Frank Andrew (September 2000:185) described burying animal carcasses found while hunting: "If we accidently happened upon a dead animal in *yuilquq* [the wilderness], we would bury it and not let it be visible. They asked us to put it in a good place, not to let it scatter. And before winter, they made it easier for gulls. They would open up their bags of fish scraps and expose them and let the gulls eat them. These gulls are the cleaners of food scraps." Nick Andrew (January 2006:276) recalled the saying that fish won't approach their dead:

> When we were at summer fish camps, they made a pit in the ground for the guts of animals and our leftovers that couldn't be eaten by dogs. They told us never to place fish guts or bones in the water. They said although the sand covers them, it stinks like rotten food and doesn't lose its smell underwater. And when fish swim upriver for the first time, they won't get close to the calm water if any fish part from the previous year is placed there. [Fish carcasses at] places where fish spawn and die are of no significance to them, since it is their way to spawn and die. But the fish won't go to those places where we placed fish that we handled.
>
> And if they sink some stinky fish from the previous summer across the Yukon River in spring, there will be no fish on the other side. After we started commercial fishing recently, some spiteful people placed old, salted fish in the places where they caught many fish, out of jealousy for people. We don't just fish anywhere. Although the Yukon River is vast, there are certain places [for fishing]. That's why those past people had us take good care of things. They treated them with care. They were thinking of their return to us. If we didn't treat the things that we caught with care, they said we would no longer be good at catching things. And although others among us were good at catching animals, they said animals won't come to us.

Men cleaned the land itself: "They were admonished not to fill the small rivers where they fished in the wilderness with anything bad. And when we went to fall camps, they let us boys clear away the grass that dangled in the streams where they fished. They would clean them, letting the river improve. They did not let us make a mess, even though it was the wilderness" (Frank Andrew, September 2000:185). Even food crumbs were kept from falling on the ground where people might step on them. Frank Andrew (September 2000:186) continued: "It was always our admonishment not to make a mess of food on the ground. And when we were going to eat dried food, they let us lay the bottom part of our *qasperet* [hooded garments] here [over our laps]. We would let the crumbs drop on top of it. That is what we were like, and I wasn't the only one." Theresa Alexie (April 2008) from Upper Kalskag told of recently seeing a beaver bone on a village trail, left there by dogs. She took it home, put it in a bag, and buried it. In the past, she noted, all bones—moose, muskrat, beaver, mink, caribou, ptarmigan, rabbit—were put in a flour sack and sunk in the lake, not left where people walked. Elizabeth Andrew from Tuluksak added that before putting land animal bones in water, one should say, "Come again."

George Billy (February 2006:338) described similar careful treatment of the land: "They really respected this land. They never discarded food waste just anywhere but always buried it, even in the wilderness. When we are going to leave our camp in the wilderness, when they were burying their bones after they collected them, some would say, 'If a small animal should get to you, do not tell it that you are the bones that I left.' He tells that to the bones." Such rites were not without humor. George Billy continued: "Those two who were teasing cousins, including my paternal uncle, ate in the wilderness when we went to herd geese. That Apassangayak was his cross-cousin. When they were finished eating and Alek'aq was burying his bones, Apassangayak said, 'He's lying! They are Alek'aq's bones!' He was teasing."

George Billy (February 2006:340) also described how people gathered human waste and discarded it in *qanitat* (dumping places): "A child would urinate and defecate in his urine container. And when it was time, his mother emptied it nowhere other than *qanitat*. They really respected the land and cared for it. They did not dispose of waste haphazardly." George concluded: "Our ancestors were like small animals. They carefully cleaned where they stayed in spring."

Aviukaqsaraq | Offerings of Food and Water

Many elders recalled how offerings of food and water were placed on the land, both to acknowledge and to provision one's dead relatives. Louise Tall (November 1999:23) of Chevak described how such offerings were made when naming a person after one's dead relative. The person bestowing the name would fill a cup with water and a bowl with food, adding a cigarette or a little chewing tobacco if the deceased smoked or chewed. The relative of the deceased then splashed water from the cup toward the ground three times and dropped small pieces of food and tobacco on the ground, making an offering to the deceased before giving both bowl and cup to the person being named. John Phillip (November 1999:34) spoke of his father offering food and water to the land itself, so requesting animals to make themselves available:

I used to accompany my father sometimes when I started traveling to the wilderness. Sometimes when I was following along, sitting back-to-back with him in the kayak, my father could not see any animals to hunt. Then when he came across a small hill, even though the little hill seemed insignificant, he would say to the hill, "We cannot see any game."

Then I watched my father take a piece from each of our provisions. And then he would take a cup, fill it with water, and walk up to that place, but I never followed him. When he came back down, that cup would be empty. I understood afterward that it was a way of asking for something, *aviukaqsaraq* [a way of offering food and water]. After that, we would finally see those [animals]. He would say that he wanted to ask for game that we could catch.

Mike Utteryuk (November 1999:36) of Scammon Bay recalled how he and his hunting partner practiced *aviukaqsaraq*: "When we arrive at *nunallret* [old village sites], my partner and I do this down by the river. We open up all our provisions before we eat, and my partner takes a piece from each. Then he takes a little of his tea and spills it there. It is said that they offer something to our ancestors. Our ancestors receive what is being offered, even though it is a small portion of something. I have also experienced this." People may also have offered songs to the land, as Mike Utteryuk (November 1999:36) recalled: "I was in the *qasgi* for the final time. I was curious and went in to see people holding drums, and they were facing our mountain and singing. I was with them and saw this."

Paul John (November 1999:70) explained how offerings were made to sea mammals and land animals alike:

> When hunters would make offerings from the contents of their bowls, they realized that, even though what they threw landed in the water, the *aviukaq* [offering] would end up in front of the seal that they would catch and that the seal would eat it. Our ancestors held *aviukaqsaraq* in the highest regard and willingly practiced this tradition.
>
> They made offerings to land animals, not just sea mammals, requesting things they wanted. And they even wanted plants to be foremost in their minds. They would go next to a fox's den and leave an offering, telling them what they wanted to catch and saying, "I am giving you this food because I want to catch something."

Not only were food offerings placed in the ground, but the land itself could be eaten. John Phillip (October 2005:7) recalled: "I was born out at Anuurarmiut, which is no longer a village. Anuurarmiut is about four miles from Kwigillingok. That's where my *al'erpak* [afterbirth, placenta] is. When a child is born, he has a placenta. When the placenta came out, they dug underground and buried it. That's why they tell us that when I arrive at my birthplace, I should put a little bit of the earth in my mouth. That was an instruction." On arriving at her birthplace, Rita Angaiak (July 2007:598) likewise ate from the land: "When I arrived there [at Arayakcaarmiut], since I heard of this custom, I took a small bit of soil from the ground twice and placed it in my mouth and swallowed it. They said that when they arrived in their home where they once stayed after a long absence, they'd take a bit of soil [and swallow it]; that's what I did there."

Michael John (July 2007:598) was given soil to eat as protection from illness: "It is said that my grandfather doctored me before I became aware. My family mem-

Upper: The coastal plain of the Yukon-Kuskokwim delta, between Scammon Bay and Black River, 2003. *Jeff Foley*

Lower: Toksook Bay at the mouth of the Toksook River, 2003. The shallow bay is less than ten meters deep where it enters Etolin Strait. *Jeff Foley*

Upper: A group photo taken at Manriq during our circumnavigation of Nelson Island, July 2007, with Simeon Agnus holding the local Bethel newspaper, *Delta Discovery*. *Seated, left to right,* Michael John, Rita Angaiak, Ruth Jimmie, Ben Angaiak, Martina John, John Walter Jr., Theresa Abraham, and Anna Agnus. *Standing,* Simeon Sunny, June McAtee, Jackie Lincoln, Tom Doolittle, Linda Joe, Joe Felix, Steve Street, Fred Joe, Edward Kinegak, John Roy John, John Walter, Joe Charlie, Ryan Abraham, Simeon Agnus, and David Chanar. *Ann Fienup-Riordan*

Lower: The sun wearing "mittens" as a prelude to cold weather, Quinhagak, December 2010. *See page 64. Warren Jones*

Upper: Lifting advection fog, formed when warm air moves over a colder surface such as land or water.
Ann Fienup-Riordan

Lower: *Amirluqtaat* (scattered clouds, stratocumulus) said to indicate good weather throughout the day.
Ann Fienup-Riordan

Upper: Here the lowest clouds are stratocumulus; above them are altocumulus at the 7,000- to 14,000-foot level and thin, wispy *cagnilriit* (cirrostratus clouds) above 16,000 feet. *See page 78.*
Ann Fienup-Riordan

Lower: Stratocumulus clouds transitioning to cumulus with a few wisps of cirrostratus above. This happens when stratus clouds move inland over warm land, causing them to lift and form cumulus cells.
See page 79. Ann Fienup-Riordan

Upper: A cold front with stratus clouds, a sign of stable air, steady winds, temperature change, and precipitation. *Ann Fienup-Riordan*

Lower: Stratus frontal activity, indicating steady winds and precipitation for longer periods. Meteorologist Jim Brader agreed with Newtok elder Michael John, who noted that elders' use of cloud types and winds often does better than the National Weather Service in predicting weather for the next twenty-four hours at certain locations. *See page 80. Ann Fienup-Riordan*

Upper: Rainbow over the Qalvinraaq River on a summer evening, July 2007. *June McAtee*

Lower: Lakes to the south of Nelson Island, 2003. *Jeff Foley*

Upper: A slough of the Yukon River. Tannin from peat moss gives it its deep brown color. *Jeff Foley*

Lower: The banks of the Kuskokwim River at Crooked Creek. *Jeff Foley*

Upper: The upper Kwethluk River in fall, 2007. *Josh Spice, US Fish and Wildlife Service, Yukon Delta National Wildlife Refuge*

Lower: The Kisaralik River, running through the foothills of the Kilbuck Mountains. *Ann Fienup-Riordan*

bers told me that my grandfather used his powers on me so that sickness wouldn't affect me, and after he was done, he had me eat some soil from the ground."

Qigcikiyaraq Yuilqumi Allat Yuut Akluitnek |
Respecting People's Possessions in the Wilderness

Elders recalled with feeling the admonishments they were given in their youth concerning respect for personal property in the wilderness. Theresa Abraham (July 2007:585) spoke from experience:

> Back when I was starting to travel to the wilderness, when I began to gather edible greens and bird eggs on the land, my father told me that when we traveled, I would see men's belongings, wooden fish traps, metal traps, various items that men had placed there. He said those items are honorable out in the wilderness.
>
> He also said never to take them. He said that people were taught that Ellam Yua would watch them, long before Agayun [God] was around. I thought, "I wonder why he's saying this although I'm a female?" When I traveled to the wilderness, I learned that although I was a woman, I was not to touch those items either, and I must not take them.

Simeon Agnus (July 2007:314) explained: "They say a man never forgets an item that he placed out in the wilderness, even if a number of years passed. If he does not find it where he put it, it does not sit nicely in his mind. Whatever the man puts in the wilderness, he puts with the intention of returning to it. What a man puts in the wilderness is to be fully respected. And if it's missing, he knows that it was stolen."

Taking personal property in the wilderness had consequences. Simeon Agnus (March 2007:723) continued: "They say a man who steals doesn't head toward prosperity. Game that he should have gotten will become scarce for one who keeps taking things that belong to others." Paul John (May 2003) concurred: "If a man steals from others, it will be as if he is placing what he would have obtained into the future of that person he stole from. The one who stole will obtain nothing. That is why our ancestors called stealing the cause of deprivation." Simeon Agnus (June 2007:87) concluded: "A male who tends to take his fellow villagers' possessions out in the wilderness will be responsible for his own demise."

Simeon Agnus (July 2007:314) also explained how one who stole would begin to find things laid out in front of him: "They say that things that he will steal start to be placed out for him. And if he thinks about that item, it will be in his path when he goes to the wilderness. The thing that he was thinking about will appear [for him to steal]." Phillip Moses (October 2002:229) told one well-known story of a man who overcame the consequences of his misdeeds:

There was a man who felt regret about his actions. He began to dislike how he stole in the past. Now objects seemed to be readily available for him to steal around the village.

At that time they never threw away bearded-seal stomachs but always inflated them and used them for seal-oil containers. There was a seal-oil container in a house, but [the thief] came upon a seal stomach filled with oil outside on the trail. It was available [for him to steal]. When he reached that seal-oil container [outside on the trail], he didn't take it, but he kicked the container hard because his mind was upset. When he kicked the seal-oil container, the one that was in the house popped without anyone touching it.

That man told about it, and from then on that became a rule. He didn't want others to be like him and take things. That was the story they told.

Simeon Agnus (July 2007:315) noted that a thief can be known through his tracks: "They will find people who have a tendency to steal through their footprints. They evidently found out that my cousin Qiurtarralek stole items through that means. Puyangun and his siblings recognized that he would take fur animals from wooden mink and otter traps through his splay-footed prints."

Peter Dull (March 2007:721) of Nightmute spoke of the importance of letting the owner know if one uses his possessions in the wilderness:

In the past, we would try to survive alone in the wilderness. And if we were desperate and found something that belonged to someone, we were told to use that and then put it back where we found it. And when we went home, we must not fail to tell the owner. When we did that, people were grateful, even though we had used what belonged to them.

And since people left those *naparcilluut* [grass storage bags] where they set fish traps in the wilderness, he could eat from that grass bag if he was hungry, but he must tell the owner when he got home.

Phillip Moses (October 2002:229) added that rather than responding with anger, the owner will be glad: "If he really is stuck and in desperate need, he will take it, and when he returns home, he will tell the owner that he used that. The owner will be grateful because he told him."

Peter Dull (March 2007:723) described how in the past when people wanted to use another's possession, they would ask and out of compassion the owner would comply: "Those people who I saw down at Qungurmiut in the fall used to be so poor. When there were about five families living here, sometimes only one person had an adze. But they would borrow that adze; if you knew who had it, this person would borrow it again. And they never failed to lend it but only told them to be careful not to break it."

Camilius Tulik (March 2007:722) sadly pointed out that times have changed. In the past people were poor but generous, while today we are becoming stingy as we get rich:

Even though those past people were pitiful compared to us, they were really fine people. Their hunting gear left much to be desired. And it was tiresome when they began to eat just one kind of food. When we think of them like that, they are so indomitable. So our ancestors lived peacefully, sharing with one another.

Accordingly, we must be the ones who are pitiful. We do not live like that [sharing with each other]. When I think about it, things are reversed. Today, even though we do not need things physically, we are indeed worse off than our ancestors.

Yuilquq Pikestaituq | No One Owns the Wilderness

At a Bethel gathering, John Phillip (February 2006:188–90) movingly described how today he uses the same land his grandfather did before him:

My grandfather took me with him when I was small by placing me on top of a kayak sled when I became tired. While we were on our way, we approached a post staked in the ground, and I didn't know what it was. When he came upon it, he stopped and shoveled snow without speaking.

Then after some time, woven grass mats became visible down below the snow. After clearing them off, he removed a wood covering, and there was a river down there. There was evidently a wooden fish trap underneath. He used that grass [mat] and wood to prevent snow from falling through. It wasn't frozen but just a thin layer of ice.

The wooden fish trap had two wooden stakes to fasten it down. At one end, there was something to prevent it from coming to the surface. After cleaning it for some time, and after picking with his ice pick and using a dipper to remove the ice, he untied that funnel. He didn't speak.

After he tilted them up, after removing their inner funnel, he told me, and I never forgot it, "These are places where you will make an effort to subsist when you become capable." Then he said, "This isn't the only place. There are other places besides this land where there are fish like this." I never forgot that.

I tell stories after that, since I make an effort to set wooden fish traps to this day. I told my grandchildren, "I probably sometimes step on the same trail that my grandfather took." That's what I say, thinking of the places that he gave me when I set wooden fish traps.

John Phillip (February 2006:190) emphasized that his use of the land and waterways did not preclude others from using them, and that many traps could be set in the same river: "My father would also show me places where there were fish. He said that those rivers are places that we must use to subsist, but we weren't the only ones. Whoever wanted to set a fish trap at that river could set it there. That's why wooden fish traps are set downriver from one another in some rivers, as no one owns places to subsist." Seasonal camps were also shared. Such sharing not only was based on kinship but was a way of forming bonds between people. John Phillip continued: "If a person wants to set up a fall camp and stay there, he can use that place. Sometimes we'd move to another fall camp. They didn't keep their neighbors from staying on certain lands. The land, the food on the land, is a way of forming kinship ties. That's what it's like. Others can use it."

Johnny Thompson (February 2006:185) of St. Marys noted that in the past land was not owned: "Back in those times, I never heard of anyone owning land. There were fall camps. Any person used [that place]. And no one said that they owned a particular place. These days, we now own land." Lena Atti (February 2006:185) of Kwigillingok described the fluid composition of fall camps: "We also lived with another family inside a home. Another family stayed across from us. They didn't always stay there, and the next year another family would be there. And after staying a number of nights, another family would replace them. Families lived in the same home. Even when we moved to Kipnuk, sometimes another family who had no home would live with us all winter, until they moved to spring camps."

John Phillip (February 2006:186) described how people in coastal communities stayed in numerous small settlements before schools were established and families began to gather at central locations beginning in the 1930s:

> In the past, some people stayed in small settlements and fall camps all winter, subsisting off blackfish. People occupied many small settlements and homes on the coast. I saw Pengurpagmiut, and downriver was another home. And right along Tagyaraq River were a number of homes reaching up to Ilkivik River. And many small settlements were downriver from there.
>
> But some people gathered in places where there were schools, and [other] villages all emptied into those places. When the village of Kwigillingok got a school, the people living in places surrounding that village all moved there and also to the village of Kongiganak.

Even when families began to spend winters in larger communities, they continued to travel to seasonal camps. Like John Phillip, many used customary sites. John (February 2006:180) noted that newcomers would ask people before harvesting in their customary places, not for permission but because the original inhabitants knew the area: "When we travel to other places thinking of harvesting foods, we first ask the people of that village since they know what type of food that area has that we can harvest."

John Phillip (February 2006:191) also described how these customary sites were shared:

Whenever people arrive, even from Bristol Bay, I bring them to my many berry-picking places, and sometimes I bring them along by boat and let them out and tell them, "Now, pick those berries." If you [Alice and Ann] come, I will bring you to a place with many berries so that I can eat your *akutaq* when I go to Anchorage.

I tell my fellow villagers that when people arrive, we should keep [the land] open to them. The place where I pick berries will no longer grow berries if I am stingy with them.

John Phillip (February 2006:193, 195) emphasized that sharing resources brings abundance, while selfish behavior results in scarcity:

Our ancestors brought anyone who arrived to their customary place.

We hear nowadays that some villages aren't happy when other people arrive, saying that particular place is their customary place. Those who don't think well of their neighbors are like that.

Indeed, no one owns the wilderness. But these days, through the corporations [and federal government], allotments have been established. And although we did that down on the coast, we kept it open for a person who arrived from anywhere. We want to have resources like our ancestors. They say a person who is compassionate and gives to others and one who takes pity on other people will be filled with well-being. A berry-picking place won't change; there will always be [berries] there. . . .

People who are stingy in my village lack things. Because they do not give to their neighbors, they can't obtain things themselves. But people who like to give things away have things to give. And their gift cannot finish. It multiplies. That's surprising to see.

John Phillip (February 2006:196) gave an example of how what one gives is replaced by more:

They admonished me that even if I have only a small amount to share, I should give that much. They said we should give a person who is asking for something as much as we can.

One day, my uncle's cross-cousin asked him to give him five dollars. Although he wasn't too happy, he gave him all that he had. When he left the next day, he caught five mink. Then he went to tell that person, "You know how I gave you all that I had in my pocket. Today I brought home five male mink." He thanked that person who had asked him for money. What he gave returned manyfold.

Though no one owned the land, harvesting activities and knowledge were often place specific: "People know the land and their hunting areas, knowing what can be harvested and where resources are located. If a person mentions that they need to use something, they will tell that person, mentioning that resource by name, and they will let them know where those things are located on the land" (Mark John, February 2006:162).

Like John Phillip, many people had places where they customarily went to harvest particular resources. Katie Kernak (February 2006:164) of Napakiak provided an example:

> We have a customary place. And for my children, during spring and fall, that place is their only wilderness and they always want to go there. They know the subsistence activities that are carried out in that place and the resources in the area surrounding that small piece of land.
>
> And since I always went fall camping there, it seems that I know only the subsistence activities carried out in the area around that land. And my children learned it and desire to go to it all the time. I'm grateful and hope that they will continue to want to do that as they are living. It was as if that small piece of land was passed down to us.

Johnny Thompson (February 2006:164) also described how his father brought his family to a place where subsistence resources were available: "When my poor father was young, he set up a settlement in a place along the Yukon River that he thought was most suitable. Various things, the plants and berries that they harvested, were available on that land that was customarily ours, the one they called Qalagpak. And when going outdoors, the place where they hunted was visible. Since they always stayed in those small settlements, they went fall camping in places that would sustain them."

Johnny Thompson (February 2006:167) described finding new harvesting places near old ones:

> When we were about to leave our fall camp, we found a place where we could have subsisted that we hadn't been aware of. We only found that place more recently when we went berry picking, and I set my fish net there. And my poor father found another place. We evidently didn't know [all] the places where we could harvest foods, and although they were nearby, we hadn't tried harvesting in those areas.
>
> When my children began to attend school, we no longer went fall camping. But I'd go to the hunting area that I was familiar with after freeze-up. A person evidently

finds his hunting places on his own only by searching for them. One who sits idle cannot obtain things. A person who doesn't sit idle and is always searching evidently obtains things after trying.

The desire to return to one's customary harvesting places was strong. Paul John (May 2003:209) told this story: "That Chefornak resident Maklaar, the *angalkuq*, since he was my mother's uncle, he called me his grandson. He said to me one time that since Uivenqeggliq [Lake] was located in the coastal region, because he did not want to quit eating what he eats now, he would go to Uiven-qeggliq when he dies. When he said that, I said to him, 'Gee whiz, why won't you go up to Heaven!'"

Nunapik Maraq-llu | Tundra and Marshland

Nuna (land) in the delta ranges between very wet and very dry, including *qaug-yaq* (sand), *nevuq* (dirt or soil), *nunarrluk* (soil mixed with sod), *marayaq* (mud), *maraspak* (very wet mud), *nikuyaq* (sticky gray mud), and *qiku* (clay). Higher ground is covered with *nunapik* (tundra). Tundra islands along the coast—often replete with berries—are known as *allngignat* (from *allngik*, "round patch on the sole of a skin boot"). John Eric (March 2008:190) explained that "[*allngignat*] are

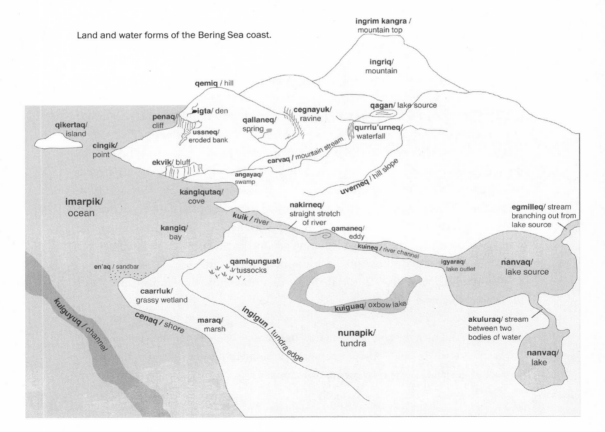

Land and water forms of the Bering Sea coast.

tundra that look like islands, and some aren't perfectly round but odd-shaped, and their surrounding area is lower. And salmonberries and blackberries can grow there." *Pengulkuut* (from *penguq*, "hill") are tussocks of grass on the tundra, and *pengunqut* are mounds. *Qalirat* (from *qaliq*, "top layer") are pieces of tundra that have broken from the bank and drifted away, landing farther along the shore.

The place where marshland and lowland tundra meet (the abrupt rise onto the permafrost plateau) is called *ingigun*.[1] Roland Phillip (November 2005:55) explained: "They call those grassy areas starting from the edge of the ocean *marat* [marshlands, muddy lowlands], and up inland is tundra, and some areas [with tundra] continue for a great distance. There is tundra, then right below it is marsh, and there is no tundra down toward the ocean. The area where [the marshland] meets the tundra is called *ingigun*."

Much coastal land is waterlogged. Nick Andrew (January 2006:35) was warned to be cautious around *puglerneret* (swamplands, from *puge-*, "to come to the surface"):

> Back when I was young, we'd hunt for muskrats in spring. They instructed me about the following, "If you are going to set a metal trap on the *puglerneq* [swampland], never pierce its surface." They said that the *puglerneq* would sink. . . . It is mossy, and berries grow on it. The area surrounding those areas is watery. . . .
>
> When you look at them, it appears as though you won't sink, as berries grow there. They told us not to use an axe to pierce their surface. I tried that twice as I was curious. When I checked it later, that [land] had disappeared.

Paul John (May 2003:209) noted: "The area underneath them is called *angayaarrluk* [swamp, bog]; water keeps [the land on top] afloat. Since the area underneath has a lot of water, the land comes to the surface." John Phillip (January 2006:36) also described swampy areas: "We call those *angayaarrluut* [swamplands]. When you step on them, they move [like walking on waves] and go down. When piercing them, a person can sink. Those are dangerous. Sometimes those occur among mossy areas. When one pierces through them, they are extremely deep and have no bottom." George Billy (February 2006:473) spoke about the dangers of bogs in the tundra region: "These *qalviryat* [bogs, quicksand] are dangerous. When one walks on the surface, the underside has no ground and it goes down. People never left anyone ignorant of those places but pointed them out."

George Billy (February 2006:401) also described how the land speaks about the season to come:

> This last summer, the land spoke to us: "This coming winter, your land will have lots of snow. It will be cold." All summer we understood by the grasses and by the surface of the land. The grasses got tall this summer.

When I saw Paul John, I said to him, "The land out there finally has snow on it!" [Paul replied,] "Absolutely! It is now cold out there as it told us it would be." The land pointed it out; the land knew how it was going to be this coming winter.

George Billy (February 2006:403) then described how a south wind and rainy weather in fall saturated the tundra and presaged abundant berries the following summer: "When the top of the ground was exposed to that wet south wind, they liked it because it was favorable to the good growth of berries. When that happens in fall, they say that it totally saturates the ground with water. So the ground, the berries-to-be, are grateful for that."

Qemit Ingrit-llu | **Hills and Mountains**

On the flat, marshy lowlands of the Bering Sea coast, high ground was physically remarkable, including the Askinuk Mountains near Hooper Bay, Kusilvak Mountain, the volcanic peaks of Nunivak and Nelson Islands, and the Kuskokwim and Kilbuck Mountains to the south and east of the Kuskokwim River. Even small grassy knolls were notable landmarks, and many were named. John Phillip (January 2006:19) remarked that berries were often abundant on high ground. Land animals also made their dens there: "When we'd travel out on the land, those very large *evinret* [grassy knolls] that were steep where grass grew were places where the land animals, [such as] mink and foxes, raised their young."

Some high points were man-made. Both Yup'ik elders and non-Native archaeologists associate prehistoric camps and settlements dating from the bow-and-

The Askinuk Mountains, with the Kun River in the foreground. Geologists note that as recent as three thousand years ago, the Askinuks were close to the Yukon River, which then flowed west to the Bering Sea through its Qissunaq and Black River paleochannels. *June McAtee*

arrow war period—seventeenth and eighteenth centuries—with high ground.[2] Paul Kiunya (October 2006:60) gave an example: "The [old village of] Pengurpagmiut [People of the Large *Penguq* (Hill)] is on the ocean side of our village. When we're on the ocean, we can see it. They say warriors made that long ago as a lookout. They say they piled up the dirt. They probably made big houses inside. There's also a small hill near Kipnuk where warriors lived."

Ingrit (mountains), *qemit* (hills, especially part of a ridge), and *pengut* (hills standing alone) served as important landmarks for travelers in the tundra lowlands. David Martin (December 2005:143) noted that in the past during winter, only the hills would be visible. John Eric (March 2008:134) observed: "My late grandfather told me to always use these Ingriik to gauge my location when I was down on the ocean. And back when there weren't any GPS devices around, when I'd bring these fuel barges upriver from [the ocean] down there, I'd record the location of Ingriik and the front of our river in my head."

Mountains also loomed large in Yup'ik oral tradition, and all the major hills and volcanic peaks along the Bering Sea coast have creation stories surrounding them. We have seen how Raven created the mountains of Nelson Island. Paul John (January 2006:217) described how Vole was responsible for the creation of Ingriik to the south of Nelson Island: "Since they appear [in two parts], they call them Ingriik [Two Mountains]. And there is another [mountain] resembling it on Nunivak Island that is also called Ingriik. They say that when that large Vole from one of the surrounding mountains carried [the mountains] on its back and was heading down, when its carrying device suddenly broke, it placed the first half there and the other half on Nunivak Island. That large Vole apparently placed them there in that way."

Vole also shaped the mountains, leaving his impression on the landscape. Paul Tunuchuk (March 2008:147) explained: "They say that one day a large Vole came and crouched down on the very top of [Ingriik]. The place where its stomach was situated is down there. They say the place where its neck was hanging over [the summit] is on the other side, and the place where its tail was situated is on this side; it's still like that now." The black rocks on the mountainside also have an origin story. Paul continued: "After [Vole] crouched down, Raven came and evidently landed alongside the top and defecated dark feces. They call those dark rocks on that side Raven's feces."

Stories also describe people lost underground emerging from Ing'errlugaat, the basaltic lava cones west of Cuukvagtuli. Mike Utteryuk (November 1999:66) told the story:

> Down in the village of Naparyaar [Hooper Bay] they were having Qaariitaaq [a fall celebration in which children went from house to house requesting food].

One of the Ing'errlugaat, the basaltic lava cones or cinder cones where "those who went below" were said to have emerged. *Tom Doolittle, US Fish and Wildlife Service, Yukon Delta National Wildlife Refuge, 2005*

Their houses were not made of wood and had [underground] entryways in the middle of the room. A person would enter through that hole down there.

Back then, when the ones performing Qaariitaaq were approaching with their noises, a married couple covered their [underground] doorway with a kayak skin cover. They could hear them down there, brushing against [the skin] and passing by. They did this one by one, and all of them continued in this manner. There were five people participating in Qaariitaaq, holding on to each other down in the ground and helping one another as they traveled. The [husband and wife] had made a mistake when they closed the [entryway].

It appears that they traveled down underneath Naparyaar for one year. You know those small mountains beyond Chevak. Their tops are cut off. Those people traveled to the highest part of that mountain. To them it seemed like they traveled only a short distance. They came out of the small mountain one by one in a place that still has an opening. But people who have seen [the opening] say that it has become small.

Apparently, when they came out, they recognized that small mountain. When they looked at their bodies and clothes, they were worn out from traveling underground. Their faces were also covered by hair. Those people who practiced Qaariitaaq told us about this because they came out of that place. They went across to Qissunaq together.

Those occurrences are very true. When they had Qaariitaaq after that, they told [people] not to close their doors because of what happened. I'm explaining that occurrence this much.

Not only did mountains have exceptional origins, but to this day many are viewed as the homes of exceptional inhabitants, especially *ircenrraat* (other-than-human persons). Although *ircenrraat* can appear anywhere, they generally prefer hilly areas, including specific places like the hilltop knob known as Qasginguaq, about five miles northeast of Toksook Bay on Nelson Island. The name "Qasginguaq" translates as "place that looks like a *qasgi* or communal men's house," referring to the belief that the knob is the semisubterranean home of *ircenrraat*.

Although people usually encounter *ircenrraat* in the human world, some stories tell how a person traveling in the hills happens on a window or door to the world of the *ircenrraat*. That person might look through that opening for what seems like only a moment but is actually much longer. Frances Usugan (July 1985) of Toksook Bay related how her older brother, Cyril Chanar, heard the songs of the *ircenrraat* while they were celebrating in their underground *qasgi* within Qasginguaq. Their songs are said to be easy to remember, and some people who heard them later used them as songs for the Messenger Feast. This event occurred when Cyril Chanar was a young man in the early 1930s. Eighty years later, Paul John (January 2006:209) recalled the story:

> They say that place called Qasginguaq up from our village belongs to the *ircenrraat*. They say it is their *qasgi*. Some people apparently heard their voices when they were singing. When they would stand, their [singing] would be inaudible, but when they put their heads down, their singing could be heard again [like listening to sound in water].
>
> And when those [*ircenrraat*] had gone to the ocean during winter, they would see their tracks leading up to Qasginguaq, and the blood that had dripped from the animals they caught was also visible.
>
> Then one time my spouse's deceased maternal uncle [Cyril Chanar] sang two songs, and they ended with *ua-ua-ua-ua*. He evidently learned them from that place when he listened to their singing. Since they say those [*ircenrraat*] are wolves, just as wolves do, they have *ua-ua-ua* endings.

A person might also visit *ircenrraat* in their homes, where the human visitor could see his hosts as people. The visitor's ability to return to the human world depended on his reception. Frances Usugan (July 1985) shared a story of a *nukalpiaq* (successful hunter) who was taken into the world of the *ircenrraat*. He was wearing a patchwork parka made of different animal skins. When the *ircenrraat* asked about them, he said that each represented the catch of a relative. Worried that the *nukalpiaq* was a powerful man with a large family, the *ircenrraat* decided to release him. Three doors led out of their *qasgi*—the lowest going farther underground, the middle door opening to the earth's surface, and the upper one into the sky. He was instructed to take the middle door back to his home.

Qasginguaq is not the only mountainous place *ircenrraat* inhabit on Nelson Island. Paul John (January 2006:210) also spoke of Qukaqliqutaat: "They say there is a village of *ircenrraat* there. When I traveled upriver to the southern part of our mountain in the evening using a sail, we saw their lights at that

Amiik, the doorway of the *ircenrraat*, is the larger vertical rock face along the ice-covered coastline of Toksook Bay. *Nick Therchik Jr.*

time. A flame would rise up there in the [*ircenrraat*] village of Qukaqliqutaat. We watched that when we passed right below." Paul followed with the story of a hunter who came upon a den with wolf pups at Qukaqliqutaat. As he traveled on, two wolves began to circle him, moving closer and closer. Although he told them he had not injured their pups, they began to hit his legs with their tails until he finally shot and wounded one. The man continued on until he met a woman who shared food with him, revealing that it was the catch of her elder brother, the wolf the hunter had recently wounded. The hunter then accompanied her to Qukaqliqutaat, where he found a village in place of the wolf den. He entered the *qasgi*, where he saw a wounded man lying in pain. The hunter explained that his shot had been provoked, and the men there agreed to let him leave. Before they released him, however, they exposed him to *nunam aaleryarai* (shrews, lit., "the land's scurriers"). They brought in a bag of shrews and let them climb on him, wriggling inside every part of his body. When they filled his nose, obstructing his breathing, the man opened his mouth, then quickly closed down on the head of a shrew that was trying to enter. Crying that he had killed *nunam aaleryaraa*, the one that cannot be killed, the residents of the *qasgi* let him exit the way that he had entered, and he returned home. Paul concluded: "I'm telling you the story that they told about the mountains down there."[3]

Ircenrraat were said to enter and exit the cliffs of Nelson Island through Amiik (Doorway), located along the coast west of Toksook Bay. Paul John (March

2008:472) explained: "Before the land and rocks in the surrounding area eroded, it looked like a real doorway from a distance. They say that noises of kayaks preparing to leave could be heard from inside the mountain. And when they seemed ready, killer whales would surface and dive along the shore, heading toward the ocean. And when they returned, they would face the shore and dive and not appear again."

The high ground near Cape Vancouver known as Kalukaat is also alive with *ircenrraat*, and care must be taken there. Paul John (March 2008:456) continued:

> Alongside the last point of Nuuget, the place they call Kalukaat is another place where there are *ircenrraat*. They said that a person from Nelson Island, when walking [along that place], just as they left, was told to say, "I am your person. I am one of your people." They say that a boulder won't roll down and fall on that person from up [on the mountain]. But they say a person who is not one of their own, a boulder can roll down and kill them.
>
> They mention that there are five kayaks underneath the place where a rock collapsed down the coast at Umkumiut. When people from Nunivak Island were sleeping there, a [rock] collapsed on them and killed them, probably because they weren't their people.

Ircenrraat were also said to reside in hills to the south of Nelson Island. John Phillip (January 2006:233) explained: "Although there are no mountains in our area, we have hills. And our elders said that some had *ircenrraat*. Now, those two steep hills next to each other called Qaurraak are visible from our village. Those before us said that those two [hills] are homes of *ircenrraat* who would try to bring people inside." John then told the story of how a hunter traveling along the coast did not return, and people suspected that the *ircenrraat* of Qaurraak had taken

Nuuget, the rocky points along the western shore of Nelson Island, with Kalukaat rising above them.
Ann Fienup-Riordan

him inside their land. Their *angalkuq* went directly to the *ircenrraat* village and entered their *qasgi*, where he found the missing man and demanded his return. The *ircenrraat* were reluctant, but the *angalkuq* threatened them. He squatted in their doorway, and as he did so, pieces of soil fell around him. Fearing the collapse of their home, the *ircenrraat* allowed the men to leave, using their original path.

Nick Andrew (January 2006:240) described *ircenrraat* associated with hills in the tundra region:

> Up between the Johnson River and Akuliqutaq is a small river, and it flows out down below the place they call Kuvuartelleq. Before it exits is a very large *qas'uqitvak* [concave bank or hillside that echoes], and there are many grassy knolls.
>
> They say Qamuutaq, a powerful *angalkuq*, slept overnight there. When he woke in the morning, there was a heavy fog and no visibility. There were many people chasing an animal down there, making loud noises. They were apparently *ircenrraat* who were chasing sea mammals. They were very loud.
>
> That man said that when it got bright and the fog lifted, [the *ircenrraat*] got quiet and there was nothing there. It so happened that he would die that fall.

Ingriik, the mountain formed by Vole south of Nelson Island, was also said to be inhabited by *ircenrraat*, who not only appeared as animals but as *kass'at* (white people). According to Paul Tunuchuk (March 2008:145): "In the past Qayuq'aq, the late older brother of Henry Kanuk and his siblings, would be gone all day. He'd arrive at night, coming from Ingriik. They say that when his parents would tell him to eat, he would tell them that he had already eaten among *kass'at*. After being gone all day long, he would return home not wanting to eat."

Many *ircenrraat* are said to inhabit the hills along the Yukon River. Nick Andrew (January 2006:220) shared the story of Ingriyagaq (Small Mountain):

> A man lived in a village right above Marshall. They say it was across from that small mountain, along the river's edge not far from the Yukon River.
>
> They say that man, when he went across [the Yukon], followed the tracks of a wolverine in fresh snow. Up above the river we call Kuik is a small mountain that resembles a *qasgi*. [The mountain] is on its own along the edge of those Ingrilukaat [Mountains] located downriver but separate from them.
>
> He continued to follow it, and not long after, one side of its footprints became human tracks. Then after a while, the other side of its prints became human prints. It was changing form like that. He became even more curious and continued to follow them.
>
> They say that when he got to that place, since it had climbed, he followed it. When it got to the top, it went inside a den. After some time, he saw that it was a window. Then he lay face down and peered inside. He saw that it was a *qasgi*. A circle

of people was inside. In front of the seated people were masks, and there were even some in the image of wolves, the faces of animals. He saw that those people, even the one sitting behind the wolf mask, was a wolf.

He circled them as he looked down there, trying not to forget what [the masks] looked like. When he had gone all the way around and looked at them all, when he got up, [the clothing] along the front of his body stayed behind [on the ground]; it had rotted. He also saw that there was no ice [on the ground]. He suddenly thought, "How am I going to return home now?"

After looking around, he saw something floating in the middle of the Yukon River. Lifting his other leg, he said, "I wish I could land on top of that down there and land right outside my home with the other [step]." He was already hovering in the air, and when he landed there, when he lifted the other [leg], he landed right outside his home. His people thought that he had died, since he had been gone for a long time. That one had evidently become a shaman, and all those [whom he saw in the *qasgi*] became his *tuunrat* [helping spirits].

They say he would use his spirit powers during winter when he returned home. It is said there were many people across there at that small hill, and that was their village.

Golga Effemka (January 2006:227) told the story of Amiigtalek (the Door Mountains) on the upper Hoholitna River south of Sleetmute and the animal people encountered there:

That mountain called Amiigtalek [Place with a Doorway] has little people. They say it is the place where these animals live, like wolves, minks, and others; they say they live inside it.

They say that the doorway along the side facing the river is very soft. [The ground] moves [when walking on it]. Since a story was told [about that place], the one who accompanied him warned him not to go there. Although [his companion] was reluctant, the other was unrelenting and went to it.

That person went inside the stones, going down, and saw many people there. Some looked angry and never smiled. Some were extremely friendly. They were apparently animals. He said they would hand each other food, including figures of animals. They say that person didn't eat [the animal figures] but placed them somewhere on [his clothing]. He got pretty far. He didn't exit through the place he entered. Some people he passed looked angry. They were evidently wolves.

He left right away, and it seemed that he wasn't there long. It had already become winter while he was under there.

When he continued on through the doorway, he came up on the south side. When he went outside, some were there wearing warm clothing. They were evidently animals. He was glad that he hadn't eaten them.

About two years ago, they went to hunt moose downriver from our village, and a number of people slept [at that place] in a tent. When they heard something outside at night, they said, "Who's that?" Those outside replied in the same way, "Who's that?" The animals were replying to them by mimicking what they said. They became frightened, as this was the first time they heard anything out in the wilderness. That never occurred back when we used to travel. Since our land has now changed, they will begin to hear of occurrences like that. They say they were extremely frightened. They looked outside, but there was nothing there.

Since I didn't want to tell you false things, I'm telling the story to this extent.

Not only were hills and mountains home to extraordinary persons; the mountains themselves were viewed as knowing and responsive. Ingriik, for example, is widely known to react to first-time visitors. Martina Wasili (March 2007:164) explained: "They say those Ingriik down there are inclined to react when seeing people for the first time. They said if someone who is not of this village ascended the mountain, a heavy fog would suddenly form around them when they went up. They still do that these days." Paul Tunuchuk (March 2008:145) gave an example:

One day our son-in-law, who lives in Anchorage, arrived during spring, and when I was about to go and get water, they asked me if they could accompany me. I agreed. After filling our bucket from the lake down there, I asked her husband if he wanted to climb up [the mountain]. He said he wanted to. The surrounding area was clearly visible.

Then we left, and right before we reached the very top, there was no visibility at all. We couldn't see in the distance. Then we returned home. When we got pretty far down, I looked up at them, but they were clearly visible. I believed at the time that they really do [react when seeing a person for the first time].

Looking out the window of the Chefornak community hall on that stormy afternoon, Paul joked: "Since they are probably reacting from seeing you for the first time, our weather has gotten bad today. [*laughter*] We are on top of [Ingriik] now."

Some inland mountains were said to behave the same way. Joe Maklak (March 2008:179) spoke from experience: "That's what it's like up on the Kuskokwim River also. Those two upriver across from the village of Napakiak, far inland, are like Ingriik. They also tend to react when they see people for the first time. They call that Cikiyakineq in Yup'ik and Lonely Hill in English. I learned what that place was like when I went there. Fog enclosed us, and animals we were hunting weren't readily available there. They say they were reacting and being wary because of seeing us for the first time. That is also true, and I'm talking about it because I experienced it."

Mountains might also emit rumbling noises, presaging a human death. Martina Wasili (March 2007:164) explained:

> Once while we were staying here [at Chefornak], Ingriik started making rumbling noises. I ran inside and told my father. Then after making rumbling noises, the land underneath this village shook.
>
> Then he said to me, "Yes, they are rumbling [lit., "kicking"] again." He said they do that when shamans are about to die.
>
> And he told me, "Since it is occupied by *ircenrraat*, when a person who was a shaman goes to their place, they kick out of gratitude." Those two small mountains up there rumbled, and since this land is their foundation, it shook. That happened only once.

Paul Tunuchuk (March 2008:144) agreed: "*Tukaralutek* [They kicked]. They will evidently do that again in the future."

Ingriik is not the only mountain that rumbles. Paul John (March 2008:457) told the story of the rumbling noises presaging the death of Edward Hooper's grandmother in Tununak: "When she started to become ill before her death, the Kalukaat Mountains down the coast evidently started to make rumbling noises. When people went inside and mentioned that the Kalukaat were making rumbling noises, she said, 'Aa, the two who will come to get me are preparing.' They say that not long after that, she died. [She said that] because she was a shaman."

Stories are also told about how in the past Ingriik tethered people, preventing them from traveling past. David Jimmie (March 2007:166) noted: "My cross-cousin told me that they'd tether a boat that they hadn't seen before and wouldn't let it pass but make it stay in that place. Although they seemed to be traveling fast, they said Ingriik seemed to be situated in the same spot." Martina Wasili (March 2007:166) explained that tying people down, like obscuring their view, was the mountain's way of reacting to people it saw for the first time: "They said that if a boat passed by [the mountain], it would just sit there right below them although it wanted to continue on. They said that happened to people when the doorway of Ingriik was facing the water."

Theresa Abraham (March 2007:167) then told how a shaman was said to have turned Ingriik around so that they wouldn't react to first-time visitors.

> Our grandmother told me that Ap'akegtaar and Luk'aq along with someone else traveled with a wooden boat upriver. Then just as they went down below Ingriik, their boat suddenly stopped, yet they were continuing to move.
>
> He said the Ingriik up there stayed in the same place, and they waited anxiously to pass them, but they were actually moving. It was probably Luk'aq who told one among them with shaman powers to use them on Ingriik.

That clever shaman turned [Ingriik] around so that the side that faced the ocean now faced the land. When he turned them toward land, they finally began to move forward.

Ingriik have an English name referring to this event. Joe Maklak (March 2008:132) explained: "We call them Ingriik. With that shaman in mind, white people call it Turn Mountain."[4]

Ingriik is not the only place known to tether people. Paul John (March 2008:462) recalled the advice people were given when they found themselves in that situation: "They say if a person doesn't move from one spot although he is traveling, he is to remove his clothing and put them back on inside out." Conversely, some areas allowed swift passage. Ruth Jimmie (June 2008:155) of Toksook Bay noted: "They said that once Acac'aq and Maacuar had their trail shortened. Although it was a long distance, they arrived immediately because *tumiignek nanilillukek* [their trail shortened]. And they say some people's destination is extended."

Ircenrraat inhabiting hilly areas could also make people sick by throwing things at them as they passed. Theresa Abraham (March 2007:168) spoke of this happening at Ingriik, when her mother (who was raised in Quinhagak) traveled past it for the first time: "They evidently came through the ocean route to the place downriver in early fall. Then she said sores grew on one side of her body because those Ingriik down there, the *ircenrraat*, had reacted to seeing her for the first time. Those Ingriik down there are extraordinary." Martina Wasili (March 2007:170) said the same thing had happened to her father: "After traveling inland, when he arrived he said that those darn Ingriik must have thrown things at him. He said that things landed on him, then [those sores] suddenly grew on his body. Those were the things that were thrown at him."

Nick Andrew (January 2006:219) mentioned how the mountain known as Cuukvaaq might react if a person pointed at it: "The large mountain they call Cuukvaaq has a pointed ridge. It is the end of the row of mountains where the village of Paimiut is located. They say that an Athabascan went downriver to sell birch-bark canoes, towing them. Those who were upriver told him not to point at that place whatsoever. That one didn't believe them. When he pointed at it, waves suddenly formed and brought the things he was towing up in the air, and it killed him. From that time on, they tell people not to point at it."

The land can heal as well as harm. Elsie Tommy (March 2009:288) described how her mind grieved following her husband's death. Sitting outside her home, she would think, "The people out there have become frost," as they seemed not to feel any sadness. Elsie's health declined, and people advised her not to submit to her mind: "They said that I should speak about the things that were ailing my mind to those whom I felt at ease with, or to things on the land, including trees, grasses, and small birds; they said if I talked to those, I would recover. During summer they constantly bring me to the wilderness to spend nights." At eighty-eight, Elsie's good health has returned.

Nuna | Land

///

akula. Land between two topographical features such as between the river and the ocean.

allngignat. Tundra islands (from *allngik*, "round patch on the sole of a skin boot").

angayaarrluk/angayaarrluut. Boggy area/s under swamp (lit., "bad *angayaq*").

angayaq. Swamp, bog. See also *puglerneq*.

angussaagviutulit. Customary hunting and gathering areas (from *angussaag-*, "to try to catch something for food").

carr'ilqaq. Clearing.

cegnayuk/cegnayuut. Valley/s, hollow/s, ravine/s.

cingik. Point of land. See also *isquq*, *nuuk*.

ciulavik/ciulaviit/ciulaviggluut. Peat moss from eroding banks (lit., "those in front, preceding")

ekvik. Bluff, cliff, riverbank.

evinret. Grassy knolls or islands.

ingigun/ingigutet. Border/s between tundra and marshland.

ingriq/ingrit. Mountain/s.

inivkaq. Mirage effect, lifting a place so that it can be seen from afar.

isquq. Point of land. See also *cingik*, *nuuk*.

kaimaq. Loose soil.

kangeq. Top of mountain, tree; summit, top part. See also *qauqaq*.

kiarrvik. Place to scan one's surroundings. See also *nacessvik*.

maraq/marat. Marshland/s, muddy lowland/s, wetland/s between rivers and firm land.

maraspak. Very wet mud.

marayaq. Mud.

nacessvik. Lookout, place to scan one's surroundings. See also *kiarrvik*.

nevuq. Dirt or soil.

nikuyaq. Sticky gray mud.

nuna. Land.

nunalleq/nunallret. Old village site/s.

nunapik. Tundra (lit., "genuine land").

nunarrluk. Soil mixed with sod.

nuuk/nuuget. Point/s of land. See also *cingik*, *isquq*.

pellatalek/pellatalget/pellaanarqellriit. Place/s where people lose their sense of direction and get lost.

penaq/penat. Cliff/s, bluff/s.

pengulkuq/pengulkut. Tussock/s of grass on tundra (from *penguq*, "hill").

pengunquq/pengunqut. Mound/s.

penguq/pengut. Hill/s. See also *qemirraq*.

puglerneq/puglerneret. Swampland/s, marsh/es, floating bog/s (from *puge-*, "to come to the surface"). See also *angayaq*.

qaliraq. Tundra that drifts away (from *qaliq*, "top layer").

qalviryaq/qalviryat. Bog/s, quicksand.

qamiqunguaq/qamiqunguat. Tussock/s. See also *ungusqunguaq*.

qass'uq/qas'uqitvak. Concave bank or hillside, producing an echo.

qaugyaq. Sand.

qauqaq. Mountaintop, summit. See also *kangeq*.

qavyuqerneq/qavyuqerneret/qavyurneret. Sheer drop/s in the mountains.

qemiq/qemit. Hill/s, especially part of a ridge.

qemirraq/qemirraat. Small hill/s, small bluff/s. See also *penguq*.

qertungucuk. Hummock, mound.

qikertaq/qikertat. Island/s.

qikuq. Clay.

tevanquq/tevanqut. Ravine/s, valley/s, deep area/s in water.

tumyaraq/tumyarat. Trail/s, path/s.

ungusqunguaq. Tussock, small hummock covered by vegetation. See also *qamiqunguaq*.

uragneq. Slope of a hill.

urr'aq. White or gray clay; dry ground that blows around easily.

uss'aryuk. Place that erodes.

uverneq. Slope.

yuilquq. Wilderness, uninhabited place (lit., "place without *yuut* [people]").

Kuiget Nanvat-llu

RIVERS AND LAKES

FOR CENTURIES THE YUKON AND KUSKOKWIM RIVERS HAVE flowed from interior Alaska to the Bering Sea coast, depositing millions of tons of sediment in their paths. In the process they created a lowland delta crisscrossed by thousands of smaller rivers, creeks, and sloughs and dotted with both freshwater and brackish ponds and lakes. Half the land's surface is water.

The topography of water is complex, and dozens of Yup'ik terms denote physical features—bends, banks, straight stretches—as well as specific hazards along the way. Conditions are ever-changing. The surface of moving water is alive with current lines, eddies, and waves. Ice cover presents new challenges and dangers. People have learned to travel riverine highways both summer and winter and to draw sustenance from the fish and animals that live there. Elders speak of rivers and lakes with respect and awe, mindful of their dangers. Water, they say, is powerful. People were admonished to carefully observe rivers and lakes to understand them. As with every aspect of *yuilquq* (the wilderness), a person must always act with caution and respect when traveling, as both rivers and their inhabitants are aware and watchful.

Anuqa Carvaneq-llu Kusquqvagmi Kuigpagmi-llu |
Wind and Current on the Yukon and Kuskokwim Rivers

Every hunter in southwest Alaska needed to understand moving water, including how to read currents and water level, use the tree line to gauge location, and stay out of the shoreline shadow to keep off the bank. A river has different current speeds all along its length and width, constantly affecting how fast a boat can travel. Some say that one can make the best time going upstream by staying just off the main current over toward the shallow side, but not too close.[1]

Nick Andrew (January 2006:37) was raised along the Yukon just upriver from Marshall. He reports:

> Back when I was a boy, when there was an abundance of snow, the Yukon had a strong current in spring. The person who raised me had a boat with an inboard engine. As he was heading upriver and reached the area below Qiuret, he couldn't continue but would go across to the south side of the Yukon and pass there. The water along the bluffs would churn loudly from the strong current.
>
> These days, since there hasn't been an abundance of snow, many sandbars and islands have grown along the Yukon. The current is no longer swift, and water along the bluffs no longer churns.

Nick Andrew (January 2006:38) explained that winter snowfall also affects summer fish runs on the Yukon:

> In my observation, when there was little snowfall, few fish [salmon] swam upriver in summer. They don't like the water's smell along the sides of the Yukon River when water stays in its place. It starts to form an odor, and since the fish taste that, since the water is where they live, there aren't many fish when water is low [in rivers].
>
> But when there was heavy snowfall and the Yukon River was full of water, the swift current cleaned out the old water that had developed an odor along the banks. Then when it was time, an abundance of fish went upriver. Fish like fresh water. When the waters are contaminated at the river mouth, they don't get close to it.
>
> Today, [the river] isn't like it was in the past. But if there is heavy snowfall once again, the Yukon and Kuskokwim Rivers will recover, and there will be [many] fish.

Wind direction at the river's mouth also affects the availability of fish. Peter Jacobs (January 2006:49) explained: "When the wind wasn't favorable along the mouth of the Yukon River, their elders said that [the wind] pushed the fish and allowed them to pass far from shore. *Tenglluki-gguq* [They say it blows them away]. They mention that the same thing occurs [on the Kuskokwim]. But they say if the wind blows directly against [the mouth], the water then pushes the fish into [the mouth]. Like herding [the fish]." Nick Andrew added: "The wind pushes the water. When a strong wind blows, the current flows in the direction of the wind. When we go to the mouth of the Yukon River, the incoming tide becomes very high as the wind pushes the water up [to land] from [the ocean]."

Wind affects the character as well as the quantity of fish on the Yukon: "In spring before fish arrive, if the wind continues to blow from the north, they always say that the Yukon River would get fish from the north. When the fish arrive, the king salmon are small, their backs look green, and they are not so big. But if the wind had been blowing from the south, they say that there will be lots of fish, that it will bring fish in" (Nick Andrew, October 2003:119).

Peter Jacobs (January 2006:45) also described the south wind blowing fish into the Kuskokwim River and the north wind blowing them away:

> Those people didn't like the north wind. They said that the wind had blown away the fish down at the mouth of the Kuskokwim. They would say that [the fish] passed [the mouth]. Here they are swimming in the water, and how would the wind reach them? That is something to behold!
>
> Once [when wind had been blowing from the north], the wind changed direction toward the south. After a while, water started to appear behind our place [in the slough behind Bethel], and in time the grass disappeared. I said to my poor wife, "You see, after there hadn't been any water [in the slough], for the first time in my presence, water has appeared behind our place later in the summer, after the ice is all gone." My wife said, "I assume that it is pushing the fish [upriver]." What she said turned out to be true.
>
> After some time, they announced that people were catching king salmon. When we fished, we saw that small king salmon had entered [the river]. We stopped fishing while the fish were still abundant. We were in awe of that occurrence. Indeed, the [availability of] fish must be dependent on the wind.

Johnny Thompson (February 2006:141) recalled a well-known adage regarding fishing on the Yukon River:

> They say when the Yukon is about to lack fish, its limbs start to degenerate first. They refer to the streams, the tributaries of the Yukon River as *ipiit* [limbs]. They say the fish in those areas start to degenerate first. You know how there are blackfish and other types of fish in lakes and rivers. They say the flesh of those fish starts to degenerate first. They tell them to be ready and to give their fullest effort in harvesting food.
>
> And they said when the fish along the Yukon were about to recover, the upper sections of the rivers start to recover first. That became believable more recently. Our Yukon River up north started to lack fish. And people farther upriver started to blame us, saying that we had caused fish to become scarce through overfishing. Then before fish started to become [readily] available, the fish in the sloughs and the blackfish started to recover.

While wind and current affect freshwater fishing, many believe that the fish themselves determine the outcome. Peter Jacobs (January 2006:40) made this clear in his explanation of how fish traps were situated when he was young.

> When I began to observe over in the marshy area [between the Yukon and Kuskokwim Rivers], they had a fishing site where they set large wooden fish traps they called *taluyarpiit* in the river. There were probably five wooden fish traps lined up behind one another. And there were mainly burbot in the river.
>
> Since those people had compassion for one another, I saw that they kept open one of the sections of the fish trap farthest upriver for the fish traps that were situated below it.

According to Peter, people respected each others' fishing sites and did not argue over fish: "Back when they depended on fish and no Western goods were available, the place where they set their wooden fish traps was respected, as they knew the owner of that place. And they could not set their traps before that person had set his, although it was merely a place where they set fish traps. That was a way in which our ancestors kept at peace, and they didn't squabble over fish." Fish, Peter said, are wary: "I also heard the following precept in the *qasgi*. They said that fish are wary and should not be the source of disagreements [or they will not allow themselves to be caught]. They would set their wooden fish traps to catch fresh fish all winter in streams. It seemed that there were no admonishments when fish suddenly became abundant in fall. They didn't keep them from others and would say that the fish run was abundant, since people in the past lived compassionately."

Aarnarqellriit Qairet | Dangerous Waves

Reading waves was an important skill when traveling on water. Nick Andrew (October 2005:188) described searching for smaller waves to cross through when traveling up the Yukon:

> When traveling by boat, the waves are large in the Yukon River and near the mouth. You have to slow down the motor when there are large waves and cross through the smaller waves. You search the waves, and when seeing a low spot, you have to go to it. That's how one must travel in summer when there are large waves.
>
> In fall when the wind hits, the Yukon River is constantly wavy. Erinvayagaq and I used to go down to the ocean, to the mouth of the Yukon River. When we returned upriver, the waves appeared like mountains. We were down on the coast in a marshy area they call Arularqurvik. Large waves were in that area.
>
> Our boat was [made of] plywood with low sides, and we were carrying fourteen seals. Since Erinvayagaq wanted to go and he was my elder, although I wanted

to wait, I didn't say anything. He said that if we didn't leave at that time, the river would freeze on us. It was cold. My, since I would be very afraid if I wasn't driving, I took over the steering.

When he sat by me, he told me that he would constantly guide me, that I should go slow. He told me to search the waves, that there are smaller waves among them that I should go to and cross. So I did as he said. Sometimes he would say, "Over there," and I would turn to the smaller waves and cross them.

When it got calm, since I had my bird gun, I took it, thinking that I would shoot downriver. Erinvayagaq asked me, "What were you about to do?" I told him that I was going to shoot downriver at the place we came from because I had been so afraid. [*laughing*] Even during summer, when trying to [travel], you have to watch the waves.

Paul John (October 2003:61) noted that river waves are more dangerous than ocean waves: "The river's waves, the Kuskokwim and Yukon waves, are more dangerous because they are shorter. Ocean waves are less dangerous and let you turn when you're on them, but river waves come one after the other." Nick Andrew (October 2005:190) agreed: "The waves upriver from [Marshall] are bad, since they are close together. Although one wants to go toward smaller waves, since waves are close together, they aren't good to travel on. Many get into accidents around our village when there are large waves. Those who were born on the coast say that river waves are more dangerous than ocean waves because they [collide with each other] near bluffs. They caution us to go far from shore if we want to pass [bluffs], avoiding waves where there is a strong current." Frank Andrew (June 2005:183) said simply: "When it starts to get windy, ocean waves are good, as they are widely spaced, but waves on the Kuskokwim River aren't good to travel on, as they are close together."

Paul Kiunya (October 2005:191) described the effects of current and wind movement: "If the wind is blowing from downriver and meeting the current, the waves build and get large. They say *makluku* [it raises (the water)]. The wind makes those [waves] large, and the current helps it because they are meeting. It makes the water rough." Bob Aloysius (October 2005:192) noted the same phenomenon upriver: "Since rivers, including the Kuskokwim, are narrow in some places and have a strong current, sometimes when a strong wind comes from downriver, they get wavy. [In rivers] with bluffs along the sides, if the wind is meeting [the current], the water rolls."

John Phillip (October 2005:191) noted that waves form when wind and current collide in a *nakirneq* (straight stretch of river):

When the wind hits a straight stretch, we refer to it as *qan'errluku* [lit., "hitting it right in the mouth"]. When the wind hits the river directly, they say there will be large waves there if the current is flowing. The Kuskokwim River is no different.

The wind meets the current flowing downriver. If they collide, that area [of the river] will have large waves, but it won't be that way where [the river] bends and the wind isn't hitting. If you face [the wind], it will go in [your *qaneq* (mouth)].

John Phillip said that the opposite is also true: "If the wind is blowing in the same direction as the current, it won't do that. It will be smooth. *Aturuarrluku* [current and wind going in the same direction, creating smooth waters, from *atur-*, "to follow"]. They [also] call it *elivqerrluku* [from *elivte-*, "to flatten a standing object"]. When the tide is going out from upriver, if the wind is following the outgoing tide, it will smooth or flatten [the waves]." Nick Andrew (October 2005:193) noted Yukon terminology: "We call it *cipgarrluni* [current and wind going in the same direction, possibly from *cipegte-*, "squeezing one's fingers or hands down something to remove the liquid"]. And [when the wind blows] from the side, they say *kepluku* [it cuts across, from *kepe-*, "to cut off"]."

Bob Aloysius (October 2005:193) said that water in front of cutbanks on the outer edge of river bends is less wavy when traveling downriver: "You have to search [for waves] when you go somewhere when it's windy. Although it's wavy, the areas around cutbanks are smoother." John Phillip (October 2005:194) agreed that the edges of rivers are calmer in windy weather.

> Although there are large waves, we are instructed that the [river's] edges are calmer and should be traveled on. Sometimes, beyond the edge of the river where there is a strong current, there is a *qisneq* [edge of the current, side of the channel]. The *qisneq* is much calmer and has less current, although it is not directly on the *qacuvkaneq* [calm water]. Although the waves are large, [travelers] can go along that edge. They call that *qairet keluarrluki*, going behind the waves, not far from shore. That's what I used to go through sometimes when boating on the Kuskokwim River.

Another strategy for river travel is staying in the deep channel. John Phillip (October 2005:195) explained: "When there are very high winds and the tide is receding, the shallow waters are wavy. It's easy to tell the middle of the river although waves are breaking in all directions; the shallow water is noticeable. The deep channel usually has smaller waves." Paul Kiunya (October 2005:196) also noted that the surface of deep water is calmer: "The waves in the deeper waters of the river are calmer because they are farther apart. That's why [Nick Andrew] mentioned that the [middle Yukon] River has waves that are close to each other. Once you go farther downriver, the waves will be farther apart in deep water rather than along the sides."

Paul Kiunya (October 2005:196) described the waves formed when currents meet: "Water where the currents meet forms pointed waves. That is also dangerous; no matter what type of boat it is, they can fill it [with water]. Places where currents meet always have those." John Phillip spoke from experience:

> We call them *cing'iktaalriit qairet* [sharp, pointed waves]. Waves like that are bad. Those suddenly [collide] on their own. When they meet, the water splashes. That's where two currents meet, and large waves also meet each other.
>
> They call those *qairrliqellriit* [places with bad *qairet* (waves)]. Some are like that in our coastal area. Those don't occur just anytime. When the south wind is blowing, it creates those waves where the two currents [meet]. The areas I know are the mouth of the Kuskokwim River, the area across from Ilkivik River, and upriver from Kwigillingok. Those are the areas that have bad waves when it's windy that I've experienced. Even though [it's wavy], if you watch them and go where it's smoother, you can go through.

Wassilie Evan (September 2003:209) noted the importance of being able to see hazards up ahead when traveling downriver: "Those going downriver with the current have to be very careful and watch out. They would stop when the sun reached a certain spot. It would make the water glow, and it would be hard to see. You probably understand *ciqineq* [ray of sun, sun's glare]. It sparkles like ice. When it did that, they would stop [traveling] and continue the next morning."

John Phillip (October 2003:62) described waves in shallow areas: "In high winds, in a shallow place, there are waves that are constantly changing. They call them *etgalqitalriit* [lit., "those that hit shallow areas"]. Those are dangerous and scary." Conversely, according to John, waves around sandbars could play a helpful role: "The waves have a job. They let you know where the sandbar is when it's windy. When waves splash, they will strike there and be very large. It's dangerous, and boats can get beached there. That's how waves work."

Nick Andrew (October 2003:63) noted that some sections of the Yukon tend to be wavier than others, concluding, "There are many explanations for waves!" Peter Jacobs agreed: "The Kuskokwim is one river, but the sizes of waves are different. Some parts are wavy, and some have small waves. They considered the area in front of Napakiak to be dangerous. They would say before someone left, 'The area right in front of Napakiak is awake.' They would always tell us because they knew about it."

Mer'em Pinirtacia | The Power of Water

Young people were taught to recognize the dangers of currents, wind, and waves as a matter of life and death. Paul John (October 2005:65) told listeners what to do when caught in a desperate situation:

While some travel it gets windy. Sometimes he thinks that he might have an accident before he reaches home. When that happens, they were instructed not to give up. When they instructed us, they used small pieces of wood that float on the river as examples. They told us to compare ourselves to them and to pretend to be pieces of wood. If we can get through it, we would get out of it like the wood.

They also instructed young people about the possibility of panicking and told them that they should always keep it in mind when they traveled. If he panics, he will have an accident. They told us that if we try to float like the wood, it can help us not to drown.

Capsizing while traveling could easily lead to hypothermia and drowning. Peter Jacobs (January 2006:268) told how the Kwethluk River, which men descended during high water each spring when returning from camping in the Kilbuck Mountains, got its name: "They say the Kuiggluk [Kwethluk River] wasn't actually called Kuiggluk [Bad River] at first. They say that Kuiggluk, during spring when it started to flow, would kill a person. Because its current would kill a person, they named it Kuiggluk, because it was dangerous."

George Billy (February 2006:473) noted a grim adage pertaining to falling into water: "When the ice has been exposed to warm weather in fall or spring, a person who falls through will not be able to pull himself out. If one is in that predicament, he can throw anything like his hat or glove on top of the ice so that they will be able to find his body." Young people were warned about the dangers of hypothermia. George Billy (February 2006:474) explained: "Also in fall when the cold has brought on the first freeze, they told them never to swim in deep water while traveling. If a person swims, though he may be strong, the backs of his knees don't have much flesh. If his [ligaments] get cold, they will cramp up, and he will drown." Teaching these lessons remains essential.

Kuigem Nanvam-Ilu Cikua | River and Lake Ice

River travel presented unique challenges both during and directly after fall freeze-up. In many places, river ice was solid and safe. As George Billy (February 2006:469) stated: "In winter, rivers are not usually dangerous. But in fall during freeze-up, some of them are always dangerous." Roland Phillip (November 2005:5) noted the advantages of freeze-up before the first snow: "It is ideal when it doesn't snow before the ice becomes thick. But if it snows a thick layer of snow before it freezes, dangerous areas cannot freeze. They refer to it as *maqaucirluku qanikcam* [the snow blankets the ice and insulates it, preventing it from freezing solid]. Although snow is cold, [the area underneath] is warm. And although these rivers have a current, they freeze [when there is no snow]. But when a layer of snow covers it, that river will not freeze all winter."

Snow-covered holes in boggy areas known as *pingayunleggluut* (from *pinga-yunlegen*, "eight") were also dangerous. John Phillip (January 2006:19) recalled: "When we'd travel during fall when it started to get cold out, my father pointed out potentially dangerous areas. They mentioned areas that had water under-neath called *pingayunleggluut* [swampy, boggy areas], which don't freeze right away and don't form thick [ice] when snow covers them, what they refer to as *maqauciqerluki* [forming a layer of insulation on top of them]. They have swamp underneath, and one cannot easily pull himself out." John (September 2009:151) also said that such areas are recognizable by the plants that grow on them and therefore "noticeable with the eyes." Paul John (September 2009:398) added: "When thin ice formed in a swampy area and snow covered it, the snow prevents [ice] from getting thick for a while. *Caarrluut* [grassy wetlands] don't freeze right away. One has to watch those during fall."

Even in cold weather without snow, some rivers with strong currents never freeze. Bob Aloysius (October 2005:145) noted: "The Aniak River is not like the Kuskokwim. Some areas never freeze even in cold weather." Nick Andrew added that the same was true of the Ecuilnguq River (lit., "clear water"), upriver from Pilot Station: "The source of that [river] doesn't freeze, even though it's very cold. It's not deep, but it has a strong current." George Billy (February 2006:469) warned, "When a river is shallow, ice becomes thin where there is a current. Those are things to watch out for."

Some areas never freeze solid. Nick Andrew (October 2005:138) explained: "Here in the [middle Yukon and Kuskokwim] Rivers, there are dangerous areas near trees, and [ice] doesn't get thick right away. Dangerous ice has a brown color and doesn't freeze. Those from the north where I'm from say that *qallanret* [springs] don't freeze completely even though it's very cold. They told us to avoid those places. Also, *quurneret* [narrow sections] of lakes and rivers as well as lake outlets don't freeze all winter."

John Phillip (October 2006:139, 154) noted that *carvat*, the fast-moving streams flowing from hills and mountains, never freeze solid. He gave an example from his experiences hunting in the mountains near Manokotak: "Around the mountains of [Bristol Bay] there are many dangers like the streams. Even in the dead of winter the ice doesn't get thick." John Walter (March 2007:1361) of Tununak also noted that the numerous *carvat* descending from the mountains on Nelson Island are a potential hazard when traveling in winter: "*Carvat* usually do not freeze solid. Sometimes we quickly come upon them [when traveling]. A person mustn't travel along the edge of the mountains with a snowmobile, especially without a traveling partner, without knowing [where they are]." Dick Lincoln (April 2009:103) com-mented: "The upper parts of these mountain streams actually freeze during fall but melt when snow has fallen on them. They no longer have ice, and the snow hovers [above the stream]. That's why some birds, especially these pintails, spend winter along those *carvat*, since they know that the entire length of that mountain stream is melted and their food is eatable and not frozen."

Paul Kiunya (October 2005:179) described eddies and areas of open water in ice, what he called *qenuilquut* and John Phillip called *kianret*:

> When rivers freeze, even though it's fresh water, some areas where there are currents have *uivneret* [whirlpools, from *uive-*, "to circle"]. When we're traveling by boat sometimes and reach those whirlpools, the boat doesn't stay still. Those areas seem to stay open for a long time, even on the Kuskokwim River.
>
> This person just mentioned the areas with eddies. *Uivneret* are like eddies. When there is a strong current, those appear from underwater. The current makes them go round and round. Sometimes, even in winter, in our river down on the coast, those open areas take a while to freeze.
>
> When there's cold weather and snow fills them, they finally freeze. But those are still dangerous, even though frozen. We call them *qenuilquut* [lit., "places without *qenu* (ice)"].

Paul Kiunya (October 2005:198) explained how peat moss along eroded riverbanks insulates ice and prevents it from getting thick: "Sometimes, the current takes pieces of peat moss from the land's edge. And when those freeze [to the ice], the peat moss insulates it. It is as if the ice is wearing a warm coat. We call pieces of peat moss taken out by the current *ciulaviit* [from *ciu-*, "area in front"]. If they freeze, the [ice] will be brown like the peat moss."

Bob Aloysius (October 2005:184) described murky areas of water that never froze: "They always told us to watch out for *ecurlirrluut* [lit., "those that are murky"]. They don't freeze, even though it's very cold." Nick Andrew (October 2005:200) described oily, brown, stagnant water that does not freeze solid: "We call that *qallanerrluk* [lit., "bad *qallaneq* [spring]"). The water looks like something is added to it. It is old water that has stayed for a while and has a light-brown color. It's gotten bad, and it doesn't freeze thick although it's cold. It looks oily and appears as though it's stuck there." Peter Jacobs (January 2006:76) observed: "In early spring when the edges of lakes and rivers start to form *imaquq* [murky water], they say that [type of water] isn't firm when it freezes and easily breaks. They would caution us about that."

Nick Andrew (October 2005:340) noted the warning to avoid any unusually colored ice: "In our area they tell us not to go on ice, even if it looks good, if the color is different. It doesn't get thick. That's why so many people run into accidents. Sometimes it's in the river, close to the bank, and when it's overflowing and becomes brown, it's not good to step on."

John Phillip (October 2005:154) noted that although coastal rivers freeze solid, they also have dangerous places, including treacherous areas where the ice cracks and shifts upward, lifting away from the water underneath:

Our rivers downcoast in marshy areas freeze all the way to the headwaters, but their outlets from lakes up there freeze later and stay open for a long while. Also where it exits [is open] for a long distance, where the current flows. When we went hunting, we avoided those places and crossed them a great distance away. Then, when our rivers froze, they froze nice and smooth. They were easy to cross.

During winter, when it gets cold, the ice that had frozen smooth [expands] and shifts upward [in the center]. They say [the expanding ice] shifts up and becomes a *pequneq* [ice ridge, from *pequq*, "upper back"]. It becomes pointed and the area underneath hovers and is not frozen. That occurs in the area [of the river] where the tide is coming in and out and also along the [river's] edges. . . .

The ice doesn't stay in one place when it freezes. Sometimes it springs up or snaps out of position. If a hole is made at the crack [along the ice ridge], the bottom [of the ice] hovers [above the water]. When we harvest needlefish, we make a hole through the crack [along the ice ridge] since it's thin and search for needlefish in those places. The area underneath the crack isn't frozen.

Nick Andrew (October 2005:182) noted a similar phenomenon on the Yukon.

In our village, when the ice freezes, the water's pressure pushes the ice, and it forms a ridge, and there is a fissure in the river's center as [the ice] rises. The pressure from the water that gathers pushes it [up]. Water is strong. Even though [the ice] was thick, when that happens, we say *culuksugtuq* [from *culuk*, "dorsal fin"].

And in spring, when the snow melts on the surface of the river, the ice is pushed upward and cracks. While the sides are attached, when the water melts in spring, the water pushes it.

John Phillip (September 2009:151, October 2005:154) also described *qas'urneret*, dangerous holes along riverbanks where incoming tides have broken through and weighted down the ice.

When the sides of rivers form *penqunret* [ice ridges] from the pressure of the incoming tide, causing them to crack and rise, when their sides fill with water [i.e., when water collects in the depressions near ice ridges], they call them *qas'urneret*. They were cautioned about the edges of the ice ridges, those *qas'urneret*, since they form holes and fill with water. That also occurs on the Kuskokwim River, down below [Napakiak] and at the mouth of the Johnson River. The area underneath is dangerous. It is filled with water, and the ice on it is thin. One cannot easily climb out because the bottom [of the hole] is very slippery. Some people fall into those inadvertently and sink; and I also see snowmobiles that have fallen into those areas along the banks of the Kuskokwim. Even though ice is underneath, the bottom of the *qas'urneq* [hole] is deep. . . .

In spring those areas become dangerous when they suddenly break off; thin ice covers them, and the surrounding area melts and becomes brown-colored. The sides of coastal rivers and the Yukon River fill with water.

Lake edges could also be dangerous before freeze-up. According to John Phillip (January 2006:101): "When these lakes freeze during fall, the [ice] edge along the north side exposed to the sun doesn't thicken right away. [The ice] along the north edge is thinner, as it is sheltered from the wind. In my village I tell children not to go ice skating there before the ice thickens. The sun shines on that location [and melts it]." Paul John (September 2009:395) explained the wind's role in creating these dangerous areas: "Before ice becomes thick, when the wind blows its water to one side, water covers the ice and it goes down. One side [of the lake] isn't covered with water, but since water covered the area along the windward side, it pushes down the thin ice and the water above it becomes deep. There would be two [layers] of ice, with water between them. A person can fall through when they happen to travel over those."

Bob Aloysius (October 2005:182) described the formation of pressure ridges in rivers and lakes: "Water expands when it freezes. That's why you get those pressures that push ice against the bank and tear it up. When the ice expands in lakes, even small lakes, it has no place to go, so it creates those pressure ridges. Rivers also have pressure ridges. The more ice freezes, the more it expands. Really powerful, there's nothing to stop it. When the ice goes up toward land, it even damages the bluffs." Paul Kiunya (October 2005:181) described *icinret*, overhangs along the ice edge: "[Expanding water] pushes the ice toward the edge of the river. We call the ice that is pushed toward the land by ice along the river *icinret* down on the coast. Because water is very strong, stronger than us, when it freezes, no matter how thick [the ice is], it pushes it up toward land."

Nick Andrew (October 2005:181) noted the admonishment to avoid ice that formed when the water was high: "They also tell us to watch lakes and rivers that had frozen during high water. When the water underneath leaves while the [ice] is thin, it does not freeze all winter no matter how cold it gets. And someone who goes on top of that will fall through." Peter Elachik (October 2005:185) noted that Kotlik residents use willow branches to mark these dangerous spots that people should avoid within several miles of the village.

Paul Andrew (November 2005:9) described dangerous ice surrounding sandbars at the mouth of the Kuskokwim: "Although the Kuskokwim River freezes in winter, these sandbars and other places filled with water from the high tide are dangerous. When the tide is extremely high, it detaches the ice [from the sand-bar], and the water piles [the ice] on sandbars upriver. They get dangerous at that time." Roland Phillip explained: "When the tide comes up, the incoming water detaches the ice. The [ice chunks] aren't large, and some will surface seemingly out of nowhere. They are cautioned not to travel on top of sandbars since it's hard to distinguish where those places are. The ice that detached floats away, and then the ice along that sandbar will not be thick."

John Phillip (October 2005:158) noted that streams with fish tend to form thinner layers of ice.

> Streams that have fish in them are easy to distinguish. When I look down at a river, although it's frozen, there will be bubbles in the ice. The [bubbles] indicate that there are fish in those rivers, and the ice is usually thin there also. And when snow fell, rivers that had fish never froze all winter. Back when there was a lot of snow, they would leave their wooden fish traps there all winter. Blackfish were available all winter. Soon, they would start to smell a little like feces, but they were tasty if seal oil was spilled over them. That is how fish in those rivers become when their water gets dark.
>
> And nowadays, it is no longer like that; the [lakes and rivers] on the coast are starting to freeze down to the bottom because there is less water and not a lot of snowfall, and lakes are emptying and no longer have fish in them.

Lake ice could also be dangerous. John Phillip (October 2005:175) noted the admonition never to walk on *mingqutnguat* (rotten or needle ice, lit., "pretend *mingqutet* [needles]") when muskrat hunting during spring: "When those musk-rats became a source of income, when we finished [spring seal hunting], we trav-eled out to the wilderness, inland from our village, to hunt muskrats. The lake ice becomes uneven, and when you step on it, it moves underneath. They cautioned against ice that had dark areas on it; when it has become *mingqutnguat*, it becomes dangerous." Paul Kiunya commented that rivers also have *mingqutnguat*: "That occurs on the Kuskokwim River also. In spring, when it starts to melt, I notice that parts of the river have already become needle ice, especially the thinner ice. The Yukon River probably does that also." *Mingqutnguat* were sometimes heard before they were seen. According to Peter Jacobs (January 2006:96): "They say that if we get close to a place where there are *mingqutnguat*, there will first be a sign of dan-ger, as our walking will produce a rustling noise; they said that the *mingqutnguat* that were hovering inside [the ice] would be making rustling noises." John Phillip (October 2005:175) explained why *mingqutnguat* are so dangerous:

> They told us not to walk on *mingqutnguat*. They said that if a person fell into the water in that [type of ice], the area above him would immediately close in. They

would not be able to surface, as those [ice needles] would poke him. They are cautioned about those dangers.

[Areas with that type of ice] are noticeable; they are rough and dark. But thick [ice] is white with a little bit of snow on it. That is how some lakes melt; all parts don't melt the same, and some have [*mingqutnguat*]. When we'd travel [pulling our kayaks] with our kayak sleds, we'd avoid those areas. When we'd come upon those as we were traveling, they immediately started to move.

[*Mingqutnguat*] form along lake edges, along what they call *menglairneq* [water's edge]; the whole lake doesn't have those. Certain areas get those first.

George Billy (February 2006:445) had also been warned against *mingqutnguat*: "They told us to be wary of those when driving dogs. They were afraid of those *mingqutnguat*." Nick Andrew (January 2006:98) warned that there was no way to survive falling through needle ice in deep water, as it was heavy and would fill all one's clothing, weighing one down.

Some conditions are dangerous to animals but not to humans. Peter Jacobs (January 2006:48) recalled the *anlut,* or exit holes, muskrats and otters keep open in lake ice during winter:

Muskrats have *anlut* on lakes that they fill with moss out from under the lake and keep open all winter.

They knew when the weather was going to kill [the muskrats]. They would start to say, "The weather outside is probably suffocating the poor, small muskrats."

When spring came and it was time to paddle and hunt muskrats, after they had said that, sometimes [hunting] all day, we'd catch only four. [The weather] had apparently suffocated them when *cikurlak* [ice formed by freezing rain] covered [their exit holes], making them airtight.

Animals knew the ice, and dogs especially helped their owners find safe trails. Bob Aloysius (October 2005:147) recalled: "My father would let [his dogs] run slowly, and sometimes they would suddenly turn, although the trail seemed okay. They avoided the places where there was no ice. [The dogs] knew the condition of the ice on their own. By hearing the current under the ice, they'd avoid [open areas]. Back then, their dogs were their partners."

Ayaruq | Walking Stick

Considering the dangers associated with river and lake ice, an *ayaruq* (walking stick) was an essential traveling tool. According to Peter Jacobs (January 2006:95): "This is how they explain *ciunrinaaryaraq* [the way of checking ahead on one's path]: we are to always check the area ahead. When they asked us not to move ahead without caution, they told us not to be without a walking stick when we went anywhere during fall and spring, as it reveals the dangerous areas." Nick Andrew (October 2005:141) elaborated:

> They never let us go to the wilderness without a walking stick and a gun. They adamantly cautioned about those dangerous places; they said a person will fall into the water and drown if he tries to travel through those conditions.
>
> Sometimes when traveling during spring when it warms up, brown spots appear on the river; one must not go on those areas. Safe areas will be blue and white. Even though it looks safe, they would always want us to carry the walking stick. In the spring, they made them with long shafts and metal points. If one accidentally fell in the water, he would push [the walking stick] down to the [river] bottom, holding the base [of the shaft], and pull himself out. [One did that] if he couldn't lay [the walking stick] on its side horizontally [on the ice] and get out of the water.

Travelers also used walking sticks to test the ice. According to Nick Andrew (October 2005:142): "In spring and during freeze-up, they would warn us not to go on the ice if the ice is penetrated on the first strike. But they said that if the ice is hit three times with great force and it doesn't create a hole, a person will stay on it [and not fall through]. They never put very sharp tips on our walking sticks because that can easily create a hole through the ice. A sharp tip isn't a very good ice tester."

In deep-snow years people were admonished to take care. Nick Andrew (October 2006:143) continued: "They told us to be extra careful when there was lots of snow because the snow will prevent thin [ice] from thickening. Also, open water [on ice] will be covered [with snow]. You can't tell that it's dangerous by just looking at it with eyes. Sometimes when we constantly traveled on foot, when we'd use a walking stick to strike an area that was covered by snow and appeared good for crossing, no ice was underneath."

Bob Aloysius (October 2005:145) told a story of falling through ice covered by new snow:

> They warned us to be careful, especially after it snowed. They said that it would melt quickly from underneath. I never believed it. [*laughing*]
>
> It snowed lightly all night. We used to go trapping by walking. My uncle told me to watch out. After we went upriver, we'd cross through the *cikullat* [newly frozen ice]. My uncle said that when he turned back, all of a sudden, I got short. [*laughing*] I fell through the ice, but my gun stopped me.
>
> I rolled to get back up on the snow. I went back to the tent since it wasn't far. You'd have to check that area where we crossed in the morning with a walking stick before you crossed it on your way back.

Although discolored ice was dangerous, ice mixed with sand was reliable. Nick Andrew (October 2005:149) explained:

> On the Yukon, ice that has sand or soil on it is solid. We travel through those areas when it gets dangerous. Even though some ice is floating, those large [ice pieces] that have mud on them are strong.
>
> Since sand always blows on the rivers, both the Yukon and Kuskokwim have [that type of ice]. If the ice is getting dangerous, a person can cross by stepping on that type of ice. Sometimes when returning from the wilderness, we see that type of ice on rivers. Although there is not much ice, they tend to stay afloat and solid. We used to cross rivers on them using a walking stick.

Walking sticks were also used to test open water, as some lakes were deep. Lucy James (March 2007:1327) of Tununak found this out one spring:

> One time [when I went egg hunting], there were two birds sitting on their eggs on an island down there. When I looked at the [lake], it looked shallow. I poked my walking stick [in the lake], and it suddenly went in deep. Then I poked around in the [lake], but it had no bottom. My goodness! I was disturbed by that for a long time. When I told my peers about it, someone told me, "It was probably trying to capture you!" They said that if I hadn't had a walking stick, I would have immediately sunk.
>
> The bottom was visible and looked like the bottom of the lake. I didn't try again and left.

Kuiget Nanvat-llu Up'nerkami | Rivers and Lakes during Spring

Peter Jacobs (January 2006:47) recalled the adage that early melting in spring created conditions in which the water would not freeze again: "Sometimes during spring, long before summer, there is a lot of water. They would start to say, 'I think [the weather] out there is forming conditions in which the water won't be chilled or frozen.' I didn't understand what that meant at that time. Evidently, they said that when summer would arrive before the water froze [again]. These were the indicators that people used to predict future conditions."

George Billy (February 2006:462) noted the adage that ice thaws from the bottom up in spring: "The dangerous parts are obvious. [Ice] along places with a current thaws fast in spring. In one day some [areas] are ice free because ice thaws from the bottom first." John Phillip (October 2005:159) agreed:

> When rivers start to melt, the snow [covering the ice] melts and forms *miiqaq* [fresh meltwater]. When that *miiqaq* forms, the lakes and rivers begin to melt from underneath.

We were cautioned about those in spring on the coast. They told us that rivers in the wilderness inland melt from underneath. Only the soft snow is above [water]. If we don't know that the ice on the river has thinned, one can fall through.

I've seen those many times when I'm traveling in the wilderness. And we are cautioned not to quickly cross rivers down along the coast when they are starting to melt. The [snow and ice] hover above water, as the incoming and outgoing tides dissolve them. They become thin and dangerous.

Following breakup, rivers brought an abundance of wood. John Phillip (January 2006:24) teasingly described how coastal residents relied on the wood their upriver friends sent them:

This person [Golga Effemka from Sleetmute] sends those who live on the lower Kuskokwim coast wood that we truly use. And when they flow out to the ocean after the river ice breaks up in spring, they beach along the shore. When I became aware, every spring, logs would flow out of the Kuskokwim River. The people always had wood to construct their kayaks, sleds, and paddles.

[The availability of wood] was made possible by snowfall; [the logs] that had been [pushed down] from the mountains near [Golga's] village drifted down, and they probably include Nick [Andrew's] logs, as he is from the Yukon River. Long ago we were given [wood] by the work of snowfall. These days this person hardly sends us any, probably because there isn't much snow nowadays. [*laughter*]

John Phillip (January 2006:24) also described portaging over the flooded tundra, hunting for birds and muskrats: "During spring down on the coast, when the land melted in the wilderness, it was covered with water although the lakes hadn't melted. When traveling in the wilderness, it was possible to paddle anywhere one wanted to along places with water, back when there was an abundance of snow. When I first started to travel by kayak during spring, I traveled wherever I wanted, using a pole to push myself along."

Kuiget Tuvingalriit Cikumek Qanikcamek-llu |
Rivers Choked with Ice and Snow

A heavy snowpack and rapid melting often translated into flooding along creeks and rivers in spring. Ice jams lower on the river also contributed to flooding and high water above them. John Phillip (January 2006:57) recalled severe flooding in the 1940s in the Bethel area:

I once saw the occurrence of heavy snow in recent times here along the Kuskokwim River. During that time, when the military arrived here, I worked across [from

Bethel] in the summer. And during that winter, there was an abundance of snow. And there was a lot of water when the Kuskokwim River ice broke up. [The water] was halfway up that building at that time.

And it floated away all of the military's equipment, including many barrels of gas, and I, too, obtained a lot of gas as they floated out to the coast. I'm not lying when I say that I never bought gas for two years. And I don't think the Kuskokwim River has had the same amount of water since then.

Peter Jacobs (January 2006:57) agreed: "At that time the water level covered Mission Road, an area that is actually very steep. I also watched that large flood. And they said it would float a house that was able to float. At that time there was *qanikcaryak* [an abundance of snow]. After that, there was very high water, and the military blew up the ice near Napaskiak to break it up when the river choked with ice."

John Phillip (January 2006:58) described the river choked with ice: "Although there is a lot of water, when the river's deep channel fills [with ice], it floods like that. When the current is flowing up at the top, then the ice piles down [inside the river], what they call *tuvvluni* [choking (with ice)]. It becomes confined as the current pushes [ice] underneath and eventually blocks the river, and the high water rises." John Phillip (September 2009:301) also described *qanisqineq* (snow in water) choking up lakes in fall: "When it starts to get cold and *qanisqineq* stays, it becomes thick and solid and safe when it freezes. That's why sometimes when it snows, they say that it creates conditions where [the snow] will choke up [the water]. When it gets cold, that snow is dense when it freezes in the lake."

Cevenret | New Channels

All across the delta, new channels known as *cevenret* (from *ceve-*, "to cut through land") are constantly created and emptying into new sources. John Eric (March 2008:62) shared one of hundreds of examples: "This is no longer a river, and this has turned into a river. A *cevvleq* [new channel] formed where there wasn't a river before. And the outlet of Kiigaq River into the lake is no longer a river today."

While most *cevenret* occur naturally, some are made by men and women to shorten travel down windy rivers and streams. In the Toksook River alone, men have cut four *cevenret* over the last fifty years. Paul John (March 2008:477) explained:

I think Nasgaum Atii and Nuyarralgem Atii made this [first] stream. After that, the people of Nightmute made this stream. And people from [Toksook Bay] and from Nightmute made this stream also. I think they mentioned that Peter Dull made this [last] stream.

I first saw this area back when there were no channels. They'd use shovels to cut across the land to make streams. Today they have become wide and have currents, and barges travel through all of them today.

The [old channels] are no longer rivers and have turned into land. They have grass growing on them.

Michael John (June 2008:42) told the story of the new channel said to have been created by a shaman along a river near Newtok:

They say that long ago, there was no river here. But when that elderly woman who was a shaman was about to die, she didn't want her grandchild to endure hardship when he portaged here. She said she wanted to be buried with a model of an *ussugcin* [land-prying tool]. She held that model and a model of a striking tool when she died. Then after she died, this evidently cut through the land and formed a new stream. When they looked at it, they saw that the edge of that channel appeared as though it had been [cut] with an *ussugcin*. And they said that became a river.

Imarpiim Ceniini Kuiget | Coastal Rivers

While inland rivers and lakes are filled with *mepiaq* (fresh water), some coastal streams and ponds are tidally influenced and brackish, creating very different conditions. John Phillip (October 2005:151) shared his experiences: "I'm going to talk about the lower Kuskokwim coastal area. Our rivers are not the same down there. Our [Kongiganak] River is freshwater, and the only part that is saltwater is the river mouth. But the rivers of the villages of Kipnuk and Kwigillingok are all salty since they are close [to the coast]. Since our river is long, it is freshwater. It freezes just the same as the lakes." John (October 2005:152) noted that ice formation on salt- and freshwater rivers differs:

Saltwater is slower to freeze than fresh water. The saltwater [ice] on the coast is much stronger and denser when it freezes, even though it's thinner. But freshwater [ice] can crack easily. It is more dangerous than [saltwater ice].

Now, before our river freezes in fall, the river will first fill with [ice] that has layered and gathered [with the tide]. When it drifts when there is a current from [upriver], it gathers and [the pieces] eventually get larger. Then they stop in a pack and block the river. The area up above [the river mouth] becomes choked with ice, and the area upriver from it stops and builds up and freezes. But the mouth of the river remains unfrozen.

But it freezes downriver. Then when the tide comes in bringing [ice] from the sandbars, it fills the river and it freezes. They used to tell us that ice is dangerous when it first freezes. The piled ice becomes safe first [during freeze-up].

John Phillip (October 2005:153) described using a *negcik* (gaff) to test the salt-water ice, similar to using a walking stick to test freshwater ice.

If I'm going to cross a river, I will test the river with my muscle. If I strike the ice forcefully, this [*negcik*] will go in the ice, and this [shaft] will enter along with it. If [the shaft] enters [the ice], that is a sign that I should not cross that place. That is our admonition in the coastal areas.

But they said if I strike with a lot of force, although a hole developed [through the ice], if it stops underneath [the wooden shaft], I would be able to stand on the ice and continue on. And although the ice in the ocean is thinner, it is much stronger. If [the gaff] comes to a halt [at the base of the shaft], although [the ice] seems thin, I am able to travel on it.

Although saltwater ice is stronger, it melts more quickly because it has a lower freezing point (28.5°F in ocean water with an average salinity of 35 parts per thousand), thus requiring colder temperatures to stay frozen. John Phillip (October 2005:177) explained: "In spring, when saltwater forms *cikullaq* [frozen floodwater, ice that freezes along open water], the sun melts it fast. But the ice on lakes seems to melt slower. When it gets warmer and the surface of the ice starts to melt, that's how it is. Although *cikullaq* forms, [the sun] melts it all, and it is less hazardous than in the fall."

Paul Kiunya (October 2005:178) used a contemporary observation to explain the different properties of salt- and freshwater ice: "Around [Anchorage], when the steps are slippery, they sprinkle that white stuff on them, and the ice melts even though it's cold outside. That's salt. Where they sprinkled it, it melts the ice. That is why saltwater melts fast when the weather gets warm. But the lake's fresh water doesn't melt right away, and the ice becomes needle ice since it melts slowly." Paul (October 2005:179) noted that on the coast saltwater lakes also melt faster than lakes with fresh water.

Saltwater ice is also more flexible. Mark John (December 2005:115) remarked, "Saltwater ice is more elastic and can bend. You can walk on it, even though it's moving." Paul Kiunya (October 2005:179) described an experience familiar to many coastal residents, past and present:

Around my village in fall, [the lake] that is close to the slough has saltwater and it would freeze. Back when we used to ice skate, although the ice was thin, we would [skate on it] as the ice was moving. The ice would never break suddenly, because saltwater ice doesn't break in the cold.

And they told us to quickly go back to land if we're skating on freshwater ice and [the ice] starts to move. Freshwater ice easily breaks. If you strike it a little, it shatters easily. Saltwater [ice] is stronger and more solid. They are different.

Nanvat (lakes and ponds) have their own typology. *Akulnguyaat* (from *akula*, "area between") are the narrow sections of lakes where women in the past regularly set nets to catch birds in spring. *Akuluraq* is a river or stream connecting two lakes. *Kangiq* (lit., "beginning, source") or *qagan* (lake source) is a headwater lake or source of a river, and *igyaraq* (lit., "throat") the outlet where a lake empties into a river. *Kuiguaq* (lit., "imitation river") is an oxbow lake, a long, narrow U-shaped lake formed when a river bend is cut off from the river on both ends. *Egmilleq* (from *egmir-*, "to keep going") is a river branching out from the lake source of another river. Roland Phillip (October 2005:318) explained: "When a river continues on, it will flow out into its lake source. And then, across there, if another river branches out that isn't an *akuluraq* but extends for a great distance from another river's lake source, they call that *egmilleq*."

Many coastal lakes surround small islands. Roland Phillip (October 2005:318) noted: "They call islands in lakes *ilanret*. They are small islands; they refer to that marshy area as having a lot of *ilanret* where birds nest and lay their eggs, especially these gulls. And they go and look for eggs around those *ilanret*."

The color and water quality of tundra lakes and ponds vary widely across the delta. Paul John (January 2006:140) explained: "The contents of lakes vary. Those with clear water have moss on the bottom, but another lake not far away is dark. I suspect the bottom of that [dark] lake to be real soil. They call the sand on the bluffs *kaimaq* [loose soil]. [Lakes] that have that kind of bottom are murky." *Urr'aq* (white or gray clay) can also make lake water appear murky (Paul John, December 2005:118).

Nick Andrew (January 2006:141) noted that brown, murky water can be undrinkable: "Some plants also color lake water. You know how the sides of the Yukon form bad, contaminated water before spring breakup cleans and flushes it out. Like that, the bad water in some sloughs and eddies is undrinkable and brown in color." Peter Jacobs added:

> Rivers that are around the tundra seem to be that way also. Their water tastes like tundra. . . .
>
> Villages on the Yukon River that are toward the mountains have wonderful-tasting water. We drink it when we take a steam bath, and it is very blue and clear. But the water is bad where swamps are located. Yes, the water is terrible.

Marie Meade (December 2005:118) from Nunapitchuk noted: "Akula [the tundra area between the Kuskokwim and Yukon Rivers] has dark water. The Kuskokwim's water is different, too. When we go up to Akula, the water gets darker." Mark John added that water color varies greatly between interconnected rivers and streams: "Part of the Yukon River flows through the Qissunaq River [to the Bering Sea], and some other lakes around there flow into larger bodies of water. At some point the water from those larger areas flows back in, giving those lakes a different color, and then they flow out. There are also lakes that don't seem to have any outlet, and they also have different colors." In fact, flying over the delta, different-colored lakes with no outlets are visible side by side. Such differences may be due to disturbance of clay-sized particles on the lake bed, algae blooms present in one pond but not another, and different water chemistry.[2]

A lake can drain out, becoming a *nanvalleq* (empty lake bed). Paul John (December 2005:113) noted: "One of the lakes near a river develops a new channel and flows to that river, drying part of the lake. [The water] leaks out to another river when it begins to flow. That is how some lakes end up becoming dry."

Water levels in lakes and ponds are tied to both summer rain and winter snowfall: "Back when there was lots of snow, the lakes were filled all summer long. At the present time, since snowfall has decreased, lakes are getting empty and some are dried up. Rains also helped in summer, and people were thankful when it rained" (John Phillip, December 2005:133). Paul John (December 2005:141) agreed: "Some summers when it doesn't rain much, lakes get shallow. If it constantly rains in summer, the lakes are full."

Periodic rain in winter also slowed the rate of snowmelt, as it created icy barriers within the snowpack. Elders considered this a good thing. John Phillip (December 2005:141) explained: "Sometimes it rained when it got warm. Our ancestors liked that because it slowed the melting process. Snow would melt slowly in spring after it rained on and off. When that happened, lakes slowly emptied out and had ice in them for a long time, even when it became summer."

People also considered fall rains helpful, as they filled lakes and were associated with good fishing. According to John Phillip (January 2006:19): "It would rain frequently in our land close to winter, starting before October when the silver salmon ran. When there was frequent rainfall, the land would start to form pools of water, and lakes became full. The people, glad for this occurrence, talked about how [lakes] would fill with water, making fish and blackfish available. Back when there was heavy snowfall during fall and frequent rainfall and it filled the lakes, blackfish coming out into the lakes and rivers were abundant on the coast."

Mepiaq | Fresh Water

Until recently, tundra ponds and streams were villagers' primary source of fresh water. John Phillip (February 2006:175; December 2005:116) explained:

> When I became aware of my surroundings, there were specific lakes where they went to get fresh drinking water. [The coastal area] is not like the Kuskokwim River area,

because when there is a flood, it covers the land [with brackish water]. But some [lakes] are on high ground, and they obtain fresh water from some places. . . .

Back then, even though the water was dark, they drank it. When we got thirsty, we cupped our hands, filled them with water, and drank from anywhere. . . .

In winter we always get [freshwater ice] from the lakes on hills near my village. The people of Kwigillingok also get their ice from higher ground where the salt hasn't reached.

David Martin (December 2005:113) described getting drinking water from brackish coastal ponds: "This top layer that froze in saltwater overnight can be pulled off, melted, and you can drink it, even though it has some salt. It is potable. This salt won't freeze. Only the [fresh water] that is easy to freeze will freeze quicker than the ocean." John Phillip explained: "It's like this—it rains and fresh water is on top. It's layered. That is how some lakes are. It is easy to tell when they aren't thick. When we get ice for drinking water, there are layers of salty ice, and it carries the salt. They have a border. I get ice that has a [salty] bottom layer. Not all lakes are reached by saltwater, but only the ones that have rivers. When it floods, those are filled."

Martina John (November 2007:457) added that before freeze-up, water was also available from freshwater springs: "Water is actually good around the mountains. Mountain water is extremely clear and good tasting. They call those places where the water comes from underground *qallanret* [springs]. When the land started to melt, the spring near Umkumiut started to flow. We used that good water all summer." Martina also drank from tundra ponds: "When we'd search for eggs or greens along the land, we'd go to lakes, not considering them unclean, and drink the water although it was dark. We never got sick. And if there was moss soaked with water, we would cup our hands and push them down on it. When they filled with water, we'd drink it. Just anywhere." Albertina Dull added: "Being someone who drank the water myself, I cannot die now. [*laughter*] And when I didn't have water to drink, when there was water along the plain ground, I'd put my head down and drink also. It would taste like land. The only water we didn't drink was one that was too dark and thick and muddy."

Kuiget Nanvat-llu Nallutaitut |
Rivers and Lakes Are Knowing and Aware

Like their inhabitants, sources of fresh water are viewed as sentient beings, responsive to human thought and deed. Joseph Jenkins (May 2003) expressed the importance of people following *eyagyarat* (abstinence practices) lest rivers respond negatively to their presence:

Girls menstruating for the first time were taught not to go near rivers and play. And if they went by boat, they were told to cover themselves completely and not to look around while traveling.

The two major rivers in our region, the Kuskokwim and the Yukon, will know and become displeased if rules are broken. These two rivers will know people's inappropriate behavior and misconduct.

Regarding the environment and rivers, the highest warning was against women going through a miscarriage. They were warned not to go near rivers or to travel on them. They'd say that if she did, the fish would leave and go to the deepest part of the river.

Smaller rivers were no less aware. Lizzie Chimiugak (January 2007) described how the Cakcaaq River on Nelson Island, a rich source of fish for the whole area, stretched out its arms to take people who fell in the water. Similarly, some interpret erosion on the banks of the Kuskokwim and its tributaries, as well as the formation of sandbars at river mouths, as the river's response to people drowning there or to reprehensible human actions at those locations.

Water itself could see. Marie Myers (February 2005:25) of Pilot Station explained: "[My grandmother] used to tell us that during spring when things were thawing, water has good eyesight. Although it was just a little tear, water would seep through the seams of skin boots and go inside." John Phillip (January 2006:275) mentioned the rule not to wade in water, as it has good eyesight: "Although they were told not to wade in water when they had their first menstrual periods, if they did so, their joints became bad. They say that *mer'em makuara takvigtuq* [particles rising from water have acute eyesight]. That's how they explained it. Since [the water] has a sense of knowing, one who doesn't follow their admonitions gets an ailment later in life."

Peter Jacobs (January 2006:42) described how Brown Slough, just behind Bethel, has a sense of knowing:

They say Brown Slough has powers of perception. My wife, through her expressions of affection, got an elderly woman who originally inhabited the land to come to our home quite frequently. She would constantly tell us about what the Kuskokwim River was like. She said that small slough has a sense of knowing. That's what she told us, since she probably observed it from the time she was small. . . .

Then when that elderly woman came inside [our home] in spring, when the water suddenly began to flow there, I said to her, "Look, the current has started to flow in Kepenkuk [Brown Slough] down there." She looked up and said, "Summer will surely come now. Before the current stops flowing, it will become summer." She used Brown Slough to determine coming conditions. She said that when the water has started to flow [in Brown Slough], it doesn't stop until summer comes.

Lakes were also home to extraordinary beings. Tommy Hooper (March 2007:1328) recalled *amllit* (lit., "things to step over") sometimes encountered when wading in lakes:

> They also mentioned that after some people had gone egg hunting, when they'd wade in the water to go to islands, they would encounter some obstacle when they returned. They would refer to those as *amllit*.
>
> They say that person shouldn't turn back, and they should try to step over it and land on the other side. They say when some people try to step over them when they are large, the inside of their legs bleeds.
>
> That's evidently what happened to some people. When one is about to [wade in the lake] desiring to get those [eggs], they said that we should check its water. They say that if it is a little white in color, opaque [like serum], one should go elsewhere, as [that lake] may have those [*amllit*] inside.

Tommy Hooper (June 2008:185) also noted deep lakes inhabited by huge pike whose approach could break the ice in winter and raise the water level in summer: "The edge of that lake couldn't freeze, as its water continually moved up and down." Tommy had seen a huge pike surface in Baird Inlet but had not tried to land it.

Lakes themselves are cognizant of human action. Irvin Brink (October 2003:182) of Kasigluk described Nanvarpak as all-knowing and responsive:

> We have a lake behind us called Nanvarpak [Big Lake]. When I became aware, we were warned adamantly that we couldn't mess around in the lake. Since our ancestors believed, they never hunted or fished there, but those who went to spring camp would only set nets along its tributaries and not in the lake itself.
>
> They said that the lake is an *ircenrraq* [extraordinary being]. If one belittles it, it always knows. It has the same rules as the ocean. That lake can cause something to happen to one's self or to one's children. They were warned against it.

Irvin Brink followed with an example of the lake's response to ridicule:

> Once after the sun rose, Qerrupalek was about to go down, so before he got down there he made a small shelter and cooked some blackfish. When they were done, his dogs began to bark. They went out to check, and there were two people from the lower coast. He invited them in and fed them just before the sun rose.
>
> When they were done and ready to leave, his companions urinated facing the lake. As they were urinating one of them said, "Gee whiz, is this the big lake? It is just a small lake." [Qerrupalek] was fearful because they say that the lake should not be ridiculed.

After urinating they left and began to quickly cross the lake. Soon they passed things on the ice that were a little taller than a person. They were perched on the ice and looked very blue. The one who hadn't said anything [about the lake] would say, "Be careful; we might reach [an opening] that has no top covering." Before long, the sun went down. The distance in front of them never changed.

Here they were going fast. Just as [the sun] disappeared, it seemed like the shore was coming toward them. Then, when they reached the other side, they felt like relieving themselves, so they faced [the lake] and urinated. The one who made a statement [about the lake] said, "No wonder they call this a big lake. It is big."

It doesn't take that long to cross; you know that. Even if you don't go fast, it shouldn't take you long to reach the other side. . . .

Today [young people] don't believe our sayings and disregard them. But like they said, since [the lake] hasn't become a *kass'aq* yet, it's not good [to break rules], and disregarding them has consequences for their children and themselves.

Elders share knowledge on where to find fresh water, how to read waves, and the dangers associated with river and lake ice. They also share an attitude of respect, both for the knowledge their ancestors had of rivers and lakes in their homeland and for the power and knowing character of the rivers and lakes themselves. Travelers must know the varied topographic and weather conditions wherever they roam, and they must also circumscribe their own conduct, as this more than anything else determines their ability to survive.

Kuiget Nanvat-llu | **Rivers and Lakes**

aciirun. Part of the river that runs under a bluff or cutbank.

akulnguyak/akulnguyurraq. Narrow section of lake (from *akula*, "area between").

akuluraq/akulurat. Stream/s between two bodies of water, something in between (from *akula*, "area between").

anlut. Exit holes in freshwater ice made by muskrats and otters (from *ane-*, "to go out").

avayaraar. Small tributary.

caarrluut. Grassy wetlands (from *caarrluk*, "dirt, debris").

carvaneq/carvarnet. Main current/s.

carvaq/carvat. Stream/s with a strong current, flowing from hills or mountains.

carvarraq/carvarrat. Small mountain stream/s.

cev'aq/cevvleq/cevneq/cevenret. New channel/s cutting across land and emptying into another source, including man-made channel/s (from *ceve-*, "to cut through land").

cikuilquq/cikuirneq. Open hole in river ice (from *ciku*, "ice").

cing'iktaalriit qairet. Sharp, pointed *qairet* (waves).

ciulavik/ciulaviit/ciulavigshluut. Piece/s of peat moss removed from the riverbank by the current (from *ciu-,* "area in front").

ecuilnguq. Clear water (lit., "one that isn't murky").

ecurlirrluut. Murky waters; waters that never freeze (lit., "those that are murky").

egmilleq/egmillret. River/s branching out from the lake source of another river (from *egmir-,* "to keep going").

ekvik. Riverbank.

elakaq/elakat. Water hole/s.

etgalqitalriit. Dangerous waves in shallow areas (lit., "those that hit shallow areas").

igyaraq. Outlet where a lake empties into a river or where a river or channel empties into a larger body of water (lit., "throat").

ilanret. Islands in lakes.

imaquq. Murky water.

ipiit. Tributaries of a river (lit., "limbs").

kangiq. Headwaters of a river (lit., "beginning, source"), open water bordered by ice or land. See also *qagan.*

kangiqaq/kangirrluk. Strait of water, bay.

kangiqutaq. Small bay, cove.

kassigluq. River confluence.

kianeq/kianret. Area/s of open water in ice. See also *qenuilquut.*

kuicuayaaq/kuigaaq/kuigaaraat. Slough/s, small river/s.

kuigem painga. River mouth.

kuiguaq. Oxbow lake, a long, narrow, U-shaped lake formed when a river bend is cut off from the main channel at both ends (lit., "imitation river").

kuiguyuq/kuiguyuut. Channel/s.

kuik/kuiget. River/s.

kuiliaq. Man-made river channel, shortcut.

kuilleq. Old riverbed.

kuineq/kuinret. Deep part of channel/s.

menglairneq. River or lake edge, water's edge.

mepiaq. Fresh water (lit., "real *meq* [water]").

meq. Water.

miiqaq. Fresh meltwater, melted snow in spring.

mingqutnguaq/mingqutnguat. Rotten ice, needle ice, shard or honeycomb ice, on freshwater lakes and rivers (lit., "pretend *mingqun* [needle]").

nakirneq. Straight stretch (of river).

nanvalleq/nanvallret. Former lake/s, dry lakebed/s (lit., "old *nanvaq* [lake]").

nanvaq/nanvat. Lake/s, pond/s.

nanvarnaq. Wide, lakelike section of river.

pequneq/pequnret. Freshwater ice ridge/s (from *pequq,* "upper back").

pingayunleggluut. Boggy areas (from *pingayunlegen,* "eight").

qaaq. Wave. See also *qaiq, yuulraaq.*

qacuvkaneq. Calm water (from *qacu-,* "loose, not taut").

qagan. Lake source, lake from which a river flows. See also *kangiq.*

qaiq/qairet. Wave/s. See also *qaaq, yuulraaq.*

qairrliqellriit. Places with bad *qairet* (waves).

qallaneq/qallanret. Spring/s (from *qalla-,* "to be boiling") .

qallanerrluk/qallanerrluut. Oily, brown, stagnant water (lit., "bad *qallaneq* [eddy]," "sort of an eddy").

qamaneq. Eddy, place lacking water current or wind.

qanglluq/qangllurpak. Deep hole in a riverbed.

qas'urneq/qas'urneret. Deep, dangerous hole/s along riverbanks where the ice has collapsed.

qenuilquut. Areas of open water in ice (lit., "places without *qenu* [ice]"). See also *kianeq.*

qipneq. River bend.

qisneq/qisneret. Edge/s of the current, side/s of the channel.

qurrlugtaq/qurrluraralriit. Waterfall/s (from *qurrlur-,* "to cascade down").

quurneret. Narrow sections (of lakes or rivers).

tevanquq/tevanqut. Deep area/s.

tevyaraq. Route one travels through, portage trail.

tugneq. Point where the river current cut the bank at a bend.

uivneret. Whirlpools (from *uive-,* "to circle").

ulerpak. Flood (from *ula,* "to rise [of liquid]").

ulevlaq. Spring (from *ula,* "to rise [of liquid]").

uniartellret. Puddles (lit., "ones left behind").

ussneq. Caved-in riverbank (from *uste-,* "to erode, to cave in").

yuulraaq. Wave. See also *qaaq, qaiq.*

Yuilqumun Atalriit Qanruyutet

INSTRUCTIONS CONCERNING
THE WILDERNESS

These instructions are because people move around a lot. People didn't always stay in one place, back then and today. They warned us about those things during fall, showing how to do things the proven way. They wanted us to learn about it.

—John Phillip, Kongiganak

O F ALL THE TOOLS MEN AND WOMEN NEEDED TO TRAVEL IN and harvest from *yuilquq* (the wilderness or uninhabited place, lit., "place without *yuut* [people]"), perhaps the most important was their understanding of how to navigate safely over the landscape in extreme and changeable outdoor conditions. Expert knowledge of tides, clouds, stars, wind, and weather in their hunting areas was essential not only for success but for survival. Elders willingly shared skills that they considered most important for traveling, especially during fall, regarded as potentially the most dangerous season. "There are many rules for the fall season," Nick Andrew (October 2003:2) began. Paul John continued: "The ones who spoke to their young people told them about what would be dangerous, starting with fall, and what would be helpful. They were always preparing for fall." John Phillip (October 2003:30) stated emphatically: "October is fall in my village, and it gets dangerous. I always warn young people when it reaches October to be observant and not to travel down [to the ocean] by themselves."

Yuum-gguq Picurlallerkaa Ikgetuq |
They Say a Person's Accidents Are Few

Young people were taught in advance about the dangers they would face. According to John Phillip (October 2003:196): "They wanted us to know about what was ahead of us before we reached that time. That was apparently a teaching taught to our elders. They taught us things concerning the wilderness before we were able to do things on our own, before we traveled, and they told us about the dangers." John continued with feeling: "I never forgot the *qanruyun* [instruction] that I first heard. Since my ears were able to hear when I began to observe, my grandfather told me, 'You dear young boy, don't live without an elder.' What he meant was that if we didn't follow the wise words of our elders, we wouldn't have a good life." Theresa Moses (May 2003) noted that uninstructed youth might have no life at all: "They tell us not to take our children for granted, but to instruct them and be especially watchful of them during spring and fall. During the spring thaw one can easily fall through rotten ice, and it is also easy to fall through thin ice at freeze-up since ice doesn't immediately thicken. Those were our precepts. They said a person's accidents are few [because the first accident may be fatal and thus the only accident]."

Simeon Agnus (July 2007:585) also recalled admonishments he received as a child: "The instructions I was given about the ocean and the wilderness are stored in my mind. When they briefly mentioned them, I listened intently. Since they always gave instructions in the *qasgi*, my mother would tell me to go and listen, that the elderly men would talk about things I would never forget." Nick Andrew (October 2003:49) observed: "They used stories like they do in Head Start nowadays, to make us children aware of life." Paul John agreed: "When we listened, we wouldn't be able to sleep anymore. And our hearts would really beat, too. When they told stories that were geared to help us, it was good to listen to and made us think."

Yuilqumi Cat Murilkelluki Ayagallerkaq |
Observing Landmarks When Traveling in the Wilderness

As boys began to travel, they were taught crucial landmarks—small and large, near and far—including rivers, lakes, sloughs, even sandbars and underwater channels. Paul John (October 2003:2) recalled: "They named what they call *angussaagviutulit* [customary hunting and gathering areas, from *angussaag-*, "to try to catch something for food"] and where they were located. When they traveled to the place where they would hunt, they also pointed out dangerous places where accidents could happen and told them what to do. Our ancestors seemed to always instruct their young people from the beginning, teaching them orally using good examples."

John Phillip (October 2003:4) described how young men were taught to know the landscape they traveled through. All places, he said, had names: "Soon when he reaches a river, that river has a name. A person will know these if he really pays attention, starting from when he travels for the first time. He will begin to travel as if he's reading his trails and recognizing them. Some of those trails have names; the sloughs also have names. The small hills have names, too. Even trees are indicators to those who travel."[1]

Frank Andrew (October 2002:191) had also been taught place-names early in life and, like John Phillip, would not get lost: "They made us learn those things and used them for reading directions, and they told us the names of the lakes and rivers or high ground with grass on top. They taught us everything we might need to know for traveling. If someone asked us, we wouldn't say, 'I was up there.' [*laughing*] When we ask some people, they don't tell us exactly where they went, but when they give us a name, we know exactly where they were. That is what they did for us. My dad used to bring me all the time and tell me the names for certain important landmarks."

John Phillip (October 2003:6) stressed the importance of being observant: "I was told to observe everything I see and hear and everywhere I walk as I live. Even if it wasn't a trail, I was told to always observe the tundra. I am able to recognize places as I travel, even if I encounter bad weather, and to tell where I am and where my home village is on that trail. A person who observes and keeps track of the ponds can reach his destination, even when the weather gets very foggy. As he travels in stormy weather and gets lost, he should go back to where he recognized a place and begin from there." Paul John (September 2007) also paid attention to local landmarks:

> When I started traveling with a sled and dog team, I used the teachings they gave. They also told us to take notice of willows that were along the land when we came upon them. Or if we happened upon a small wooden post sticking up, they said to take notice of where it was located. These various landmarks, including small hills and wild celery that grows along the edge of lakes—if we paid attention to those things, although the weather suddenly became stormy while we were out, they said we wouldn't veer too far from our destination when returning home.
>
> And if we happened to miss our destination, if we had taken notice of a small thing that was erected, or if we had taken notice of the direction that the grasses were flattened, when we recognized those markers, we would suddenly realize the direction of the village.
>
> When continually trying to pay attention to the teachings, since they are useful

when traveling out to the wilderness to avoid getting lost, one has to know the direction of their village when traveling.

John Phillip (January 2006:23) was admonished to find his way by means of hills and grasses, as lakes would not be visible in winter.

> They instructed us to pay close attention, as one tends to lose one's sense of direction while traveling in the wilderness. And they told us to place in our memory those steep areas and places with grass growing on them. After observing their location, they told us to note the location of our village. I later learned that they gave us the teaching to observe the small hills and bushes because the lakes that we were once able to locate became concealed when there was snow. If one lost one's sense of direction or the weather became stormy, they could recognize that small hill or clump of grass and use those to locate their village. That was especially a teaching for us young people who traveled in the wilderness.

John Eric (March 2008:135) noted the importance of keeping one's destination in mind: "I also used the mountains [to gauge my location] because that's what my grandfather told me to do. He said that if I was alone, I should first record my destination in my mind [before I left]. He said that *ella* isn't always good." Peter John (March 2007:1215) noted the importance of keeping one's relative position in mind while traveling: "Although I kept my destination in mind, I was always familiar with landmarks before reaching that place. When I'd reach those markers on the land, I'd know how far I was from the place where I was heading. When I reached a [landmark], I also knew my position when I traveled."

Peter John (March 2007:1214) emphasized the importance of not taking one's knowledge for granted: "I'm so glad that my father reminded me to always know where high places are located every time I traveled. Although I knew where they were and constantly saw them, he told me to observe them closely. He told me that although I had previously traveled through that place, I should study it well and not think, 'I know that place.'"

Yuilqumi Ayagalleq | Traveling in the Wilderness

They instruct you so that you will know how to avoid accidents, so that you will walk with caution in the wilderness.

—Nicholas Tommy, Newtok

Men rose early when traveling in the wilderness. Paul Jenkins (March 2007:238) recalled:

In fall, we left in the early morning in the dark over at Cuukvagtuli. We always left before the sun came up.

Now, this is what it is like to hunt without sleep. People who sleep long lose their opportunity to hunt and fish. But if a person wakes up before daybreak, when he reaches that place where an animal is before others reach it, he catches it.

Simeon Agnus (July 2007:99) noted the adage: "They say *pitarkat* [animals, lit., "those to be caught"] in the wilderness are easily approachable in the early morning. The morning isn't daunting. Only the evening is intimidating. That was an instruction given by our parents." Simeon (July 2007:250) admonished his fellow travelers: "When out in the wilderness, one mustn't sleep long, but one can sleep when in the villages. Although we had nothing to do, our parents told us not to sleep long out in the wilderness."

Once up, men worked hard. Wassilie Evan (July 2000:58) recalled: "Our ancestors would walk in the wilderness from the time they woke, all day long until nightfall." Peter John (October 2007:1226) commented: "We were also told that even if we rested, we shouldn't lie down. Since I walked for a long distance twice following that teaching, I tried not to stop for too long." Wassilie Evan continued:

These women store their needles and threads in containers. Even men in the wilderness had these in their pockets as they traveled on foot. In those days, traveling was slow. They would start in the morning and travel to try to catch animals. Even without provisions, men traveled on foot. When they hunted for red foxes, they walked from their village in the morning without eating. They set traps on the tundra and walked to check their traps. Sometimes as they walked, the wind picked up and became very cold. The part on his parka where the man was breathing became white [with frost] because he was walking against the wind. The visibility would get bad, but since it was their land, they knew where to go.

Irvin Brink (October 2003:12) noted the importance of checking one's trail: "Those people made trails on good land. They made trails by [frozen] lakes that had the shortest route. It was also a rule that if we used a trail for the first time, we should keep an eye on the place where we first went down [the bank] by looking back." George Billy (May 2003) had also been taught to look back while traveling: "When my uncle brought me out to the wilderness for the first time by sled, when he was going out to hunt red foxes, he told me as we got to the lake, 'Now, look back.' We looked back and saw where we came from and also where we were. [He said], 'If you do that, you will be knowledgeable.'"

John Phillip (October 2003:6) explained the advice to go back to a place one recognizes when lost: "A person who observes and keeps track of ponds can reach his destination, even when the weather gets bad. As he travels in stormy weather and gets lost, he should go back to where he recognized a place and begin from there. Then he can go when the weather improves by following the trail. He can travel where there is no trail, but he can return by observing markers on the tundra and remembering."

To this day some areas on the tundra are known as *pellaanarqellriit* or *pellatalget*, places where people tend to lose their sense of direction. John Phillip (October 2005:240) explained: "Bethel is close to Kasigluk and Nunapitchuk, and their lights are almost visible. But many times, those coming from up there get lost and veer north, and some [people] are never found. They say that place between Bethel [and the tundra villages] is a place where one can lose one's sense of direction. And there is a place upriver where people get lost around the trees. That's why they instructed us that if we lose our sense of direction, not to continue but to stop in a place we know."

Many view such experiences as the result of willful interference by *ircenrraat* or other extraordinary beings when people travel near their homes. George Billy (February 2006:368) explained:

> The area between Atmauthluak and Napakiak has started to claim lives recently. These *ircenrraat* cause one to get lost in some lake up there. Those who travel through *pellatalget* get lost.
>
> Those elders said that Nanvarnarrlak [Lake] had a *pellatalek*. Apassangayak told me that while he was going across, someone would shoot a light on him.
>
> And one time we heard those *ircenrraat* at Vegcuaralek [Hills] across from Nanvarnarrlak beside the lake. When we listened, there was a droning from across there. They said that the *ircenrraat* were kicking.

Encountering *pellatalget* can be counteracted by cutting one's trail. Bob Aloysius (October 2005:234) recalled: "When we come upon those, they say that we have arrived at *pella*. We seem to be moving, but we're still there. They told us that if we realize that we are in that situation, we should get off our snowmobile and cut the snow behind us with a knife so that we can pass." Paul Kiunya agreed: "Apparitions did that [to people]. If they followed you from that place, it is said that if you took a knife and made a cut, it would slow down if it arrived at that place. Although you leave, [it stays] behind you." Lucy Sparck (October 2005:235) described her brother's experience traveling from Chevak: "My younger brother, John Jones, went down to Scammon Bay after Christmas, and he said the weather was good. It takes one hour to go down to Scammon Bay. Two hours passed while they were traveling fast. John got out and told Loddy, 'Don't look back.' He took out his knife and cut the snow behind them, and when he got in and left, the village was there."

Cutting one's trail was not the only solution. Paul Kiunya (October 2005:234) noted: "Also, they have an instruction in the Kuskokwim River area that if you're traveling through trees and get lost, you should remove your clothing and put them on inside out, and you will reach home from there when you leave. There are places where Kuskokwim River people are told not to go. They say there is a *pellatalek* there. Those invisible ones who aren't human can do that to people, causing them to stop." Nick Andrew (October 2005:235) noted a Yukon admonishment: "Where I'm from, they refer to that as being tethered to a place. If they continue to go but don't arrive at their destination, if they don't have a knife, they say if they urinate, they will go without being held back by anything."

Elders emphasized that stopping when lost was of paramount importance, as continuing to travel could take one farther and farther away from one's destination. Mike Utteryuk (November 2000:208) emphasized this instruction: "They told us that if we didn't know where we were, we should stop immediately and stay there as long as the weather was bad. They said that if we tried to go in a hurry, we would travel far away and could not be found."

Tim Myers (September 2003:289) of Pilot Station recalled losing his way: "I know this through experience. I tried to follow the wind direction, but I began to travel in the wrong direction, as wind constantly changes direction. Fortunately, I recalled what they said, that I should not continue to travel but stop when I lost my sense of direction. When I woke the next day, when the sun came up, I found that I had gone in the opposite direction from my village." Peter Jacobs (October 2003:37) lamented that today this advice is rarely given: "Nowadays, since young boys aren't instructed, they don't stop when they get lost but continue to travel by snowmobiles, and many die from hypothermia and are found far from home."

In upriver wooded areas, one could light a fire and wait out a storm. Tim Myers (September 2003:296) explained: "If you're around spruce trees, some usually have dry branches. After gathering some, if you light a fire under a live tree, the surrounding trees will block the wind and the area underneath will start to get warm. Then you can put something around it [to block the wind]. If there isn't any material [like canvas], you can cut branches from live trees and place them on the windward side, staying there until sunrise." Tim added that inland people were admonished to immediately go into the trees to a place with dry wood and try to warm up if they fell into water.

Men sometimes made snow shelters to wait out stormy weather. Paul John (October 2003:21) explained that shelters should be made on the windward side of steep areas, as the leeward side tends to accumulate snow, which can lead to suffocation. He also noted that a person should always make a breathing hole in the snow shelter, using a piece of wood to either poke a hole in the shelter's ceiling or to scrape through the snow. Once inside a shelter, a person should always push the snow away from one's parka as well as sit on something, such as grass or branches, to protect one's body from melting snow. Tim Myers (September

2003:303) recalled the fate of those who take shelter on the leeward side of snow-banks: "They said that if we stayed on the leeward side, that snow would cover us and we would eventually suffocate. And when they finally found us, there was a saying, 'If you sleep there, you will become the end of the snowbank.'" Wassilie Evan then told of a young man found suffocated in the snow. Tim Myers concluded: "That's why our elders told us not to feel that we could conquer the weather, even if it's just the weather. They also told us not to belittle it. They say that the weather could easily take a person's life, causing one who belittled it to die of hypothermia, to get wet, or to become part of the snow. That was our teaching for those of us who traveled, not to belittle *ella*."

Elders recalled other admonishments intended to keep one safe and warm while traveling. Peter John (March 2007:1216) noted: "They said that if I was cold while sitting, I shouldn't put my head down and breathe inside my clothing, moistening my body with my breath. They said if I did that and went out in the cold, I would be responsible for my own death, as my body would freeze. It was an admonishment." Paul John (October 2003:22) said, "If he gets cold, he is to continuously stiffen his muscles, and he will survive." Irvin Brink (October 2003:13) agreed, "They told us not to stay still but to move around to keep warm." Wassilie Evan (September 2003:294) added that when he started to get cold while traveling by sled, he would get off and run until he warmed up.

Hypothermia can be accompanied by hallucinations and confusion, including feelings of apathetic peacefulness, and elders shared advice to deal with them. John Phillip (October 2003:38) recalled the instruction that when lost and cold, one should never accept a parka that presents itself: "They say that when one is extremely cold and chilled, one sees a parka that is very thick and warm. Even if it appears, one should not accept it and put it on. Once someone experienced that. When he remembered, even though it seemed to be very warm, he left it alone. When it disappeared, his body began to feel better and got warmer." Simeon Agnus (June 2009:66) agreed that if a person suffering from hypothermia put on a parka presented by Qerrum Yua (the Person of Death by Freezing), he would surely die. Peter John (March 2007:1218) added: "Those who put on that parka would be wearing no clothing when they found them."

Peter John (March 2007:1219) recalled his father's instructions regarding cold weather: "After my father brought me traveling with him, when we'd enter the porch, he would tell me to remove [my parka]. Here it was cold. After removing my parka, I'd just stay for five minutes until he told me to go inside. If I went inside right away, he said that I wouldn't be able to get warm for a while if the warmth suddenly entered [my body]."

Peter Jacobs (October 2003:94) was admonished not to immediately warm his hands after being out in cold weather: "When I entered from outside, the one who raised me instructed me not to get my hands close to the heat. Once when

I was warming my hands over the stove, she told me not to do it, and I got mad and threw myself on my small bed. She made an '*ugg*' sound and said, 'You are a young boy. Since you will travel, I always tell you not to go to the heat because you might always have cold hands. In spring you can warm up with the sun's heat as much as you want.' I finally understood." George Billy (February 2006:507) was also trained to endure cold weather: "When they slept in the *qasgi*, when a boy was about to go to sleep, his grandfather tucked him in warmly. They covered them, but once they fell asleep, their grandfather would take off their blankets. Sometimes when they woke up, they would be so cold, and their grandfathers would say, 'You have kicked your coverings off when you were sleeping.' They did that so that when they started going to the wilderness, they would be better able to fight the cold."

People were advised that drinking their own urine and, if available, seal oil, could help them stay warm. Paul John (October 2003:22) noted: "A mouthful of urine will make you warm. If one happens to drink urine and has seal oil handy, if he mixes them they will really warm him up." Nick Andrew (January 2006:11) said that eating the raw breast meat of ptarmigan could warm a person, as could frozen whitefish: "When we were out in the wilderness, they took along raw frozen broad whitefish to eat in cold weather. When we ate it, we got cold, but when they melted inside our stomachs, the raw foods made us feel warm." Nick noted that some foods had the opposite effect: "They usually cautioned us against eating animals that we caught from the wilderness while they were warm, and they especially told us not to eat swan meat while it was still warm and raw. Only raw ptarmigan meat would warm a person."

Travelers were also admonished not to drink too much water or eat soft, new snow when thirsty. Peter John (March 2007:1225) explained: "Although we were extremely thirsty, they told us not to constantly eat snow. Eating snow can make a person tired. But they said if the snow had packed down, the bottom layer has small pieces of hard snow; they told me to eat a little bit of that. They also told me not to swallow [the snow] before it melts. And if our mouths couldn't form any saliva, if we chew on a piece of wood, we will develop saliva. That was what they told us to do when we were walking." Why this is true is unclear. Although new snow has more air in it than ice, ounce for ounce, it takes as much human energy to melt snow as ice.[2]

Ellminek Ikayuayaraq Pellaalriani | Helping Oneself When Lost

Elders shared a number of time-tested means of orienting themselves in the wilderness. Travelers throughout the region used *iqalluguat* (snowdrifts formed by the north wind) as guides. Paul John (September 2007) explained:

When weather gets stormy, those *iqalluguat* form as mounds on top of the snow. Starting along the direction of the wind, they extend in length toward the sheltered side. They are obvious, as their windward side is a little concave after it's windy, but their sheltered side extends out.

When we'd travel to the wilderness, they would point those out to us. They said before we got too far from the village, we should stop and look back and also ahead toward our destination. After observing, we should determine if we were about to travel directly cutting across the *iqalluguat* or at a slant.

If we traveled using *iqalluguat* as guides, they said if the weather suddenly got stormy along our trail, we would not miss our village. Although the wind direction turned, they said those *iqalluguat* that won't move would allow us to reach our destination.

Nick Andrew (October 2003:25) noted: "When I get lost, I look at the *iqalluguat* made by the north wind and walk alongside them. I still use them today." Irvin Brink (October 2003:26) added, "They are our compass."

Grass bent by the wind and frozen in place by cold weather will point in a fixed direction. Even if the wind shifts, an observant traveler can use the grass to orient himself in whiteout or foggy conditions. Paul John (September 2003:221) explained: "Grasses in winter face the way they do after it was windy and wet from the south during fall. That is when they fall to the ground toward windward, and they won't change direction all winter. They face the way they froze. If someone is lost and knows that the land has grass, he can dig [in the snow] and check to see how the grasses were flattened by the wind during fall. By looking at the grass he can travel toward his home." Simeon Agnus (December 2007:63) concluded: "Grasses bent by the wind and frozen in place were used like compasses long ago, when they didn't have GPS devices."

Paul John (September 2003:222) described traveling on the muddy lowland bordering the coast and digging under the snow to locate *evget* (grasses): "If we get lost and we don't see high tundra, we can dig and find the short grass, which will be flattened toward land. When it floods in fall the water goes over them and flattens them facing the land. By looking at that we will know where the land and ocean are." Grass was not the only plant useful in navigation. The windward side of plants such as willow and wild celery also freezes following wet weather. Travelers could tell the direction of their home if they knew which direction the wind had been blowing from.

In clear weather with snow on the ground a person could travel in safety. Frank Andrew (October 2003:191) observed: "When there is snow, it's never dark, even at night. And when the moon is visible, it's very bright and the whole trail can be seen. It's very dark when there's no snow and the moon isn't out." Stars were used to navigate in clear weather. Paul John (October 2003:23) said: "They told us to observe the stars. If one goes out at night and looks up at the stars, he can understand where he is and which direction his village is."

Hunters used the sun's reflected light to find their way on a cloudy day: "If one has a knife, he can open it and go around a tree watching the knife. They say that when [the blade] faces the direction of the sun, it gets bright. That will tell you the sun's direction" (Irvin Brink, October 2003:34). Paul John (October 2003:39) said that the same advice could help a man disoriented in the ocean. A lost hunter was advised to watch the blade of his knife or lick his thumbnail, paddling in a circle until the wet surface began to shine. Knowing the direction of the sun, he could judge which way the land lay. Frank Andrew (October 2003:123) added that when lost in fog on the ocean, a man might also watch the tiny ice crystals in the ocean water, called *makuat*, sometimes referred to as the ocean's eyes: "The *makuat* were also indicators down on the ocean. If the sun wasn't visible, we would check the *makuat* and understand the direction of the land. They said that if we lost our sense of direction, we should watch the area around the edge of floating ice and go around it. Nothing would be visible on the shaded side of the ice. They say that side is the direction of the land, down on the lower Kuskokwim coast, and the *makuat* will begin to sparkle if we get to the side in the direction of the sun. That side is the direction of the ocean."

Navigating in wooded areas also had rules. Nick Andrew (October 2003:35) reported: "When you're in the trees, you can get lost. When you look at cotton-wood, one side has moss from the bottom up pretty high, but the other side is clean. That [clean side] is the south side. Then you can tell where you are. South wind cleans the south side when it rains hard, but the other side is dirty and has moss growing on it." South-facing branches also tend to grow longer. During a rainstorm, the windward side of trees would be wet and the leeward side relatively dry. If all else failed, observant elders ruefully noted that jet trails over southwest Alaska always run east to west.

One thing that was not dependable when navigating in the wilderness was wind direction. Tim Myers (September 2003:289) explained: "Wind should not be used to determine direction when weather suddenly gets bad during winter. The wind is constantly changing direction. After blowing in one direction, it moves from there and then goes back toward the same direction. The wind isn't a good guide."

Paul John (September 2007) concluded with an apt comparison between modern travel aids and the traditional *qanruyutet* that not only helped travelers but guided people in every aspect of their lives:

> These days, they have started to travel by navigating with a compass. And they have the GPS, which some people call *angalkucuaraat* [lit., "little shamans"]. When I think about how those people asked us to observe the weather and our surroundings

as we traveled, I realized that those people who gave instructions about what to do when traveling evidently allowed me to have a GPS from the beginning.

When I started traveling out to the wilderness alone, I never got lost. Sometimes the weather got stormy while I was out, back when we traveled with dogs. By constantly paying attention to their teachings, I knew how far the village was although the trail was covered [by snow].

They allowed me to have a GPS. If a person pays close attention to any sort of moral advice, even if it doesn't have to do with traveling, when encountering a particular moral teaching he will think, "Oh yes, this is how those past people asked us to handle this situation." Since those are true, a person who follows the advice he was given will evidently always lead a good life.

Aarnarqellriit | Dangerous Places

They gave qanruyutet in the past to keep people safe. These instructions prevent a person from getting into danger.

—Peter John, Newtok

Frank Andrew (October 2003:52) declared, "There are many potentially dangerous areas on land." People traveled armed with knowledge. Peter John (March 2007:1207) noted:

These places that are typically dangerous are usually mentioned. My father gave me good common sense from the start, and I knew the dangerous rivers, and I also knew the grassy wetlands that are dangerous. We were instructed from the beginning that the *akulurat*, narrow streams between lakes, stay dangerous [unfrozen] for a long time.

And they told us to be careful around small tundra lakes and lakes located along sloping land, [saying] that those are the deepest spots.

Nick Mark (May 2003) of Quinhagak noted the admonishment not to quickly cross watery areas in spring: "If someone is going to cross a body of water as the snow melts in the wilderness, they should cut tree branches and cross using snowshoes. We tell them not to quickly cross the [melting] snow since we grew up in the wilderness." Similarly, *qallanret* (springs) were dangerous in spring, when snow covered deep holes that might trap the unwary traveler.

Wet areas covered with wheat grass were dangerous in fall. John Phillip (February 2006:177) explained: "Wheat grass grows in old lakebeds. They are dangerous during early freeze-up, when there is a thin layer of ice over them and water underneath. Once I fell through ice around wheat grass. The area underneath the ice wasn't frozen since wheat grass is hollow."

As noted, swampy areas covered with snow can be dangerous as well as river outlets on lakes: "During fall freeze-up, these larger rivers were void of ice a good distance downstream in places where water flowed out of lakes. The headwaters of rivers and sloughs at the lake outlets didn't freeze right away and were dangerous; we could cross them only by going a great distance downstream" (John Phillip, January 2006:19).

John Phillip (January 2006:22) noted that lake edges were another potential hazard: "During fall and winter, we boys were instructed, 'When you go out to the wilderness, be careful when you are about to go down into a lake.' Sometimes willow bushes are visible alongside lakes and not covered [with snow], but during winter when there was snow, those willows were no longer exposed and everything became level. One would not know that he was on a lake, and it was difficult to determine their locations." Camilius Tulik (March 2007:711) noted that these places were not as dangerous before snow covered them. After snowfall, however, travelers should be wary. The same applied to both cliffs and bluffs. John Eric (September 2009:141) cautioned: "Bluffs are dangerous when suddenly coming upon them in a blizzard. Although it isn't a very steep place, a person can get hurt there, as those bluffs are difficult to distinguish during winter." John's warning was prophetic: a man died falling from a bluff on Nunivak Island in whiteout conditions during winter 2010.

Paul Tunuchuk (March 2007:89) remembered the admonition to travel quietly and make noise only in an emergency: "They were told not to be noisy on the land. They would know that someone had encountered danger by the noise he made. Or if for some reason they couldn't let their voices out, they were told to erect something in the air." George Billy (October 2006:7) noted how silent people were when floating downriver in spring so as not to disturb those steering the boat: "Back when they went up to the mountains along the rivers to hunt beavers, when they were about to go downriver in the swift current, they gathered together and stayed very quiet. Then, following the leader, they drifted downriver in complete silence. Once someone whistled because he was so joyful, and they scolded him."

Men were also instructed on how to treat one another in life-threatening situations. Paul John (October 2003:81) explained: "Back then they told them things to give them more courage so that they wouldn't become hopeless. They also instructed young people about the possibility of drowning. When he hurries excitedly and panics, he ends up having an accident." To make his point, Paul (September 2009:273) shared the story of a young boy who saved his companions by keeping a cool head:

> They used to tell about those many kayaks that suddenly encountered strong winds. Since they couldn't paddle, they gathered together along a thick piece of ice and set up their windbreaks.

They say those great hunters expressed how overwhelmed they were. There was a boy among them, and he was the youngest in their group. He looked at them and when he turned back and faced his kayak, he said to those who were despondent, "My goodness, if we happen to die, there will be so many husbandless females!" [*laughter*] Those who had been despondent said, "Or, when it becomes calm, we will head up to shore."

They say by not panicking, that boy brought his many companions to their senses.

People were admonished not to give up when they found themselves in dangerous situations. Camilius Tulik (March 2007:734) declared, "Never think that you will perish or resign yourself to death." Frank Andrew (October 2003:250) recalled: "They told us that if we got into a grave situation, we should not show fear around our young people, to make their minds stronger. They say that if they follow their feelings of hopelessness, they can let the young people around them have an accident." Simeon Agnus (July 2007:585) added: "Out in the wilderness, it is an admonishment for one not to speak sternly to his son although he was defiant, since he won't always be that way."

While keeping a positive attitude was critical in an emergency, men were admonished not to help others at the cost of their own lives. Simeon Agnus (March 2007:798) recalled: "If one of us is going to perish and take us along with him, they told us not to help that person. Even if he is going to perish as they watch him, if he is going to take him along, it was okay to ignore him." Phillip Moses (January 2007:522) spoke of what men did following a death: "If one of their companions died in the wilderness, they were told to pass the village when they arrived. Then they went back up to the village. Through this, people understood that one of them had perished, but they did not know who. That was their way so that they were not shocked."

George Billy (February 2006:366) made the wise suggestion that knowledge of hazardous situations be shared not only with young people but with everyone: "When *kass'aq* teachers come, we must not hide these dangerous places from them, these places where one can get lost or places where one could fall through ice. Because it is not their land, when they arrive these *kass'at* do not know dangerous ice or how to negotiate overflows covered by snow. Those are also things that they should know about."

Ayagassuutet Caskut | Tools for Traveling

People traveled to the wilderness with little gear, but they considered some tools essential. Perhaps the most important—for both men and women—was the *ayaruq,* or walking stick. Peter John (March 2007:1224) declared: "In fall, they told us to always use a small walking stick during freeze-up. A person will find dangerous spots with that walking stick; they will know if the ice is dangerous or if grassy wetlands [are dangerous]." Nick Andrew (October 2006:333) declared: "Even when snow is deep, it's not good to walk without using a walking stick. [It's better] to keep testing the snow with a walking stick. It doesn't freeze under snow, and it doesn't freeze where beavers move about in those *carvat* [mountain streams]. Streams are usually shallow but could be dangerous, even in the dead of winter, because beavers make their dams in those areas."

Women also traveled using a walking stick. Theresa Moses (June 1995:23) explained: "We women always used the walking stick when we went out to the tundra. We also used walking sticks without points to check the dangerous areas on our trails when we gathered greens and plants. Down in the Canineq area we women used walking sticks at all times. Ice may look solid, but if you poke it with your walking stick, needle ice is easily shattered."

Frank Andrew (June 1995:17) vividly described the life-saving properties of the walking stick. A properly used walking stick, he said, would lead a person on the right path, allowing safe travel by pointing to hazards along the way. He concluded by comparing the walking stick to the *qanruyutet*: "The *qanruyutet* work exactly like the walking stick. A person who merely hears the teachings and does not live them is only holding the walking stick and has no idea how to use it properly on his path. When a person masters the teachings by listening and observing and applies them to his life, they will work like the walking stick for him."

Irvin Brink (October 2003:12) described essential tools for traveling in the tundra area: "Before we traveled we were instructed not to leave the tarp, shovel, axe, and walking stick and to always have them with us. Also this seal-gut rain parka may look small when it is stored away, but when they get wet they get larger. We were instructed not to leave those whatsoever, even though we only went a short distance. We also had large knives with us, hanging on our belts. They would tell us not to leave them." Tim Myers (September 2003:300) noted the importance of bringing water: "To those of us who lived inland water was most important when ice and snow became thick. They told us to take a little bit of water with us. They told us to place our shovels in our sleds along with an ice pick. They mostly made tarpaulins out of tall cotton grass in my village, and they told us to place our provisions under those."

One of the most ingenious pieces of equipment was the *qasperrluk*—a wide, roomy, fish-skin parka that could double as a tent. Frank Andrew (February 2003:706), who had both worn and slept in *qasperrluut*, described them at length:

> They used them for traveling out on the land and never left them behind during win-
> ter. My father had one as well. When we were going to travel, when he started bringing
> me along, he would fold it over the back of the sled so I could go inside. When it was a
> little blizzardy, he would let me put my head down inside in the back of the sled.

When we were going to sleep, after staking the ice pick into the ground, he would scatter grass for matting inside and tuck in the garment and put weights around the bottom edges inside. And we put our few belongings in and slept inside. They put the sleeves on the inside, and placed the hood over the end of the ice pick, tying the fastener onto it but not closing it tightly up there, so the air could come out. They aren't like cloth but are very airtight. It would not be cold.

Peter Dull (March 2007:720) noted the equal importance of the *imarnitek* (seal-gut parka) in coastal communities, both for warmth and because the breathable gut prevented one from sweating: "If a person is going to go anywhere, he was told never to leave his seal-gut parka behind." According to Peter John (March 2007:1227):

When I began traveling, my mother always had me bring my seal-gut parka, even in winter. And my father told me that if I had a seal-gut parka with me, if I was cold, I should wear it inside my hooded cloth garment and the wind wouldn't enter.

One day when we went dipnetting, the weather suddenly became stormy and we were weather-bound. We stopped and tried to build a snow shelter. Since I had a seal-gut parka, I put it on and put my hooded cloth garment over it. Because wind can't get through, it felt as though I had put another parka over my parka. I realized that they are indeed warm.

A shovel was another essential tool. Nick Andrew (October 2006:333) said: "When we used dog teams and then Snow-gos, we were told to always bring a shovel with us, no matter where we were going. We'd pile snow across the stream, even if it was wide." Peter Dull (March 2007:1224) noted: "If we were walking on the land, they said the shovel was a big help if the weather suddenly got stormy and we were weather-bound, that we could use it to dig." Simeon Agnus (June 2009:59) said simply, "Although it isn't human, a shovel is a partner in the wilderness."

Nick Andrew (October 2003:16) also mentioned the importance of dry kindling when camping: "When we traveled in winter, they told us that when we were going to sleep, we should always have kindling ready so we could light the fire. Even though it didn't seem like it would rain, I got the dry kindling ready and put it under the bedding. When I used it, the wood would quickly burn."

Food was also essential. According to Peter Dull (March 2007:720): "Though they were going close by and planning to return the same day, they were told to take along some food since no one knows what the future may hold." Frank Andrew (October 2003:47) added: "They never let them go without provisions, and they always had them take seal oil, even though they were traveling close by." *Meluut* (aged fish eggs) and *sulunaq* (salted fish) were considered particularly sustaining, as was rich, oily *nin'amayuk* (fatty herring aged in seal oil). John Phillip (September 2009:423) observed: "They prevent the body from getting weak and tired."

People were cautioned to watch what they ate in the wilderness. Simeon Agnus (July 207:584) explained: "While we are traveling in summer, during the time when fledglings and their mothers are molting, they told us not to eat a variety of things. They told us not to eat just anything. We were especially cautioned [against eating] swan meat. They told us not to supplement our food with other things when eating them."

Although provisions are important, Peter John (March 2007:1226) noted that one could also rely on the land: "As for food, I didn't take too much. When I came upon Ungusraq's fish trap, I took just three blackfish. When I returned, I told Ungusraq about that since it is a teaching that if I take something that belongs to someone else, I should tell the owner about it."

Canegteggun Anangnaqsaraq | Way of Saving Oneself with Grass

Wassilie Evan (September 2003:299) said that travelers always carried grass, especially during fall. Frank Andrew (October 2003:47) agreed: "They never left grass behind." Peter Dull (March 2007:720) said, "They really valued grass. Grass kept us warm and took care of other necessities." Many recalled bedding down in grass. Peter continued: "In fall when night overtook us, we would collect lots of grass and put them in a depression and crawl inside them like a mouse. There is no cold at all inside them, and they are warmer than sleeping bags!"

Theresa Anthony (November 2000:233) of Nightmute recalled using grass as over-clothing to keep out wet weather: "When we traveled to *yuilquq*, if it seemed that we would get wet, we pulled out lots of grass and braided them. We then put them on and continued. When wetness hit, it would not immediately get on our bodies. This is extremely true." Peter Dull (March 2007:715) emphasized the importance of grass for survival: "They used it on all kinds of things, and if they were going hunting, they took grass to use as canvas. If inclement weather came or if they spent a night, they could use it as protection against the blizzard or as a mattress. So grass was their survival material before [modern] materials became available."

Like many before him, Tim Myers (September 2003:288) had used grass as mitten liners: "When it was cold out, they filled their mittens with grass that they had flexed and softened, or they gripped grass in the palms of their gloves. Their gloves might be wet, but they will dry and their hands will no longer be cold." Peter Dull (March 2007:716) described using grass not only to keep one warm but to save one's life:

At that time they never let us go without grass boot liners. When I was a small boy, Aanilaniq used to carefully wrap my feet with grass. And when my boots were on, I wasn't cold at all.

And grass under the sole wears out when we keep stepping on it, becoming pulverized. When our ancestors lived on the edge of fear, sometimes those wolves surrounded them. When [wolves] approached them from downwind, when they poured out the crushed grass, some of those [wolves] started sneezing. When they sniffed around and that [crushed grass] got into their nostrils, they panicked and stopped going after them. People used grass to save their lives.

Falling in water was particularly dangerous in cold weather, when a person could quickly succumb to hypothermia. Irvin Brink (October 2003:14) explained: "If we fell in water in fall, they told us not to give up but to keep the possibility of death alongside us. It is very cold in fall. To have been in water is like being tightly tied up." Travelers were admonished to use grass to save themselves. Irvin Brink continued: "When we get out of the water, we are to undress very quickly and wring out [our clothes]. Then we put on our underclothes. Then when we put on our clothes, we would put grass in between them. Grass is very warm. A person can save himself with grass."

Peter John (March 2007:1215) had also been admonished to use grass in an emergency: "If I fell into water, after quickly filling my clothing with grass along my body, they told me not to just stand still but to flex these joints that we usually bend before [my clothing] froze. Even if [my clothing] froze, those areas wouldn't get stiff and I would be able to flex the clothing along my joints." If two fell in, they were told to stuff grass inside their clothing and then wrestle with each other to warm up. Simeon Agnus (March 2007:718) shared his experience: "When my snowmobile fell through the ice, my boots were really taking in water, and it was cold. When I got up on land, I took off my boots, my coveralls, and pants and told [my son] to collect grass. I stuffed grass into my boots after softening them, and wrung out water from my pants and coveralls before they froze. When I put on my boots, my, there was no cold whatsoever. I did not put on my socks but used only grass, and I stuffed my wet coveralls with grass. So everything was warm on me, including my feet."

Qimugtet Usvituut | Dogs Are Wise

No discussion of traveling on the land would be complete without mention of the teams of dogs that accompanied travelers. Men spoke with affection about their dogs, praising them not only for their strength and stamina but for their intelligence. Wassilie Evan (September 2003:305) declared: "Back when they always had dog teams, hardly anyone got lost. Although there was a blizzard and no visibility, their lead dogs brought their owners home. The dog won't get them lost, but only the owner would [get them lost]." Tim Myers (September 2003:307) described how dogs used landmarks to find the trail, just like their human companions: "They use the places where they come out from lakes, and when there is no visibility, if there are trees or small hills, they see those and use the same trail every year." Peter Dull (March 2007:505) recalled:

When we came of age to go by dog team, they warned us that when we did not know where we were [in a blizzard] and were on our way home, even though we had doubts about which way our lead dog was going, to let him go his own way. They say that they take them home where they usually stayed. We always fed dogs in the evening. They say that they always want to go to places where they ate. So when the weather gets inclement, they taught us that in the wilderness, since the dog knows its home and the outlying areas, they always go to their home. They know by their keen sense of smell, and they do not rely on how they have gone.

George Billy (February 2006:372) recalled his own experience:

Lead dogs were intelligent. Some dogs were something else. Several times I inadvertently tried to make my dog get lost when I couldn't recognize my trail. Although it was snow-covered, my dog [knew the right way to go]. When I commanded it to go north because I thought that he was lost, that dog would just look at me [and not go].

So after trying to make him turn in vain, I let him go on his own; so he turned one way and went on. Though snow was blowing hard, [my dog named] Kavirli-rralek took me home. These dogs are intelligent, and some of them know the dangerous spots.

All were in agreement: dogs don't get one lost. Elders were also unanimous in declaring that snowmobiles are not like dogs. Tim Myers (September 2003:292) said with feeling: "It's not as good to have a snowmobile as a dog during winter when lost. Although it seems like they turn just a little, since snowmobiles are so fast, sometimes we divert a long way from the trail when these tracks are covered, and we lose our sense of direction."

Wassilie Evan (September 2003:292) observed that snowmobiles create their own wind:

These days when he goes by snowmobile in calm weather, it gets windy, and as he increases his speed, the wind gets stronger and blows his clothing as he continues on his way. Also, when it's snowing, when the snowmobile starts to go, it seems windy, and they will not know what the weather is like.

That's why if a person is using wind to tell direction, when he thinks that he is going the wrong way, he stops the snowmobile. Since he used the wind, he would realize that although he had not changed course, his speed had gotten him lost.

From the time they were young, most contemporary elders were given instructions that allowed them to travel through the wilderness and harvest its bounty. Nick Andrew (October 2003:15) told a story testifying to people's self-reliance and independence:

> I just want to tell a story. The barge used to arrive not too far from the ocean downriver from us. After it arrived in spring, it returned in the fall. One fall it didn't arrive. It probably had some difficulty. Since it wouldn't arrive, a *kass'aq*—the one who took care of the barge—happened to come by dog team. He gathered the people and asked the elders, "The barge has not arrived. What are you going to do all winter?" .
>
> One of the elders replied slowly, "Tell that one that even if the barge doesn't arrive, it is the same to me." Maybe [the *kass'aq*] thought that they were going to go through desperate times. It was the same to them, even though the barge wouldn't arrive.

Elders combine empirical observation of the present with past experience to help them travel safely. They must know local conditions, named trails and landmarks, and their position relative to key geographical features.[3] They must also know the traditional instructions to survive dangerous situations in a knowing and responsive universe. Paul John (January 2006:66) declared: "It is no wonder they said to us: 'If you pay close attention and heed those who are giving instructions during your short life, you will live like one who speaks for yourself.' It so happens that when one recalls his oral instructions while carrying out tasks and continually makes an effort to follow the *qanruyutet*, they won't allow one to come upon unfortunate circumstances. They told us to pay close attention and heed their words." Failure to follow the rules, like failure to pay attention to changes in wind and weather, could have disastrous consequences, both for individual hunters and for their families and communities. Following the *qanruyutet* allowed one to travel safely, even in extreme and potentially dangerous conditions.

Qanikcaq

SNOW

YUP'IK PEOPLE'S UNDERSTANDING OF *QANIKCAQ* (SNOW) IS AS vast and deep as their own multilayered landscape. Snow covers the land, and often fills the sky, as much as two-thirds of each year. Once on the ground, winds shape it and thaws followed by cold weather harden its surface. The Yup'ik relationship with snow is nuanced, involving emotions non-Yupiit more often associate with interpersonal relations, including love, fear, humor, and respect. People drive on it, mix it with berries for food, shelter in it, and die from disrespecting it. Interactions with snow continue to be an intimate part of Yup'ik life today.[1]

A common myth holds that Inuit languages have hundreds of words for snow, while a scholarly rebuttal declares that these myriad terms are variations on only two root words.[2] Neither generalization reflects reality in southwest Alaska,[3] where elders use a rich vocabulary, including more than seventy unique terms to concisely describe their intimate experiences with snow.

Qanikcaryak | An Abundance of Snow

Contemporary elders agree that the abundance of snow in the past no longer occurs. John Phillip (October 2005:118) declared: "Sometimes when I traveled at night, I only knew I had arrived at Kipnuk when my dogs stopped. A chim-

Homes covered with snowdrifts by the strong coastal winds. *Augustus Martin, Martin Family Collection, Anchorage Museum B07.5.1.A2*

ney would be sticking out. That's how much snow there was. Or when I'd reach a home out in the wilderness, I would only know by that chimney that it was a house." Peter Jacobs (January 2006:51) remembered:

> There would be lots of snow shoveled from the entryway of the *qasgi*, and they would construct stairs [out of snow] to go down [to the entryway].
>
> Before their houses were Western-made, the boys in the *qasgi* who were willing would shovel snow from outside the sod homes. They wouldn't just shovel snow for their relatives but for all the homes. There truly was an abundance of snow back then.

Elders remembered snow covering the land in winter in the past. Golga Effemka (January 2006:27) noted: "There was a lot of snowfall when I was a child, and the moose that roamed about would drag their stomachs when walking upriver." David Martin (December 2005:143) recalled: "Back then the snow was so deep that only hills would be visible on the land." John Phillip (December 2005:143) agreed: "Truly, the land that we saw back when it snowed abundantly was smooth. Trees along riverbanks and the shores of lakes would be gone, and it looked level. People could no longer recognize the land, and when they reached lakes, they wouldn't know that it was a lake."

The snowy landscape also had distinctive sounds. Paul John (December 2005:144) explained: "Back when I was young, when a person walked, his footsteps made a crunching sound because it was so cold. These days we don't hear that sound 'qerriuq' anymore because the coldness isn't strong." John Phillip

(October 2005:25) added: "When I became attentive back then in winter, when a person walked, the crackling of his feet was apparent although he was far away. Because of that, when some people hunted for red foxes, since our first ones were so ingenious, they placed *qaliruat* [caribou-skin boot outsoles] fur-side down [under their feet]. Before they made a crackling sound, they'd reach [the fox] and shoot it." Bob Aloysius (October 2005:32) had a similar experience:

> When it used to get cold, you could hear everything, especially animals around trees. When it was windless you could hear, and the old men would know what was walking, "Ah, the large, male moose is going to cross by the Aniak River." Some would say, "A female with two calves is walking."
>
> When we walked on top of the snow, we could be heard for maybe two miles. The sound would get louder as they approached, and people on dogsleds in the dark could be heard from far away.

One could not only see and hear cold snow but use one's nose to sense it. John Phillip (October 2005:24) said: "Sometimes, as soon as I went outside, ouch, these two [nostrils] would [sting] very hard. They say *pupengqualukek* [those two sting from the cold]." Bob Aloysius agreed: "We didn't know about temperature, but just used this [nose] to measure the cold."

Years of heavy snow were marked by both high water and abundant wood. Paul Kiunya (October 2005:3) recalled: "In spring, when the snow melted, there would be so much water on the land." Rivers ran high, and wood was abundant. John Phillip (October 2005:21) noted: "When I became aware of life, there was always wood down on the shore. During spring, when there had been lots of snow and it melted and the ice broke and flowed out to the ocean, many logs came out through the Kuskokwim River for our use. When the logs stopped, the shore was lined with wood." Deep snow made wood hard to find in winter, however. Frank Andrew (June 2005:9) said: "At that time we lacked wood for stoves because snow covered it. Down along the coast we poked a large stick down [in the snow] to search for wood, constantly poking. And when we found a log, we dug it up, and sometimes it would be as deep [as this house]."

Bob Aloysius (October 2005:30, 33) also described the quick freeze-up and extreme cold associated with past winters:

> In fall we moved to our fall camp, and it wouldn't take long to get cold. Sometimes it froze overnight upriver around the trees. The streams that had current would take three nights to freeze, but the sloughs without current would freeze overnight and soon were good to travel on. It never snowed until after the ice got thick.
>
> When it finally became winter, we wouldn't travel anymore because it would be too cold. All the animals quit walking around, too. From the middle of December

until the end of January we would mostly stay at our winter camp gathering fire-wood, packing water, ice fishing, or checking our snares for snowshoe hares.

We followed the animals' movements at that time. When animals began to move, we would follow them and begin to travel. When they stayed, we stayed.

Cold weather created its own diversions, including a unique game similar to hockey enjoyed by children in coastal villages. Paul Kiunya (October 2005:3) explained: "In our village, when dogs started to eat needlefish, they defecated very hard feces. When their small feces became round like balls, they wouldn't break although you hit them with great force. We used that *akakataq* [small round feces] as a puck and hit it with wooden sticks and let it move. We also had opponents. We made a hollow spot in the snow and piled snow [around it] forming a barrier, so the feces could land inside."

Qanikcam Caliara | The Work of Snow

An abundance of snow in winter was viewed as a good thing for a number of reasons. John Phillip (December 2005:114, 143) recalled: "When snow was abundant, some of the houses would be covered, and it would insulate them and make them warm. . . . They wouldn't get cold inside, even though they had no stoves. Snow is very airtight when it gets damp and hardens. And since they were sod houses, they would put a piece of snow [by the skylight] facing the sun, reflecting inside the house so it could be bright."

Nick Andrew (October 2005:276) spoke of the benefits of abundant snow for fish habitat: "An abundance of snow is good for those of us who subsist from the wilderness. It is good for blackfish and also good for fish in the Yukon and Kuskokwim Rivers. If those two rivers are full, the current is very strong and will carry away the water that was along the bank all winter and clean it out. If fish taste the water below and don't like it, many fish don't go upriver. But if water is abundant, since they want fresh water, then many fish will go upriver. When water is high, there are lots of fish."

John Phillip (January 2006:80) spoke of how melting snow helped the land: "Before winter, sometimes when it snows a little, they say that the weather is providing conditions that will melt [the snow]. After it snows the conditions get wet and it melts [the snow]. That snow is beneficial for the surface of the land and growing things. And sometimes when you shovel, there are distinct layers of snow." John (October 2003:103) also recalled fall conditions conducive to a plentiful berry harvest the following summer: "In fall when it constantly freezes [after thawing], they were thankful. Snow itself can't stay on the ground for a long time, but if it periodically gets wet and freezes, it will help the salmonberries when it is time for them to grow. They say that when there is too much south wind, the snow melts fast and the land doesn't get damp. But if there is south wind followed by snow, the snow would stay longer during spring and toward summer."

People, John Phillip (January 2006:25) concluded, were sustained by the work of snow: "The snow also helps things that grow on the land by watering them. When there was little snow, those people said that the berries would be scarce because they didn't have anything to water them. But when there was snow, they were grateful for the plants and berries that would grow on the land. We eat from the land, and it sustains us. It was our source of sustenance when Western goods were scarce. We people of the lower Kuskokwim coast were sustained by the work of snow."

Qanillerkaan Nallunailkutai | Signs of Coming Snowfall

As winter drew near, people looked ahead to a more or less snowy season. Peter Jacobs (January 2006:51) recalled: "Back when there was an abundance of snow, they knew of coming conditions, and they would say that there was going to be *qanikcaryak* [a large quantity of snow]. . . . The elderly woman who cared for us would keep an eye on the grass. When grass grew abundantly, she said that there would be more than the usual amount of snow. They called it *maqarqucir-luku canegnek* [grass insulating (the ground)]." Nick Andrew (October 2005:367) recalled another well-regarded predictor: "If it snows in June, there won't be much snow, the usual amount, all winter long. But if it hasn't snowed during that month, then there will be more snow. That's how it is." Nick (October 2003:72) also noted the admonition not to pull the feathers from gray jays during winter or snow would follow: "Even in extremely cold weather, if a person pulls off the feathers, it will begin to snow before dawn. . . . When they caught it, they plucked the head and let it fly. When they woke up the next morning there was already lots of snow."

Peter Jacobs (January 2006:75) recalled being taught to recognize conditions that indicated approaching snow: "If the horizon at daybreak was split when we woke from sleep, although the weather was calm, our dear father would come indoors and say, 'The daybreak out there indicates that heavy snowfall is about to arrive, along with wind. Do not travel far.' What he said would come true." John Phillip (January 2006:82) also recognized the weather conditions that foretell approaching snow:

> Sometimes we wake in the morning and look along the horizon and see that it is very red. They say that before the end of that day, the weather will turn bad. That is also a sign of coming snow. When the sky starts to form clouds and when [clouds] come in and cover the sky from the direction of the ocean [south], snowfall ensues.
>
> When it will eventually snow down on the ocean, the area along the horizon

becomes dark periodically, and sometimes it starts to change appearance because of the sun and looks as though there's a downpour from the sky. When that front approaches, it arrives as bad weather from the direction of the ocean.

Also, when wind is blowing from the north, when it brings the clouds above us, that is a sign of impending bad weather with snowfall.

When it began snowing, the size of the flakes also predicted what was to come. Nick Andrew (October 2005:208) mentioned what many knew from experience: "Sometimes in winter there is suddenly a very large snowfall [with large flakes]. When that happens, it will pass. It will stop snowing. But if it slowly starts to snow small snowflakes, it will snow for a long time."

Qaniqerraarqan | When It First Snows

Peter Jacobs (January 2006:73) recalled the signs of winter: "Before there is snowfall, we know winter is approaching. At some point when we go outdoors, we start to smell the scent of cold weather, before there is snow. One day we wake to find that *qakurnaq* [frost] has formed. That is the first sign of snow. Since small villages had different sayings, when they saw *qakurnaq*, they would mention that it is a sign that winter would come in the next month."

Peter Jacobs (January 2006:75) explained that the freezing of the surface of the tundra before the first snowfall was considered a good thing: "In fall, sometimes the top of the land freezes before there is snowfall, an occurrence they call *ciulivigluku*, 'placing a covering over it.' Those people would start to say, 'It is a good thing, as there will be berries.' They were grateful for that occurrence." Paul John (January 2006:64) added: "If there is a south wind and wet conditions that soak the land's surface before there is snowfall, when the weather gets cold, it freezes the land's surface. Since they viewed that condition as a means to water the berries that would grow when it melted, they were grateful and said that berries would grow in abundance."

The first snowfall, Peter continued, also had a saying: "They said that the new snow of the first snowfall isn't good to drink when one is thirsty, that fresh snowfall is harmful. They said that if there was nothing else to drink, we could just place [the new snow] in our mouths for moisture and then spit it out and not swallow it."

John Phillip (January 2006:79) spoke of fall freeze-up following the first heavy snows: "Also sometimes just before winter, it starts to snow. When it snowed a lot when the weather was calm and it had started to get cold out, those elderly men would say that the snow that settled on lakes, in water, and in the ocean *tuvqau-tekiurtuq* [was providing conditions that would freeze and become solid, from *tuve-*, "to cake up"]. When it starts to get cold, the snow doesn't melt, and the [snow's] surface immediately freezes and creates conditions that are safe to walk on, as [the snow] freezes thick."

John Phillip (January 2006:80) also spoke of the wind's role in solidifying snow: "When a light wind is blowing and [the snow stops falling], the snow hardens. When it starts to get cold out, one can stand on top of the *nutaqerrun* [new, fresh snow] and use it. They say the wind solidifies that [snow], making it solid. Since I've used that many times when I walked out to the wilderness, I know that occurs." Snow scientists call this process sintering. Wind blows and tumbles newly fallen snowflakes about, driving one grain into another. When the wind stops, the broken snow grains become glued together in a superstrong mass. Anywhere two particles touch each other, they exchange molecules and bond into a single mass. The thicker the bonds, the stronger the snow.[4]

Nutaqerrun wall' Qaninerra'ar | Newly Fallen Snow

Yup'ik people cultivate a rich vocabulary surrounding snow, vividly describing different conditions. Newly fallen snow went by several names, including *nutaryuk* (fresh powder snow), *nutaqerrun* (new snow on the ground, which melts right away in the mouth, from *nutaraq*, "new one"; also, new ice in fall), and *qaninerra'ar* (newly fallen snow, first snowfall). John Phillip (October 2005:271) recalled: "They call that white and clean [snow] *nutaqerrun*. After it snows, that *nutaqerrun* stays for a while. Sometimes, when it gets cold, it makes a crunching noise, and it has a lot of friction when it isn't damp. And a sled makes noise as it's moving on that *nutaqerrun*. It tends to stick to sled runners and won't let them glide." George Billy (February 2006:406, 415) noted: "*Nutaqerrun* is airtight, and it was so warm when we used to pitch a tent in the wilderness. . . . If it snows this morning, that is *qaninerra'ar*. New-fallen snow provides much warmth when used as a shelter. One will not freeze."

Kavtakuaq wall' Kavtagaaq | Granular Snow Used for Akutaq

Of all the types of snow discussed, among the most enjoyable was *kavtakuaq* (lit., "pretend *kavtak* [hailstone]"), also known as *kavtagaaq* or *kavtayagaat,* the sandy bottom layer of fresh-fallen snow used for *akutaq. Kavtakuaq*, or depth hoar, is formed when dry snow metamorphoses into large, ornate grains when subjected to a strong temperature gradient inside the snowpack.[5] John Phillip (October 2005:266) recalled: "In fall when it first snowed, the ones I watched always made *akutaq* [a mixture of fat, snow, and berries]. My grandmother would try to make *akutaq*, gathering the snow. After putting oil [in the bowl], she mixed it and added snow to it. She then added caribou back fat. Then she finally put in the supplements [berries]. My grandmother, Cingarkaq, never went without making *akutaq* when it first snowed."

Peter Jacobs (January 2006:53) recalled: "There are different terms for snow. There is what they call *nutaqerrun* [new, fresh snow]. And they call the old [snow] *kavtakuaq*; the *kavtagaaq* [granular snow] that is inside snow. The *kavtayagaat* [lit., "little hailstones"] in the middle [of the snowpack] are good. When they were collecting snow to make *akutaq*, they obtained *kavtayagaat*." Lucy Sparck (October 2005:267) noted that her great-grandmother used this middle layer of metamorphosed snow, which was not only coarser but often colder than the top layer: "When she told me to get [snow] for *akutaq*, I would remove that [snow on top]. She told me not to get the ones on top." Nick Andrew observed: "That snow that is soft is easy to remove. Then there is *kavtakuaq* underneath."

According to Paul John (January 2006:63), *kavtagaaq* is preferred: "And if the snowfall is damp, [it will form] snow called *kavtagaaq*. *Kavtagaar* isn't extremely soft; it isn't *nutaqerrun* [new snow]. They call that very soft snow that isn't supplemented with anything *nutaqerrun*. The snow that contains *kavtagaar* is preferred for making *akutaq*, by adding it to seal oil." David Martin followed with an apt comparison: "Think of our food. Fresh ones have a different smell, and the next one has a different smell. The fresh snow melts right away when bitten, but snow that is from below is sandy and doesn't melt right away and is cold when swallowed. That's the kind of snow used in *akutaq*." George Billy (February 2006:406) enthused: "They could wait to make *akutaq* using [the bottom layer of] new-fallen snow. But if they do not have new-fallen snow, they use older snow, too. That snow used to be so cold on their hands. Those females are hard acts to follow! Some were in such a frenzy when the first snow fell and would make *akutaq* with gusto."

Kavtak | Hail

Kavtak (hail, hailstone) was another well-known phenomenon. John Phillip (October 2005:267) explained: "Before freeze-up, *kavtiit* [hailstones] fall, the frozen rain that comes from above; they are white and resemble snow. They hurt when they hit. Sometimes, even in summer, after the weather is extremely warm, it thunders. And after thundering, we see *kavtiit*." Nick Andrew (October 2003:97) called hail the "partner of snow," while Irvin Brink jokingly declared: "*Ellalluum-gguq ilurai*" (They say they are the rain's cross-cousins).

When he was young, Roland Phillip (October 2006:170) watched a shaman use a hailstone to cure his father: "When he was about to work on [my father], he told his wife to get a *kavtak* [hailstone] from outside. I was curious, thinking that when he got the *kavtak* he would do some sort of incantation. When she brought it inside, without putting on his seal-gut rain garment, he told my father, 'Now put that inside your mouth. Don't bite on it, but when it melts, swallow it along with your saliva.' Then [my father] put it inside his mouth and swallowed that. [My father] probably didn't feel it [inside his mouth] because that *kavtak* was just one small piece of snow."

Qakurnaq Kaneq-llu | Frost and Hoarfrost

Peter Jacobs (January 2006:87) recalled the adage that frost was said to chill the water as winter approached: "When *qakurnaq* [frost in fall] started to form, they would say that it was chilling the water for eventual freezing, saying *kumlacirluku* [it chills it]. They say that is the purpose of *qakurnaq*; frost periodically forms to chill the water. That is the instruction they gave about *qakurnaq*." John Phillip added: "That occurs when winter will soon come. Frost is a sign that the water is about to freeze and cold weather is near."

When moisture content in the air is high, intricate crystals called *kaneq* (hoarfrost) form on trees and plants. Since warm air carries more liquid in suspension than does cold air, *kaneq* often forms overnight when temperatures drop. Although this frosty coating is lovely to look at, Nick Andrew (October 2003:96) noted its negative association: "During winter, when it's foggy in my village, the trees get frosty. They say that it makes people sick. Some want to cough, too, because it is damp. Our ancestors were told to observe the trees and bushes when they got frosted. They say that sicknesses develop from that." Irvin Brink added: "Sometimes in winter trees get a parka of frost and snow when there is no wind." Paul John noted that *kaneq* developed during cold temperatures, and John Phillip (October 2005:23) observed that on the lower coast Kanruyauciq (Time of Frost) is the name for January: "Sometimes during winter, those wild rhubarb and wild celery plants would appear like tall people. *Kaneq* makes that happen when there is no wind." In windy weather, however, *kaneq* could cause blizzard conditions. *Kaneq* formed inside sod homes as well. Many recalled how small fires were periodically built inside in winter to remove the heavy frost that formed along the ceilings and to dry the inside.

Taqailnguut Kitulget-llu | Ones That Don't Stop and Ones That Pass

Taqailnguut (small snowflakes that do not stop falling, lit., "ones that are persistent") are often distinguished from *kitulget* (large snowflakes, clumps of flakes that fall for only a little while, lit., "ones that will pass"; also called *qanugpiit*). One reason *kitulget* stop so fast is that large flakes require a lot of moisture in a cloud, and in Alaska in winter such moist clouds quickly run out, returning to the slow snowfall rates typical of everywhere but coastal mountains.[6] Elders noted that *taqailnguut* often come in fall before there is a lot of snow. Peter Jacobs (January 2006:74, October 2003:97) explained:

> When winter came and snow first arrived, the snow was a different type. Sometimes, it will snow very large flakes. And the people where I was raised, since those snow-flakes were too large, would [pretend to] say to these very large snowflakes, "The snowflakes' buttocks are visible." It is said those very large snowflakes stop falling right away [when spoken to like that]. But they said these small snowflakes don't quit falling right away and are never-ending. . . .
>
> They don't place importance on the large snowflakes. They call them *kitulget* [from *kitur-*, "to pass"] because they will quit right away. The weather will be good after the snowfall.

Nick Andrew (October 2003:107) added: "When it suddenly snows huge flakes, it doesn't snow for very long and the weather gets good; but if it doesn't begin to snow right away and snows small flakes, it gets thick. That snow doesn't quit right away." John Phillip (October 2005:209) noted the same phenomenon: "It is said those *qanucuayagaat* [small snowflakes] don't pass right away. They call that *pellukainayuilnguq* [one that doesn't pass right away]."

Paul John (October 2003:107; June 1995:23) added detail:

> They call those big snowflakes *kingukegglit* [ones with a good outcome]. The weather will be good when it is over. *Kingukegtuq-gguq* [They say the outcome is good].
>
> When it's not really windy and it snows light snow with little ice particles, they called that *kaimlenguuyutaq* [from *kaime-*, "to make or drop crumbs"]. It does that even though the skies are clear. . . .
>
> When it's overcast and *taqailnguaraat* [small *taqailnguut*] begin coming down, the cloud where the snow is coming from is called *mamtulria* [one that is thick]. But when the cloud where the snow is coming from is thin, the snow will come down rapidly and suddenly stop.

Mecungnarqellriit | Ones That Make You Wet

Men also described wet snow called *mecaliqaq*, *mecungelriit* (ones that are wet), or *mecungnarqellriit* (ones that make you wet, from *mecunge-*, "to be wet") and sleeting conditions known as *mecaliqerluni*. John Phillip (October 2005:209; January 2006:83) explained:

> In wet weather, those moist ones are *mecungnarqellriit*. That kind of snow is moist, and those large [flakes] make you wet. They also melt fast. When it snows that type

of snow, they say *mecaliqerluni* [it forms sleet]. The surface of one's body gets damp, and one will make tracks when traveling. . . .

They also refer to [the snow] as *yukutanirluni* [damp]. They use two terms for it; they refer to [the snow] as *yukutanirluni* and *mecungnaqluni* [causing one to get wet].

Nick Andrew (October 2005:210) recalled the hazards of wet snow: "It's [dangerous] because it causes one to get wet when it melts. And when moist [snow] falls, the [snow's] surface freezes. When that happens in my village, they call that *patugluni* [from *patu*, "cover"] because it covers the snow and prevents it from blowing away." Nick (October 2003:110) noted the admonition against traveling in wet-snow conditions: "When [snow] lands on the body and melts, they say not to travel. *Tegulallugnarqeqatartuq-gguq* [They say handling things will become difficult] when it melts as it lands on the body. It doesn't take long to get clothing wet. That happens when it's not cold. Snow that is dry is better." Paul John added: "They say *urunglugnarquq* [it is such that bad melting conditions tend to occur] when the snow melts on whatever it lands on, even though it's not a person's body." Wet snow, George Billy (October 2006:47) concluded, is powerful: "That was our teaching. They say that during winter, when *mecaliqaq* gets on a person's clothing, a person won't be strong for long. They will become weak although their destination is close. They [were concerned] about one getting cold when the water soaked one's clothing. They say *mecaliqaq* is overpowering. They told people that they wouldn't be able to travel for long."

Nevluk | Clinging Snow

Clinging snow was known as *nevluk*. According to John Phillip (October 2005:273): "*Nevluk* sticks to one's body. When it snows wet snow they'd say *nevnarquq* [(the snow) has a tendency to cling]. It [snows] the type that sticks and is damp." Bob Aloysius noted that clinging snow could be hazardous: "When we traveled and that type of snow came, they told us to stop and somehow pitch a tent or to stay. Even though a hole was small, the [snow] would quickly [plug it]. The first snowmobiles had carburetors that were exposed, and [the clinging snow] tended to enclose them and we weren't able to go."

Qaninerra'ar Tengyukaaralria-llu |
Fresh Snow and Snow That Easily Blows Away

Other snow to be wary of was *qaninerra'ar* (new-fallen snow). If it came to rest on the ground and a wind came up, it might blow and become blizzardy. John Phillip (October 2005:215) explained: "They call that snow *tengyukaaralria* [one that easily blows away, from *tenge-*, "to fly"], and they say it is dangerous. They told us that if it snowed *tengyukaaralria*, a blizzard would occur right away if a light wind started to blow from any direction. We call it *tengqeryuka'ar* [(snow) that easily flies], as it easily blows like feathers and tends to scatter." Nick Andrew added: "When it has just snowed, it is dangerous. But after the snow stayed a while, the blizzard wouldn't be as bad." Frank Andrew (June 1995:23) observed the same conditions: "Snow that came down to the ground on a calm day was called *tenganaviaraat*. They'd say it was dangerous and could bring a blizzard as soon as the wind started to blow."

John Phillip (October 2005:215) noted that care should be taken when traveling in fresh snow, even after the surface had hardened:

> If we were traveling far away, we were cautioned about *tengyuka'ar qanikcaq* [snow that easily blows away]. That immediately starts to "smoke" and scatter after the wind starts to blow. There is no visibility, and there is no way to recognize your surroundings although close by.
>
> If the wind doesn't blow that *qaninerra'ar* and it stays for a while, sometimes the surface becomes wet and then hardens, and a small layer of ice forms on it. But when the wind starts to blow, it can penetrate [the frozen layer], and [snow] underneath can still blow since the area under [the frozen layer] isn't soaked and is snow.

Bob Aloysius (October 2005:36) described whiteout conditions and their aftermath:

> When it got windy, especially up on the mountains, things that were close weren't visible because the wind was so strong. Since the snow was so dry, it would really blow away. Sometimes we would stay in one place for two days. After the wind stopped blowing, we'd attempt to travel again.
>
> When the wind died down in the morning, we would go out of our tent, and our dogs would be nowhere in sight. Everything would look [level]. My paternal uncles would whistle, and all of a sudden the dogs' noses would come out of the snow. They stayed put and didn't move when there were severe blizzards back then.

Natquik Pirtuk-llu | Drifting Snow and Blowing Snow

The treeless tundra of the Bering Sea coast is both a snowy and a windy place in winter. *Natquik* denotes drifting snow on the surface of the ground, while *pirtuk* (blizzard) is blowing snow. John Phillip (October 2005:217) explained the difference, expertly scaling drifting and blowing snow from mild to severe depending on the wind: "*Pirtuk* is worse than *natquik*. The area above is visible when there is *natquik* and [the snow] blows close to the ground. But when it is windy and there is a blizzard, it blows high up. They refer to it as *qulvagguarluni* [going high above the ground] when the snow blows high. *Pirtuk* is the same. There is no visibility, and one cannot recognize the surrounding area, and even a house that is very close isn't visible. It gets very blizzardy in our area on the coast where there are no trees."[7]

Paul John (October 2003:110) noted slightly different terms for drifting snow in different areas: "Down in my village [on Nelson Island] and in the tundra area, drifting snow is called *natquiggluni,* but at Chevak and Hooper Bay they say *natquissaraarluni.*" Peter Jacobs added: "They look at the drifting snow and, if it's worse, they call it *natqugpak* [lit., "big drifting snow"]. If the blowing snow is low, they refer to it as *natquik,* but if it is higher in elevation, it is *natqugpak.*" Paul John continued: "In my village, if [drifting snow] is high, it is called *agerpak* [large snowbank]. If the drifting snow covered [something], they say *agellrullinia* [it went over it]." Conversely, a small snowbank is sometimes called *agneq.*

Irvin Brink (October 2003:110) noted the admonishment not to travel through an *akuurun* (thick snowstorm covering the ground, from *aku,* "lower part of a garment"): "The people of Bethel instructed others not to travel when the weather was bad and an *akuurun* was visible. They would tell them to stay." Sometimes blizzards started while men were on the trail. John Phillip (October 2005:218) described his experience:

> Back when we traveled with dogs, we depended on them; some lead dogs are wiser than us people.
>
> When there was a blizzard, when my wise lead dog's eyes became covered [with snow], sometimes it would stop and clear them by itself. Then we'd stop, and I would clear their eyes and paws [of snow], and we'd continue. And while we were traveling, although there was a blizzard, when the trails had many bends, those wise dogs could travel using a shortcut and reach their destination without using those trails. My dogs would take me, and I never told them where to go.

John Phillip (October 2005:221) also shared advice he was given on what to do if lost in a blizzard. He was told to stop where he was and hollow out a place in the snow to wait out the storm, never allowing the snow cover to become too thick. Once inside the snow, he was told to use his forearms to push the snow away from his torso. Then, using his walking stick, he should create an air hole: "If we stay in the snow, we won't get cold. Our breath will moisten the snow [inside], making it airtight." John concluded: "That's the instruction they give about stopping during a blizzard. I haven't experienced that, but when I'd stop sometimes at night, I would hollow out the snow and sleep there, and it was very warm in the snow. But I'd place my sled so that my dogs wouldn't be covered [by snow]. Sometimes snow covered some dogs, and we'd take them out from deep in [the snow], when it started to snow without our knowledge. My dogs were covered like that a number of times, but I'd get to them before they died, shoveling them out by stretching out their harnesses since we had them connected."

John Phillip (October 2005:224) noted the term for allowing oneself to be covered by drifting snow: "They call that *agevkarluni* [letting the snow drift around you to make a shelter, from *agevkaneq*, "snow that gathers on the sheltered side of an elevated area"], sitting down and pushing the surrounding [snow] outward. They have a lot of room and are able to stand." Paul John (January 2006:66) added that if the snow got too thick overhead, a person should create an airway: "If blowing snow covered us, if we got to the point where a stick could not pierce through the top [of the snow shelter], we would scratch [the snow] above and along the sides with our hands and place it underneath us to gradually move up, following the thickening snow. If we just stayed put underneath the snow, we would suffocate. Those are good instructions they gave us." George Billy (October 2006:406) warned: "One is told never to fall asleep if a blizzard causes him to stop. Though snow appears light, it is heavy, even in small amounts."

Sometimes travelers encounter a blizzard's edge or border. John Phillip (October 2005:238) explained:

I've experienced *pirtuum engelii* [the blizzard's edge]. Last year, we went [north] to Chefornak, passing through Kipnuk. Just as we left [Kongiganak], snow began drifting along the ground's surface, and there was low visibility, and the mountain [near Kipnuk] was no longer visible. As we continued, we suddenly came out [of the blizzard], and the mountain was visible. That area was actually clear.

When I arrived, I asked people if the weather had been like that all day and not stormy. Only the area between [Kongiganak and Kwigillingok] was blizzardy. Sometimes, some of these large lakes are like that. There is a [blizzard] line, and you can get out [of the blizzard]. That isn't the only place, but that happens to me from time to time upriver from us, right above the village of Tuntutuliak; I come out of the blizzard like that. It has a border.

Since the wind has many different strengths and has a line [where it ends], that seems to determine the [blizzard's] edge. The ocean is like that, too. When it's wavy, there is an edge [where it becomes calm] in some places.

Drifting and blowing snow were problems not only while traveling. Blizzards filled villages with snow and covered houses. John Phillip (October 2005:224) continued:

Since there used to be a lot of snow, even in villages, [snow] would fill their porches. When it started to get blizzardy at night, they'd bring the shovels inside the house where they were easy to get to.

One morning when we woke up after a snowstorm, they opened the door, but the porch had apparently filled with snow with only the edge of the upper section open a little.

Since those elderly people couldn't do it, they asked me to try to go through [that opening], bringing the shovel with me. I complied, and after they cleared it, even though [snow] went inside the house, I'd go a little farther, and eventually I went outside. When I got out, I cleared [the snow], and when an adult could fit through, he went out and finished clearing it.

Back when it used to snow a lot, we'd pile up the snow that we constantly cleared from porches, and the piles became steep outside. The doorway also got covered by snow and developed an [adjoining snow porch]. They'd make a door for that and make stairs on the snow [going up from the doorway to the outside of the snow pile].

Patuggluk wall'u Nungurrluk | Ice Fog

Ice fog was another well-known weather condition, referred to as both *patuggluk* (ice fog, lit., "bad *patu* [cover]") and *nungurrluk* (ice fog, fog in cold weather). Peter Jacobs (January 2002:48) recalled: "A man would come in and say, '*Patugglilliniuq keggna* [There is ice fog outside].' It's ice fog when it's cold and freezing. The sides of logs are covered with ice." Paul Kiunya (October 2005:226) explained: "It's *cikuya-gat* [small ice particles]. It's like fog, but it constantly turns into ice when it settles on things. It freezes everything it touches when it settles on the ground. It coats airplanes, too." Paul John (June 1995:23) noted the difference between *taituk* (fog) and ice fog: "Fog in summer is called *taituk*. And when it's very cold in winter it's called *nungurrluk* [ice fog]. In the Nunivak dialect fog is called *nunguk*, as those Nunivak people speak a little differently from those on the mainland."

John Phillip (October 2005:226) associated ice fog on the coast with an east wind:

That especially occurs when there's wind from the east, along with cold; it tends to freeze. If the windward side of the upright grass freezes, it will bend toward [the leeward side]. They tell us to use that as a guide. If we knew which way the wind was blowing, the windward side [of the grass] will freeze and bend. The windward side of wild celery will also freeze. We can tell the direction of our home through those if we know which direction the wind was blowing from. . . .

When it's moist, it freezes. Sometimes it doesn't freeze when there are small flakes. That type [of weather condition] approaches as a cloud that has sun rays. They say *kuvuarnirluni* [lit., "it appears as though it is pouring"]. It's a reflection. That doesn't occur all the time.

Nakunarqellria | Condition That Makes You Cross-eyed

Snow and fog sometimes combined to create whiteout conditions, characterized by poor visibility. Nick Andrew (October 2005:230) explained: "It's like the snow and sky are alike when looking around at your surroundings. When the sky is cloudy and the snow is white, they become identical." Paul Kiunya (October 2005:230) elaborated: "When it's like that when traveling, you can't see anything. The *kass'at* call it whiteout. Even though the weather is really good, you can't see anything around you. That only happens when it's cloudy in the snow." John Phillip (October 2005:231) recalled its evocative name:

> They call it *nakunarqellria* [lit., "one that makes you cross-eyed"] down on the coast. If you walk with your eyes crossed [*everyone laughing*] you won't see the ground. It's hard to tell what is dangerous ahead of you, and deep areas are hard to see.
>
> *Nakunarqellria* is indeed dangerous when you're traveling by snowmobile. Even though you get to sloughs, you don't see them. Once I suddenly fell into [a small slough] when conditions were like that.
>
> Also, once I was traveling by snowmobile with the wind. To me I was going fast. I went when the conditions were such that it made you cross-eyed. I looked at that small dark tussock over there as I went fast, and I thought, "Why isn't that changing?" Everything around me appeared the same. Since that [tussock] didn't seem to change, I put my [foot] down and found I wasn't moving at all. [*laughter*] I'm talking about my experience.

Qengarut Iqalluguat-llu | Snowbanks and Snowdrifts

When it settled on the ground, blowing snow took on distinctive shapes depending on weather conditions. John Phillip (October 2005:105) explained the difference between *iqalluguat* (snowdrifts formed by the north wind, lit., "imitation dog salmon," so called because they appear like fish wakes) and *qengarut* (snowbanks, from *qengaq*, "nose," for their shape).

> Only the north wind creates the ones we call *iqalluguat*. They aren't good to travel over because they make us jolt.
>
> But when the south wind is blowing snow, it creates those smooth [snowbanks]. They are smooth, but they periodically become steep if there is something [on the ground], like a tree stump or clumps of grass. The south wind will pile snow onto its sheltered side. It places a *qengaruk* [snowbank] next to it. The [snow] that hits the sheltered side will gather at that elevated place, and then it forms smooth [snowdrifts] a great distance out from that spot. Those are not as high, but they are steep.

Qengaruut could also be referred to as *agevkanret* (whaleback dunes, because of their shape; lit., "ones that have gone over"): "They call ones that are smooth like this and periodically go up and down *agevkanret*. Only the south wind forms them" (John Phillip, January 2011:72).

Paul John (January 2006:66) noted the warning pertaining to *qengaruut*: "When the weather suddenly became stormy while we were traveling, they instructed us not to build a snow shelter along the sheltered side of the *qengaruk,* as the snow would cover us. Since it is the calm side of a steep area, another snowdrift will grow along its sheltered side. But they told us to go along the windward side; thick snow won't accumulate on us although we form a snow shelter there." Paul Charles (February 2010:448) of Newtok noted that *qengaruut* usually formed toward spring, when snow was constantly blowing along the ground.

Travelers used hard-packed *iqalluguat* to orient themselves. Nick Andrew (October 2005:235) explained: "When I get lost, I look at the drifts called *iqalluguat* made by the north wind. They are very recognizable. Those drifts have overhangs where the wind has eaten away [snow]. I still use them today." John Phillip (October 2005:105) explained how *iqalluguat* were used to guide people traveling along the coast: "Along our coast the *iqalluguat* form following the shore. Later those *iqalluguat* will be covered [by snow]. While a person is traveling, it will get foggy, and he will lose his sense of direction. Even though those *iqalluguat* are covered with snow, you can tell where they are. You can clean it and see which way it faces." Nick Andrew (October 2003:115) added: "There is a big snowstorm when there is wind from the northeast. That is the only wind that gives us a strong snowstorm. That's why we use the *iqalluguat* that it creates to determine our direction, even when visibility is low."

John Eric (March 2008:197) also described *tenguguat* (steep snowdrifts spaced at a distance from one another, from *tenguk,* "liver"):

> If it had been windy, a *tenguguaq* forms, gradually descending on top of the snow. If a snowmobile travels [across it], it travels okay, then suddenly drops off. And if it goes [across] in the other direction, it will suddenly go up, and then it will [drive smoothly] on the other side. And a snowmobile traveling along it will slant, and then it will pass. *Tenguguat* are wide.
>
> When driving over an *iqalluguaq*, the snowmobile will jolt up and down. *Iqalluguat* are close together, and there are many of them. If you try to count them, the numbers will run out.

Frank Andrew (June 2005:185) observed: "*Iqalluguat* aren't good for traveling. The surface of the snow is rough, and it makes traveling slow."

Peter Jacobs (January 2006:75) recalled the admonition that those who did not listen, and thus did not know how to act in dangerous situations, would be found dead on the edge of a snowbank: "When they were imparting instructions, those people even used snow, *qengaruk*, as a lesson. They told us boys that if we tended to ignore instructions, we'd be found with nice teeth along the edge of a large snowbank. What they said about the snowbank's edge made us very afraid. They even used snow as a lesson."

Aniguyaq | Snow Shelter

Travelers let snow cover their bodies in blizzard conditions, hollowing out a space to keep warm and dry. They might also build an *aniguyaq*, or snow shelter, dug into the snow and provided with a door. Peter Jacobs (January 2006:53) described their construction: "Back when they weren't wealthy, they used snow to construct *aniguyat* for shelter. You probably see the [igloos] they construct up north. But for us, when there was a lot of snow, we'd just dig and hollow out the snow to make [shelters]. They are very airtight and warm." Nick Andrew (October 2003:115) described how to create a layer of ice inside, making it even warmer: "When we sleep inside a deep snow shelter, we can make a small opening and light a little fire. When it starts to get damp inside, you put out the fire, and it will ice up inside. When you close the opening, it will get warm from your body heat, even when the weather is very cold. Because of this, they always say to bring a shovel when we travel so, if we have problems, we can dig." Frank Andrew (October 2003:47) added: "When they were close to rivers, when they were done making those, they opened a hole in the river [ice] and spilled water on top [of the snow shelter], letting it freeze. Then they went in and fixed the inside. They were very warm and airtight."

Paul Kiunya (October 2005:223) described building an *aniguyaq* for emergency use:

> Once when we started to get wet, my partner wanted to stop where we were because the village wasn't that close.
>
> After we dug the snow and hollowed it, we covered it with a tarp. We placed snow along the sides. We went in there; it was comfortable, like a house. But while we were in that snow shelter, it was damp because the south wind was blowing. The wind got strong, about a sixty-mile-an-hour wind at that time. We survived and didn't die of hypothermia, even though the weather was bad.

People also constructed *aniguyat* to use for harvesting activities. John Phillip (January 2006:84) explained: "If a person is staying in one place in the wilderness, building an *aniguyaq* can help, when one is trying to stay on the land without traveling. They are told to build a snow shelter before they get wet." Peter Jacobs (January 2006:53) also spoke from experience:

In the place where we fished with lures through holes in the ice, the people of the tundra area built many *aniguyat*. They would hollow out the snow.

That's how we used the snow, and I used it many times. When my ancestors weren't wealthy, they would make snow into small shelters, and when we'd hunt for foxes, we'd use the snow to construct *aniguyat*. We always had them and never spoke of tents, but they would travel by depending on the snow.

As with emergency snow shelters, it was critical to bring a stick to create an air passage inside the *aniguyaq*. Peter Jacobs (January 2006:54) explained: "They told us to keep a stick inside the *aniguyaq*. The inside of snow can be suffocating. It is said if they poke a hole through the area above, they save themselves. A snow shelter must always have air inside, as it is dangerous [if there is no airflow]." John Phillip (January 2006:84) shared his experience: "During winter, a blizzard suddenly came upon us as we were traveling and covered us, and eventually our tent started to get wet and became airtight. The next morning when we woke and lit the camp stove, just as it got warm, the stove went out. We then created a hole as we started to feel suffocated; we ripped [the tent] up at the top through a thin section, and when we lit the stove, it stayed on. That is one of the dangerous situations."

The *qanikciurun* (snow shovel) was an essential traveling tool. Peter Jacobs (January 2006:55) continued: "Back when there was an abundance of snow, they always told us to keep the snow shovel with us when we were going to travel. I would also bring it inside [the snow shelter], relying on their teaching. The first snow shovel that I saw, probably because they took such good care of it, was made from a caribou shoulder blade. It was a good snow shovel. And some had wooden snow shovels."

Ella Yugmek Pitangyutuuq | The Weather Hunts for a Person

Like all aspects of the natural world, freezing temperatures and blizzard conditions were viewed as a social response by *ella* to human actions in the world. The weather, they say, can hunt for a person. John Phillip (October 2005:241) explained:

Ella-gguq yugmek pitangyugtuq [They say that the weather is eager to catch (a human)]. . . . They also say that the weather *ellangcarituniluku* [puts a person through a life-threatening situation, causing him to gain awareness]. Those two seem to be true.

They say that when a person doesn't heed instructions, sometimes the weather goes after a person and allows him to go through a situation to make him gain awareness without killing him. Or when it is eager to catch someone, there is a blizzard. It isn't just [the weather] that does that, but they say our ocean does that, too.

They say when it caught someone, it suddenly got calm. It is said that when it catches someone, the weather is satisfied with its catch and becomes very likable and pleasant. They say it is satisfied and suddenly improves when it has caught [a person].

Nick Andrew (October 2005:242) recalled a similar understanding among those living on the Yukon: "Some elders would understand by looking at the weather and say, '*Ella* out there looks like it wants to catch a person.' Then when it suddenly improved, they'd say, '*Ella* out there has caught a person.' It would turn out to be true; it had caught a person." Frank Andrew (October 2003:115) added: "After the weather got bad and impossible for travel and then suddenly changed, they would say *ucuryagaarrluni* [it was proud] once again. It was proud of its catch." Paul John noted: "They call it *ciriyagaarrluni* [from *ciri-*, "to have an abundance of things"]. *Ella ciriyagaarrluni* [The weather (is satisfied) with the abundance]."

Then, after catching a person, they say *ella* suddenly gets cold. John Phillip (October 2005:243) explained: "They say *kumlacirluku* [it's cooling off the (body)]." Paul Kiunya added: "After a funeral in winter, the weather doesn't fail to get cold; they say that it's cooling off the deceased's body."

Muruanarqelria | Snow That Causes One to Sink

Winter was also characterized by deep snow, especially upriver. According to Paul John (October 2003:37): "When snow is so deep that you sink in, they call that *muruanarqelria* [(snow) that causes one to sink at every step]. When snow is deep and easy to sink into, they use their snowshoes." Nick Andrew noted: "It's good to use snowshoes when the snow is about a foot and a half to two feet deep. If you don't have snowshoes and walk quite a ways, you get tired. You will keep sinking into the snow. But if you wear snowshoes and walk on the soft snow, you won't get too tired."

Bob Aloysius (October 2005:34) described hunting in deep snow in February and March:

> At the end of January we would go back to our winter camp and go beaver hunting for two months. It was hard work. Sometimes the snow would be [waist] deep. It was so soft. The dogs pulled the sled, and we led them with snowshoes. We would travel all day, starting from early morning, going up on the mountains and crossing over to other rivers. There was lots of snow around the mountains and the trees where it wasn't windy.
>
> We never traveled without snowshoes. Snowshoes were our only means of travel. In the evening, when we'd pitch our tent, we used them to shovel the snow down to the ground. After pitching the tent, we'd place spruce branches down for matting, and we made our sleeping area thicker.

Nep'arrluni Qanikcaq | Blowing Snow Getting Moist and Sticking

Snow and wind did not automatically translate into blizzard conditions. John Phillip (October 2005:225) explained: "After heavy snowfall, if the wind immediately starts to blow from the south and a blizzard forms, the snow will gradually get moist. When the [snow] becomes wet, the blowing snow will stick [to the ground], as the moist [snow] covers it. That is what they call *nep'arrluni* [(blowing snow) suddenly sticking]. And although it's windy, there won't be a blizzard, and it suddenly becomes clear. Snow blowing along the ground will [suddenly] stick when weather conditions become wet."

George Billy (February 2006:401) also described how the snow "crouches down" in wet weather following a south wind: "This fall there was a lot of snow, but then a strong south wind came [melting the snow]. There was wet weather for a long time, without much snow. The south wind creates wet weather, and [the snow] becomes water and 'crouches down.' The snow has *ekiaq* [crust] when there is wet weather."

John Eric (March 2008:200) associated sticky snow with the end of winter: "Sometimes when days are getting long, and it's moist and there is a blizzard and falling snow, then [the snow] suddenly stops blowing and sticks, what they call *nep'arrluni*. In the afternoon, the weather is very clear. That's why they mention that toward the end of winter, the weather gets stormy during the morning and then gets good in the afternoon."

Qetraq | Crusted Snow

Hard, crusty spring snow was known as *qetraq*. John Phillip (October 2005:272) explained: "It's called *qetraq* [snow frozen overnight, crusted snow]. Then in spring, after the snow crusts in the morning, even though the snow is thick all the way down [to the ground] and had been hard, when the sun softens it they said *aniullinguq* [snow on the ground is softening and melting]. One sinks in deep, and sleds sink down low, too, when traveling by dog team." Just as backcountry skiers today look forward to crust skiing in spring, Yup'ik hunters enjoyed the advantages of traveling on crusted snow. John Phillip continued: "When we are going to travel, we sometimes say, 'When the snow crusts, I will go.' When *qetraq* hadn't formed and I had to travel a long distance, I wouldn't go. The trail is better when crusted snow forms in the morning. It enables us to travel faster, too, and doesn't cause us to sink in deep."

Conversely, wet snow that froze into an icy, slippery surface was not good for traveling. Again, John Phillip (January 2006:81) explained: "Sometimes during winter, after snowing wet snow, real raindrops fall and trickle down through the layers and wet the snow underneath. Sometimes when it hardens, one no longer falls through the deep snow. But sometimes, after snow followed by wet conditions, the surface freezes. Snow with a frozen surface is not good to travel on, as sleds tend to continue gliding forward and are unable to turn until [the icy surface] melts [or] until another layer of snow covers it."

Urugcuun wall'u Tumliranarqellria | Thin Snow Layer Showing Tracks

Bob Aloysius (October 2005:273) described the thin layer of new snow that sometimes covered crusted snow in the morning in spring: "When crusted snow formed in the early morning, even though it was cloudless, we'd wake to find [a small amount of] snow on top of the crusted snow. We call that *urugcuun* [lit., "something that helps the melting process," from *urug-*, "to thaw or melt"]. The sun quickly melts it, and it helps the [snow] underneath melt. At night, even though it is cloudless, there will be a little. As soon as the sun rises, it melts." Nick Andrew recognized the same condition: "He seemed to call it *urugarun* [snow that helps it melt quickly]. When [the crust] is covered with that light snow, [the light snow] quickly melts the *qetraq*. But if it didn't snow, and the snow crusted, the [crusted snow] didn't melt right away although it was warm."

John Phillip (October 2005:274) described comparable conditions on the coast: "They refer to that thin [snow layer] as *tumliranarqellria* [(snow) where *tumet* (tracks) easily form]. That snow is thin. It disappears when it melts, and tracks show. That happens, even though it isn't spring in our area. When the snow is hard, these fur animals like red foxes can't make tracks on it. When it snows a little, they say, 'It is now able to show tracks out there' on that snow. When we went out, our footprints were visible."

Qanikcaq Up'nerkami | Snow during Spring

John Phillip (January 2006:82) recalled a saying accompanying the thick, soft spring snow: "When it gets to a certain point during spring, it will snow a little, and the animals down in the ocean will give birth to their young; but they aren't seen as they give birth far from shore. Since those people knew, sometimes when it snowed a little, they said the following about the snow, that it has given the seal pups that will later stay on top of the ice some thick padding to lie on when they come out. It is thick, soft snow." David Martin (January 2002:48) added detail:

> Down in the coastal area during spring when the sun gets warmer, sometimes in the morning when it's snowing dry snow, they'd say, "They are making bedding for the sea mammals." They'd say that it was making bedding for the newborn sea mammal by snowing on the ice.

Also when referring to land animals, when it's snowing both wet and dry snow, those wise elders would say, "Since the newborn animal is wet [newborn], the snow is moistening the ground it will step on." Those had two different sayings for land animals and sea mammals. *Ella* does that when there is a newborn, making it bedding. [That snow condition] is usual at the end of February when walrus and bearded seals have young.

Qanisqineq | Snow in Water

John Phillip (January 2006:86) described another well-known snow condition— *qanisqineq* (snow in water, slushy snow):

> Down on the ocean, they were also afraid of the formation of *qanisqineq*. They told us to immediately head toward [the shore-fast ice] as soon as it started to snow. They told us not to head to our intended destination when the weather started to get cold. The possibility that [*qanisqineq*] might freeze in place when it started to get cold was dangerous. A kayak cannot glide forward through snow. I know that because I've tried it. It isn't like [regular] ice. Snow is very abrasive when it's thick.
>
> And these sea mammals also [get stuck] if *qanisqineq* is thick after it snowed, even if they are very heavy. One time I saw an adult bearded seal approaching, and it surfaced periodically along that [*qanisqineq*] although it was thick. When it swam down below us, when the one who claimed it shot it, when it suddenly jerked [from the gunshot], it stayed there stuck in an upright position [in the place where it came out of the *qanisqineq*]. Then using a kayak, they pushed [the *qanisqineq*] and helped him harpoon it and take it. Usually when we shoot these [bearded seals], they immediately sink.

Qanikcam Urullra | Melting Snow

The warming weather and melting that accompanied the return of the sun in spring posed as many hazards as freezing conditions in fall. John Phillip (January 2006:26) explained: "During spring, when *miiqaq* [fresh meltwater, melted snow] started forming, it was dangerous to suddenly sink into snow when it became soft, when there was water underneath. We were instructed that when the snow and ice melted and formed water, the areas underneath [snow covering] sloughs and rivers are eaten away and collapse. Those areas are open and dangerous because of the current."

John Phillip (January 2006:90) also described *qiliqaumakait* (places that [snow] merely covers):

> Sometimes the snow conceals an area, and the [ice] underneath is thin. They say

that the snow just merely covers those places, and the area underneath [the snow] is open down to the bottom. They refer to those *qiliqaumakait* as dangerous areas.

I sank into the river some months ago, when I wasn't paying attention, as I came upon the river where I had set my wooden fish trap. I had actually stopped right on the lake outlet that flows into the river. The snow had evidently covered that place before it froze. When I checked the place, I saw that it had only a thin layer of ice. Since snow is very warm, [the water] underneath was unable to freeze.

The freeze/thaw cycle that created *kavtakuaq* (granular, transformed snow) was desirable not only for making *akutaq* but to slow down the melting process in spring. Nick Andrew (October 2005:269) noted: "The south wind makes it have ice particles. It suddenly develops *cikurlak* [ice formed by freezing rain]. When snow doesn't have that, they say it won't melt into water, even though there's lots of snow." John Phillip explained further:

When the weather warms every so often, [the snow] is called *kavtakuaq*. Then when it gets thick, the melting process is slower. When it doesn't have that [kind of snow] in it, when it doesn't have [layers of] rain in it, the snow melts quickly. Sometimes when it rains occasionally in between [snowfalls], soaking the snow, it hardens and gets thick. They like it when that happens.

That snow with occasional rainfall helps [berries grow]. We call that *ungalilleq* [snow crust formed by *ungalaq* (south wind) accompanied by warm, wet weather]. When there was no south wind, they said *ungalillritniluku* [the snow doesn't have crust created by south-wind weather conditions], that the snow would melt right away.

Paul John (January 2006:63) also commented on conditions conducive to slow melting in spring:

During fall, sometimes the first snowfall is dry and doesn't melt right away. If [the dry snow] doesn't become moist, and another snowfall that is damp ensues when there is a south wind, it is referred to as *ungalillengluni* [melting conditions forming a layer of wet snow that later refreezes].

After that, if another dry snowfall ensues, and if once again it is followed by wet snow, they would say that the *ungalilleq* has formed another layer of snow over [the dry snow].

When it has snowed without forming *ungalilleq* all winter, it melts quickly in spring. But if there is *ungalilleq*, the snow [melts] slowly, even in spring when it gets warm.

Layered snow also prevented flooding. Bob Aloysius (October 2005:268) noted: "Back when it snowed a lot, they had a few [layers]. The first [snow layer] was when the wind blew from the south and packed it down and the surface froze. Then another new layer of snow fell, it warmed up, and packed again. Then another, and it froze. These days that doesn't happen. The snow is all soft, just fluff. And that person would say, 'Oh my, it's going to flood up there.'" Nick Andrew (January 2006:70) commented: "Heavy snowfall on its own doesn't determine the amount of water that will form [during melting]. If the heavy snowfall doesn't contain *ungalilleq*, [snow] will melt quickly and the rivers will flood."

Nick Andrew (January 2006:70) noted that constant cold weather in spring could cancel the benefits of *ungalilleq*: "Although [the snow] contains *ungalilleq*, if the wind constantly blows from the north, bringing cold weather during spring when water starts to melt, it will dry up the water that was to form. Water will not form [on the land], and [rivers] will not be full. They don't favor that condition, as it is not beneficial to these fish and other living things."

Peter Jacobs (January 2006:76) noted that heavy snowfall in spring was associated with rapid melting: "At some point in spring, close to summer, very heavy snow will fall. They would start to say, '[The heavy snowfall] is creating conditions that will cause [existing snow] to rapidly melt and release water.' Right after that, since that snow was soft, it suddenly developed water and started to flow; it caused that [existing] snow to rapidly melt and release water. That [heavy snow] isn't good to walk on and causes one to get wet."

John Phillip (October 2003:106; December 2005:141) remembered an adage pertaining to melting, recalling the personhood of both land and sea:

> Down on the coast, if the ocean melts earlier, they say that the land and lakes will melt slowly. They say that the ocean has outdone the land at reaching summer. [When the ocean] becomes summer earlier, they say that the land is jealous, and ice on the lakes stays for a while longer. Also, if the land melts first, sometimes the ice on the ocean doesn't melt right away, and it's not summer right away. They say that they switch intermittently, and they call it *kipullgutaarluni* [going back and forth] in my village. . . .
>
> This world has a saying: when our land melts first, then the ocean's ice doesn't go out right away but stays for a while. When that happens, our river breaks up later. The lakes hold in the water. It lets it lose water slowly along with the ice.

Paul John repeated the adage succinctly: "When the ocean melts before the land, the land will melt slower than the ocean. That was the saying from those people."

Qanikcakun Ayagayaraq | Traveling in Snow

As one might surmise, detailed instructions guided those traveling in snowy conditions. First and foremost was staying warm and dry. Men were admonished never to travel without their seal-gut or fish-skin rain gear. John Phillip (October 2005:217) recalled:

> I had a *qasperek* [seal-gut rain garment] back then, and I never left it behind. If I was going without a sled, I would hang it over my shoulder since it was light.
>
> We always brought our *qasperek* [or] *imarnitek* [seal-gut rain parka] with us. When we were about to leave, even though the weather was very good, they told us not to be of the same mind as the weather, *umyuallgucirluku pisqevkenaku ella*, thinking that the weather would continue to be good. They said that we should not leave concurring with [apparently good] weather, that although it seemed good, we shouldn't leave behind the seal-gut rain garment but bring it along for safety. . . .
>
> When weather conditions were wet, we dampened the surface of those [seal-gut rain garments] and put them on inside our clothes, placing a white-colored garment over us. They are very warm and dry.

John Eric (March 2008:190) noted that snow goggles were crucial when traveling over snow in spring: "Since the sun is powerful, people who were outside became snow-blind. The sun caused their vision to go bad. And back when there weren't a lot of glasses around, they constructed snow goggles. They used a piece of wood and made small slits, and then they painted the inside with charcoal." Wassilie Evan (September 2003:316) said: "They used them to prevent snow blindness, and they didn't call those *ackiik* [glasses] but *nigaugek*. It's almost the same as the noise that ptarmigan make, '*Negauq, negauq*.'" Tim Myers added an apt comparison between *nigaugek* worn by hunters and the dark feathers around a ptarmigan's eyes in spring: "You know how the ptarmigan have glasses during spring when the days are longer, they get red sunglasses; that's what they're like." Wassilie Evan concluded: "When you get snow blindness when there's snow, you can't keep your eyes open, and when trying to open them, they shut and get red. Since the snow's surface is white and has a very powerful glare, that causes sore eyes."

Peter Jacobs (January 2002:95) described using the sun's reflection on snow as a guide in stormy conditions: "My father used to tell me not to be scared in blowing snow if the sun is still visible. We used the sun as our compass when we traveled by sled." Bob Aloysius (October 2005:109) added: "If the sun is shining on any surface, it will reflect, no matter how small the amount of light might be. If we get caught in fog while traveling, they always tell us to find a high place and go around it. You'll see little reflections on the snow on the south side of the snowbank. It means the sun is there. The other side will be all shade. That's how we find our direction."

Paul John (January 2006:71) explained that although thirsty, one should not eat new snow: "When traveling to places that aren't nearby, although we were thirsty, they cautioned us not to periodically eat snow. Snow isn't actually poisonous, but it could cause the body to weaken and become extremely tired. But they said that if the snow contained *ungalilleq* [snow hardened by south-wind conditions], we should dig into the dry surface snow and take pieces of ice from the *ungalilleq* underneath and not chew it but let it melt in our mouths. If we kept some ice inside our mouths when we were thirsty, they said that our bodies would not weaken too badly." Susie Angaiak (March 2007:1448) was admonished to eat *cikurrluk* (icy, metamorphosed snow that has melted and refrozen): "When a person is walking along the snow and gets thirsty, they told that person not to take snow from the surrounding ground and eat it but to have some *cikurrluk*, [snow] that is icy; they say that is better. After melting it inside his mouth, he should swallow it. They say that if he reacts to his thirst and eats that snow out there, he will get more thirsty." Tommy Hooper added: "Our mother used to tell me that snow is like foam inside the stomach, and it isn't good. It's better to melt *cikurrluk* inside one's mouth and swallow it when it turns to water."

Dangers also beset those traveling through snowy mountains. Nick Andrew (October 2006:329) said that snowbanks are to be avoided: "[If we are] in the mountains, we were told not to go below high snowbanks in winter, in case the snow starts sliding down. They'd tell us to walk a safe distance from it. Sometimes that happens when there's a lot of snow. Like the mountain behind our village. The snow creates a *elluugaq* [avalanche]. Some mountains upriver from my village are steep."

Men were taught not to approach avalanche areas. Nick Andrew (October 2006:335) recalled: "They say not to go near it. It is easy to detect. The snow looks pure and white, and nothing shows through it. It is beautiful to look at. They say it is dangerous if nothing is visible on the mountain side. The deep snow covers everything." John Phillip (October 2006:336) recalled advice to move diagonally if one was caught in an avalanche: "They say if a person moves with the avalanche, he'll only get into more danger. And it is hard to go against the avalanche, because a lot of snow is sliding down. They say not to go straight down or up if caught in an avalanche but to go diagonally. That's how a person can survive."

Ice-covered snow could also be dangerous. Nick Andrew (October 2006:329) recounted his experience:

> I was hunting a moose, and I was wearing short boots. In winter, when it freezes after a thaw, the snow on the mountains would be covered with ice. There was an area covered with snow, and above it was a rock.
>
> I saw a moose on the other side of a small valley. My mind was somewhat wary, and I could have shot it from where I was, but I said, "I'll go and shoot it from over there."
>
> As I started across that valley, I slipped and fell on my back as soon as I took a

step. I looked and saw my gun far from me, and at the same time I was sliding down the mountain. I tried rolling toward my gun, and I finally got it, at the same time I was sliding faster and faster.

Good thing the gun strap was toward me, that's when I grabbed it. When I got hold of it, I used the stock to slow myself down. When I finally stopped sliding, I saw how far the bottom was. I might have gotten hurt if I hadn't caught my gun.

Caribou [and moose] in our area usually stay on the mountains on the north side. That's why they tell us these things, and that sometimes we should wrap dog chains around our boots when going up a mountain.

Nick added that men were told to watch for steep drops when traveling over snow: "Today they chase caribou on the mountains with snowmobiles. When we started using Snow-goes, our elders told us not to speed, even when it looked smooth. They warned us in case we might suddenly come upon a *qavyuqerneq* [cornice, curling waves of snow frozen in place over the edge of a bluff or gully, from *qavyurte-*, "to frown"]." John Phillip (January 2011:74) explained with a teasing scowl, "If you suddenly frown, *qavyuqerneret* are reminiscent of these eyebrows [coming over one's eyes]."

Nick Andrew (October 2006:331) also observed that mountain travelers needed to be attentive to wind direction: "They told us to observe and base our calculations on the east wind when we were on a mountain. [They told us] to be aware of the wind if it's coming from the east. Snow accumulates when the wind is from the east, and snowdrifts build up. They say to avoid those areas because the snow could cave in." Travel in snowy mountains, Nick emphasized, required vigilance: "They'd tell us always to be wary in the mountains, even when we were excited about the animals we were hunting." Nick concluded: "There are dangers everywhere on the land, and they have sayings concerning them."

Qanikcaq | Snow

acaluruaq. Snowshoe. See also *pupsugcetaaq, tangluq.*

agerpak. Large bank of drifted snow.

agevkaneq/agevkanret. Snow/snowdrifts that gather on the sheltered side of an elevated area (lit., "that which has gone over"), whaleback dunes, because of their shape.

agneq. Small snowbank.

agqercetaar(aq). Snow fence.

akuurun. Thick snowstorm covering the ground (from *aku*, "lower part of a garment").

aniguyaq/aniguyat. Snow shelter/s (from *aniu*, "snow on the ground").

aniu-. Snowing. See also *qanir-.*

aniu. Snow on the ground, snow cover. See also *apun, qanikcaq.*

apun. Snow on the ground, snow cover. See also *aniu, qanikcaq.*

cikurlak. Ice from freezing rain (from *ciku,* "ice").

cikurrluk. Icy snow, metamorphosed snow that has melted and refrozen.

cikuyaaq/cikuyagaat. Small ice particle/s.

ciru-. To be covered with snow. See also *qanikcir-.*

elluugaq. Avalanche.

iingir-. To be snow-blind.

ikamraq. Snowmobile, dogsled. See also *snuukuuq.*

iqalluguaq/iqalluguat. Snowdrift/s formed by the north wind (lit., "imitation *iqalluk* (dog salmon)" so called for their fish-wake-like surface).

itrugta. Powdery snow that enters through cracks in the house (from *iter-,* "to enter").

kaimlenguyutaq. Light snow with little ice patches (from *kaime-,* "to make or drop crumbs").

kaneq. Hoarfrost developed by cold temperatures.

kanevvluk. Light snow or rain.

katautaq. Snow-beater. See also *kavcircuun.*

kavcircuun. Snow-beater (from *kavtak,* "hailstone"). See also *katautaq.*

kavtagaaq/kavtagaar. Granular, coarse-grained, transformed snow with larger crystals, known in English as depth hoar (from *kavtak,* "hailstone"); bottom layer of snow, used for *akutaq.* See also *kavtakuaq, kavtayagaq.*

kavtagaraq. Snow on the ground after a warm spell close to spring (from *kavtak,* "hailstone").

kavtak/kavtuk/kavtiit. Hail, hailstone/s.

kavtakuaq. Granular, transformed snow (lit., "pretend hailstone"), depth hoar; bottom layer of snow, used for *akutaq.* See also *kavtagaaq, kavtayagaq.*

kavtayagaq/kavtayagaat. Granular, transformed snow (lit., "little hailstones"), depth hoar; bottom layer of snow, used for *akutaq.* See also *kavtagaaq, kavtakuaq.*

kitulget. Large snowflakes or clumps of flakes that stop right away (lit., "ones that will pass"). See also *qanugpak.*

mecaliqaq. Wet snow, sleet (from *mecunge-,* "to be wet"). See also *mecungelria, mecungnarqellriit.*

mecungelria/mecungelriit. Wet snow (lit., "ones that are wet"). See also *mecaliqaq, mecungnarqellriit.*

mecungnarqellriit. Wet snow (lit., "ones that make you wet"). See also *mecaliqaq, mecungelria.*

muruanarqelria. (Snow) that causes one to sink at every step (from *murua-,* "to sink into snow at every step").

nakunarqellria/nakunarqellriit. Whiteout conditions with no visibility (lit., "condition/s that make you cross-eyed," implying distorted vision).

natqugpak. Severe drifting snow, higher than the ground's surface (lit., "big *natquik* [drifting snow on the ground]").

natquik. Drifting snow on the surface of the ground.

navcaq. Snow cornice. See also *qavyurneq.*

nevluk. Clinging snow.

nungurrluk. Ice fog in cold weather. See also *patuggluk.*

nutaqerrun. New snow on the ground, new ice in fall, which melts right away in the mouth and is not good for making *akutaq* (from *nutaraq,* "new one"). See also *nutaryuk, qaninerra'aq.*

nutaryuk. Fresh powder snow. See also *nutaqerrun, qaninerra'aq.*

patuggluk. Ice fog (lit., "bad *patu* [cover]"). See also *nungurrluk.*

pekutaq. Snow shovel. See also *qanikciurun.*

pirtuk. Snowstorm, blizzard, blowing snow.

pupsugcetaaq. Snowshoe. See also *acaluruaq, tangluq.*

qakurnaq. Frost in fall, first frost.

qanikcaq. Snow on the ground, snow cover. See also *aniu, apun.*

qanikcaqegtaar. Flawless *qanikcaq* (snow).

qanikcaryak. A large quantity of snow.

qanikcirluni/qanikcir-. Snow-covered. See also *ciru-.*

qanikciurun. Snow shovel. See also *pekutaq.*

qaninerraaq/qaninerra'ar. Fresh or new-fallen snow, first snowfall. See also *nutaryuk.*

qanir-. Snowing. See also *aniu-.*

qanisqineq. Snow in water, slushy snow.

qanucuar/qanucuaraat. Small snowflake/s. See also *qanucuayaaq, taqailnguut.*

qanucuayaaq/qanucuayagaat. Small snowflake/s that do not stop falling. See also *qanucuar, taqailnguq.*

qanugkaq. Coming snowfall.

qanugpak/qanugpiit. Large snowflake/s that stop right away (lit., "ones that will pass"). See also *kitulget.*

qanuk. Snowflake.

qavyuqerneq/qavyurneret/qavyuqerneret. Overhang/s, cornice/s, curling wave/s of snow frozen in place over the edge of a bluff or gully (from *qavyurte-,* "to frown").

qengaruk/qengarut. Snowbank/s (from *qengaq,* "nose," for their shape).

qerretrar-/qetrar-. To form a hard snow crust during a cold spring night.

qerrute-. To be cold (of humans and animals).

qetraq. Crusted snow; snow frozen overnight.

qiliqaumakait. Places that (snow) merely covers, potentially dangerous.

snuukuuq. Snowmobile, Sno-go. See also *ikamraq.*

tangluq. Snowshoe. See also *acaluruaq, pupsugcetaaq.*

taqailnguut. Small snowflakes that do not stop falling (lit., "ones that are persistent"). See also *qanucuar, qanucuayaaq.*

tenganaviar/tenganaviaraat/tengyukaaralria. New, still snow on the ground; snow that falls to the ground on a calm day; dangerous because wind can lift it and cause blizzardy conditions (from *tenge-*, "to fly"). See also *tengqeryukaaq.*

tengqeryukaaq/tengqeryuka'ar. Snow that easily flies (from *tenge-*, "to fly"). See also *tenganaviar.*

tenguguaq/tenguguat. Steep snowdrift/s spaced at a distance from one another (from *tenguk*, "liver").

tumliranarqellria. Thin layer of snow over crusted snow that shows tracks (from *tuma*, "footprint, track"). See also *urugcuun.*

ungalilleq. Melted, refrozen layer of snow formed by *ungalaq* (south-wind) conditions.

urugcuun/urugarun. Thin layer of new snow over crusted snow (lit., "something that helps the melting process," from *urug-*, "to thaw or melt"). See also *tumliranarqellria.*

utvak. Block of snow.

Imarpik Elitaituq

THE OCEAN CANNOT BE LEARNED

*I hope that no one says down on the ocean, "I've learned the ocean." Before we have
learned [to predict its conditions], we have reached this age.*

—Simeon Agnus, Nightmute

C OASTAL RESIDENTS EARLY LEARNED AN ATTITUDE OF HUMILITY
and respect for the ocean which sustained them—an attitude that
went hand in hand with the practical skills of ocean hunting. John Eric
(March 2007:52, 101) reflected this high regard: "As someone who has lived along
the ocean, I have always viewed it as the most important element of our environment. We cannot live without the ocean. Our ancestors sustained themselves
mainly from the ocean. It was their source of clothing, it provided skin boots,
seal-gut rain garments, kayaks, and pants. Also, all species of fish enter the rivers
from the ocean. They were evidently food for us to eat when winter came again."
Earlier, John Eric (January 2007:114) commented on the ocean's name: "It's no
wonder that the ocean has the name *imarpik* [from *imaq*, "contents"], because it
holds everything."

Just as *qanruyutet* embodied time-tested rules for interacting with the land,
instructions likewise guided one's interaction with the ocean. John Eric (January
2007:117) said: "It seems that there are many *qanruyutet* for using the land, and

the ocean also has its own distinct *qanruyutet*." George Billy (October 2006:7) reminded his listeners: "We heard [of Atertayagaq] from time to time. They said by having listened to the teachings concerning the ocean when elderly men spoke, instructions for what to do when facing life-threatening situations, he followed the teachings when he became desperate and didn't lose his sense and lived through that." The ocean, they said, was variable and quick to change. John Phillip (October 2003:239) noted, "They say that we cannot depend on the ocean or all the bodies of water."

Many contended that a person could never fully learn all the potential hazardous situations out on the ocean. According to John Jimmie (March 2007:75) of Chefornak: "I'd hear the cautionary lessons of the ocean when I'd go to the *qasgi*. They would mention that a person won't learn the ocean. *Imarpik-gguq un'a ayaperviituq pinirliinani-llu* [They said there's nothing on the ocean that one can lean on for support, and there's nothing stronger than the ocean]." Paul John (December 2007:69) explained: "They said that when we are on land, we can lean on something with our hands to avoid danger. But they said there is nothing on the ocean that one can lean on for support when encountering danger. This person just mentioned that the ocean cannot be learned; indeed, because it cannot be learned, they referred to it as *ayaperviilnguq* [something that a person cannot lean on with his hands for support]."

Paul Tunuchuk (March 2007:257) recalled: "They said if someone says, 'I have learned the ocean,' they will be lying. They said a person can't learn to predict conditions on the ocean. That's true indeed." John Eric (March 2007:264) added: "The ocean indeed cannot be learned. One person said that before he learned to predict ocean conditions, he stopped going there." Simeon Agnus (December 2007:56) wholeheartedly agreed: "Let no one say that they've learned the ocean. There is no one who has learned the ocean during spring when lots of ice is around. Although he follows the ways of traveling when lost in fog, when ice obstructs his path, he cannot continue."

Hunters were warned against overconfidence. John Eric (March 2007:268) recalled: "I used to hear that a person shouldn't underestimate the ocean because it is daunting." John Jimmie (March 2007:75) agreed:

> The ocean is a worrisome place, even though it is a joyful place to be. I know because I was in a desperate situation twice, and we were down there and felt desperate, even though this person was with me.
>
> I think that it's less frightening to face danger on land, but the ocean is extremely distressing. And I thought, "I probably won't see my family again."

Coastal hunters emphasized the need for quick response to changing conditions. John Eric (March 2007:113) stated: "They say that the ocean doesn't delay things for a later time. It won't say, 'Wait,' because it does things at that moment. Since ice detaches and breaks when the tide goes out, although you say, 'Wait a minute,' it won't listen to you. The conditions will change right away." Roland Phillip (November 2005:65) added: "They would say that the ocean doesn't announce what it is going to do. When it is going to do something, it isn't hindered by the fact that [the ice] is stable and safe. [It doesn't say], 'I am stable, I will stay.' But these elders who started to do things before our time would say, 'The ocean is getting dangerous.'"

Hunters often emphasized differences between land and sea. John Eric (March 2007:113) recalled: "Since there's nothing down on the ocean, it isn't like the land. Although one attempts to do something right away, because it's water he will fall behind. That's why the people of Nunivak Island are fast in carrying out work. Because they are ocean people, they are very fast and efficient. But because we mainly live here on land, we are slower." Camilius Tulik (March 2007:570) agreed: "The ocean is not like land. Those on land do not lie. Though it is extremely foggy and dark, we can travel on land and come back home. But the ocean keeps changing. They may go fearlessly knowing where they come from by this [GPS], but that is worthless if the ice pack cuts off their way."

Hunters approached the ocean with respect and fear, but they also experienced it as a joyful place. John Eric (January 2007:114) recalled:

> In the 1950s, when I brought my grandfather down to the ocean, how very serene and joyful it was along the ice edge. Listening to the birds, the long-tailed ducks, it was extremely windless and calm. I told him that since I wanted to go with him so badly, "I'm going to tie our dogs on the ice up there, so that you can paddle with me back-to-back." He told me that people don't sit back-to-back in a kayak while paddling out on the ocean. [laughter] I wanted to go with him, as I was so joyful because of the surroundings.
>
> He was elderly at the time, and he paddled down along the ice that extended up to shore. I felt such pleasure over there along the mouth of our river. I headed back to shore, and the sound of long-tailed ducks gradually quieted, and then they got far.

Imarpik Nallutaituq | The Ocean Knows

Qanruyutet are not the only guidelines in one's dealings with the ocean. *Eyagyarat* (traditional abstinence practices following birth, death, illness, miscarriage, and first menstruation) must also be taken into account. John Eric (March 2007:101) explained:

> My father and those before him mentioned those other customs that we followed—our *eyagyarat* having to do with the ocean.

He said that if someone died during winter, their son or his siblings have to wait until someone brought a seal pup up to the village before going down to the ocean.

They also said that the occurrence of a miscarriage is something that the ocean doesn't like. Someone who experienced it said that when that occurs [in a family], they have to wait for the appearance of a red-necked grebe [typically in late March] before a person can go down to the ocean.

Since the ocean has a sense of awareness, it will sense a person who has experienced something that prohibits them from being down on the ocean; the ocean indeed has always been aware and knowing.

Tommy Hooper (June 2008:159) added that contact between the ocean and a miscarried fetus is particularly serious: "They say that's the most reprehensible circumstance that causes a scarcity of resources. There won't be anything in the ocean and the ones that usually swim won't come around if a miscarried fetus is discarded there. They say that the ocean gets angry."

Paul John (October 2005:74) further explained the admonition not to go down to the ocean until after hearing the call of a red-necked grebe: "Regarding the grebe making noises, there is a teaching that states that if a person's wife had a miscarriage, he was not to go down [to the ocean] whatsoever until all the birds arrived. But they say that even though all the birds hadn't arrived, the grebe makes its call for a reason. When it yells and we hear it, it is said that it is opening up the possibility of that person going down to the ocean."

People's safety during transformative events rested on their invisibility in a sentient universe. Recall Frank Andrew's (September 2000:32) advice to the girl who had experienced her first menstruation—to climb a hill and to throw moss or dirt at *ella* to blind it. The ocean also had eyes that must be closed. This requirement to stay out of view explains the reasoning behind restrictions regarding ocean hunting. Frank continued: "They said that men practicing *eyagyarat* should not go down to the ocean until the red-necked grebes and ringed seals arrived in spring. The grebes come first to the ocean and move on to land. When they started to defecate and when the blood of the ringed seals soaked the ocean down there, *makuara-gguq mer'em cikmirtetuuq* [they said the *makuat* (eyes of the ocean) close and become blind]. Nothing would happen to hunters when they went down to the ocean, and they would no longer risk developing physical ailments."

Frank Andrew (October 2001:106) noted that "the ocean knows everything that is going on through its *makuat*, and they recognize who is practicing *eyagyarat*." These *makuat*, which play such an important part in a person's well-being, can be seen as dust in the air when the sun shines inside a house. If allowed to settle on a person, they cause sickness: thus, the admonition to rise early before the *makuat* begin to fly. *Makuat* in the ocean can appear as tiny glittering particles around icebergs. When lost or disoriented, hunters used them like a compass to get their bearings.

Frank Andrew noted the consequences when those restricted from hunting or traveling on the ocean disobeyed *eyagyarat*:

> The ocean knows when a child dies. If those who have a duty to abide by *eyagyarat* go down to the ocean before the family's time of abstaining is over, large waves break the ice apart. But if they had not gone down, large waves would not develop until after the sea mammals down there have their offspring. That is the *piciryaraq* [way] of the ocean.
>
> The red-necked grebes blind the ocean's eyes that it uses to look around. After the grebes arrive, the ocean won't react if they go down. The ice cannot be broken and the weather is calm as in the past. The ocean down there *nallutaituq* [knows everything that is going on].

Walter Tirchik (April 2001:150) of Chefornak recalled that his uncle told him the same thing—when the red-necked grebes arrive and defecate in the ocean, the ocean's eyesight dims, and hunters practicing *eyagyarat* can safely approach the sea at that time. To ignore this admonition was to invite disaster. Walter continued: "Avegyaq's mother told her son, 'They have not spoken of grebes; don't go down.' He replied, 'Those things are not true.' After he went down, large waves as high as this ceiling developed, and we returned home. After that, I think he started believing it."

Imarpik Naklegtaituq | The Ocean Is Pitiless

They say the ocean down there, when feeling displeasure, doesn't feel pity for people.
—Paul Andrew, Tuntutuliak

Large waves and windy weather were visible and immediate penalties for failure to follow *eyagyarat*. Frank Andrew (August 2003:75) described his understanding of ocean swells known as *qairvaak* (lit., "two large waves"), the long, high-amplitude waves generated by storms in the open Bering Sea and North Pacific:

> During winter, the ocean is quiet and very calm, and the *qairvaak* are not there. And when sea mammals begin to have their pups, the *qairvaak* begin to appear. They call the smaller of the two swells the wife, *ulcuar* [small swell], and her husband is the larger one, *ulerpak* [large swell]. They exist only down in the ocean and don't get to the Kuskokwim River. . . .
>
> Those who talked about them said that they are a married couple, a female and a male. Then in spring when the waves begin to get stronger in our area, they constantly

make "*engg, emmm*" noises like a person in pain. When the wave lifts the shore ice, something makes noise like that, "*Eeeeee-mmmm.*" It seems to be the ice. They say it's those two, the noise of the persons of the water. The female is higher pitched.

Sometimes, however, the *qairvaak* appear before their usual time. Frank (October 2003:66, 75) continued:

> They say one must not try to make them angry. They will know one who is breaking the laws they were told to follow concerning them. They say when they're upset, they are pitiful. They make that noise because they are suffering over that. Our area down there is not wavy during winter, but if a person is disobedient, if they go against their admonishments, the ocean down there will know. . . .
>
> They say those two [ocean swells] have awareness, and they don't like it when a person breaks a law. Even though we don't see them as human, they become upset when a person who must follow abstinence practices offends them and will appear before their usual time, breaking up the shore ice.

Walter Tirchik (April 2001:150) described what happened when his uncle went hunting on the ocean soon after his uncle's father's death: "From that moment it was like those animals they sought were cut off, and the hunters did not see them anymore. The sea ice came in fast with constant large waves. After he quit going hunting, the ice on the ocean's edge melted quickly." John Phillip (November 2001:152) also spoke about a man who lost a close relative and went seal hunting before the red-necked grebes arrived: "After that, the sea was very wavy, and the next time the hunters approached, the bearded seals and walrus disappeared. King eiders that used to come right by the shore were hardly seen. The ocean won't become a white person. Those things are still true."

Eyagyarat are an abiding expression of the personhood not only of humans and animals but of the land and sea. As Frank Andrew said, "The ocean has eyes" and "The world knows."

Imarpigmi Pissulriim Qanruyutai |
Rules for One Hunting on the Ocean

Although hunters did not presume to know everything about the ocean, they did not travel away from shore uninstructed. As when traveling on the land, time-tested rules guided them. Paul Tunuchuk (March 2007:90) noted: "There are actually many *qanruyutet* concerning the ocean. We also know the hazards and the instructions for when there's a lot of ice." Young men were carefully admonished before they traveled to the ocean, where experience continued to teach them. Paul continued: "The potential dangers we encountered in the past, if our children experience them firsthand, they will learn like I did. We learned only through direct experience. That's how we all are."

Ocean hunting began as early as February. Frank Andrew (October 2003:19) recalled:

The month they called February was their time of hunting since the beginning. They say that when the weather was in its normal state [before it changed], the wind would weaken in January. And the wild rhubarb would get as tall as people because there was so much frost. The shore-fast ice would be far out there, down near the *iginim engelii* [edge of deep water]. That's why those people were successful catching animals when they hunted. They say that it was like the coastal people were at spring camp in February, starting to catch sea mammals.

They say that it used to get warm in February, and lakes would fill with water. And even though it got cold in March, it began to turn into summer before it got too cold.

We never paddled down on the ocean when it was windy, but we floated on the ocean when it was calm. And we would climb on top of floating ice to have lunch and sleep. It was never windy.

The wind removes the salt from the ice down there. They would melt the ice to drink, and it would be very clear with no salt in it whatsoever. When the ice began to melt, its surface had deep places filled with very clear water. We would constantly fill our water containers.

Men were admonished not to travel down to the ocean alone and to make sure other boats were in view in case of accidents. Those who did go alone were considered "big hearted" (fearless), but this was not an admirable quality. Roland Phillip (October 2006:102) explained:

[Albert Beaver] said that those they call *cacetulriit* [ones who are brave and fearless] especially encounter danger. But he said those who are cautious and fearful don't encounter danger as much.

You know, when dangerous conditions are just about to occur, they stay on the shore-fast ice. Some who are fearless go out, even when conditions are hazardous, and sometimes when ice drifts in along the shore, they are unable to go up to shore.

Since Angutevialuk [Albert Beaver] was fearful, he never caused me to encounter hazards when I accompanied him, and I followed whatever he did.

Hunters were taught to be cautious and to travel with their customary hunting partners. Roland Phillip (November 2005:95) recalled: "Back when we hunted by kayaks we were told to always accompany our usual partner. There is also the following *qanruyun*. They say if someone changes partners too often, since that person hadn't gone with his usual hunting partner, at a later time when conditions are hazardous, one will ignore his [plea for help]."

Hunters were admonished to stay together when traveling. John Walter (July 2007:357) said: "Since the ocean is vast and there is nothing a person can lean on for support, it is of utmost importance not to leave each other behind out on the ocean." Theresa Abraham (July 2007:363) cautioned: "When you young men travel, you should keep an eye on your hunting partners. They say we should watch out for each other when traveling as a group. And we cannot leave our traveling companions behind."

Simeon Agnus (July 2007:369) noted that the one exception was when helping one person put others at risk: "Back when our ancestors hunted by paddling, the following was their teaching. If a person in their group was about to encounter peril in windy conditions, if helping that person would cause the others to encounter a mishap, they said it was okay not to help that person. But if there was a way to save him [without putting themselves at risk], they could help that person. I used to hear them mention that inside the *qasgi*."

Imarpik Kiarnaitaqan Tumkarcuryaraq |
Finding a Path When There Is No Visibility on the Ocean

They went down to the ocean without a compass but with words of wisdom.

—John Phillip, Kongiganak

Although windy conditions were considered hazardous, knowledgeable hunters did not fear fog or low visibility. Roland Phillip (November 2005:87) recalled:

> My father mentioned that although nothing is visible, those who know how to navigate in those conditions will hunt and travel. But they say that those who don't know the ocean will not leave the shore-fast ice.
>
> I was cautioned about fog down on the ocean. When fog was about to form, I would head along the shore to a safe place. One day, I was with an elderly man, and the weather was very calm. The tide went out, and there was no visibility, as there was heavy fog. When the large amount of ice that had been beached [along the ice edge] went out, that elderly man said to me, "Now, it is time for us to go." I told him, "My, I don't want to go while it's foggy."
>
> Since he knew his location, although the sun was gone he went. I didn't go with him because I was afraid. [*laughing*] They say those people who pay attention aren't afraid of fog.

Although hunters today use the GPS to find their way in foggy weather, Simeon

Agnus (July 2007:370) noted: "The GPS isn't the only device available. There are also old teachings they used—the ocean swells that flow toward shore or the sea mammals that swim northward. Their bearings are animals and waves." Simeon (January 2007:107) elaborated: "These walrus don't lie whatsoever, even though it's extremely foggy. Walrus don't swim in just any direction during spring; they always swim north. Back when they didn't have compasses, they used those walrus to learn their location during heavy fog. When it's extremely foggy, the ocean appears as though one's eyes are closed, although his eyes are open."

Simeon Agnus (July 2007:370) also remarked on the reliable flow of the *qairvaat* (ocean swells) mentioned above. Dangerous in spring, when these powerful, long-period waves from afar can break the ice with little warning, they are potential lifesavers during the season of open-water traveling: "Near summer, although a man doesn't have a compass with him, he won't get lost. These *qairvaat* that don't break head toward land. They don't head anywhere else but only toward land."

Qairvaat were also recognized as directional indicators in the Canineq area. Roland Phillip (November 2005:89) noted: "My father told me that when there was no visibility, if I followed *qairvaak* [two swells] I would reach shore. When I'd watch them, I saw that when they started to form, they always flowed toward shore. These *qairvaak* aren't always visible, but they appear, then smooth out and appear again." Frank Andrew (June 2005:179) noted that *qairvaak* head straight toward the mouth of the Kuskokwim, hitting the lower Kuskokwim coast at an angle: "These *qairvaak* don't face straight toward the land, but they tilt toward the mouth of the Kuskokwim River down below our shore. And if there's low visibility and fog, if *qairvaak* form, they know which way to go."

Paul Andrew (November 2005:90) said that in the Canineq area, the sound of long-tailed ducks can guide a hunter to shore: "When I heard one of those long-tailed ducks, I headed toward the sound, and as I went I found the shore-fast ice. There are many indicators down on the ocean. The birds they call *aarrangit* [long-tailed ducks] don't leave the shore-fast ice when it's foggy." Paul John (October 2003:40) noted that men lost on the ocean were admonished to always pay attention to Pacific loons and red-throated loons: "They said those two birds fly northward. They said to follow those when they pass because they knew they were flying north." King eiders were also reliable indicators. Simeon Agnus (July 2007:683) added: "When these king eiders arrive, they head in our direction [north toward Nelson Island] along the lower Kuskokwim coast. Then from Nuuget [points on the west end of Nelson Island], they turn toward the open ocean, facing Russia." Gulls indicated that land was near.

Other well-known directional indicators are the glittering ice crystals that reflect sunlight, even in foggy weather, clearly indicating the direction of land to those who know how to read them. Paul John (July 2007:278) explained:

We may look down along the edge of the small ice sheet, but if we aren't along the side where the sun is located, we won't see *mer'em makuari* [particles in the water along the ice edge]. But if we circle it, when *makuat* become visible when the sun reflects on them, they find the land's direction.

If we don't see an ice sheet, sometimes we have a shiny object in our possession, those we call small pocket knives or even a butter knife inside our provision box. Those are also like ice, if there is no visible ice in the water.

If one's knife is shiny, one can wet it, and after going in a circle, the knife will suddenly glimmer when the sun is behind him. When it becomes shiny, he will find where the sun is located. Or one can wet one's fingernail and use it in the same way. It's like going around a sheet of ice. That's what a person can do if he loses his sense of direction.

Joseph Patrick (January 2007:108) described using a paddle to view *makuat* in ice-free water: "When the ocean is clear of ice, he could just place his paddle on top of the water, and those *makuat* will be visible on top of his paddle or any wooden implement. Those *makuat* will always be visible in the sun's direction." John Eric (January 2007:110) added: "Evidently those *makuat* don't disappear whatsoever because they're in the ocean. Without placing an object down, I always see those *makuat* when looking over the side of the boat." *Makuat*, the ocean's eyes, not only allowed hunters to find shore but were the means by which the ocean could find them.

Frank Andrew (October 2003:86) noted: "There are some people who are skilled. They say those who were good at finding direction by determining the direction in their mind would know where they were going, even though it was dark and very foggy." John Eric (January 2007:110) described how experienced hunters used their knowledge to bring their companions safely home.

Sam Alexie told that story a number of times. He said that the ice detached
and floated away at night while they were sleeping on top of the shore-fast ice,
along the mouth of the [Chefornak] River, in complete darkness.

And when he alerted [Billy Yupayuk], after putting on his sealskin boots
[Billy] placed his kayak down in the water. He said that sheet of ice wasn't large.

He paddled and went around the place where they had slept, and when
he'd gone all the way around [the ice], he said to them that he had found
a way to travel, and here it was completely dark. He said to him, "Follow
behind me."

[Sam] said that they traveled for a long time, and the area ahead began
to turn white. They arrived at the shore-fast ice. I wonder how that person was
able to see that? It's true that our fathers who hunted with kayaks were awesome.

The shallow coastline of the lower Kuskokwim coast—only five meters deep over a mile from shore in some places—is characterized by extensive mudflats during low tide and an intricate system of sandbars and channels. John Eric (December 2007:241) described potential hazards: "The area down below my village is too flat. Although the land is a great distance away and not in view, a boat can get stuck in shallow water." Camilius Tulik (March 2007:758) warned: "Although there is absolutely nothing wrong with us, if we happen to hit the shallows [and ground the boat], it will be our demise." John (March 2008:137) also advised caution: "When the ocean is calm, it is blissful during summer. It all appears deep when there are no waves around, but here some areas are shallow. One cannot just travel anywhere. Those before us evidently used a pole to dip in the water and check the depth [when they traveled]." Land travel, John concluded, is much easier: "Since this land is visible, it is not difficult to travel on. If a person goes somewhere looking at a map, he will arrive at his destination. But some areas of the ocean turn into sandbars and dry out when the tide goes out, blocking the path he is planning to take."

Just as men needed to know place-names to travel safely on land, so they named sandbars and channels. Frank Andrew (October 2003:80) named eight sandbars on the lower Kuskokwim coast between the mouths of the Kipnuk and Kwigillingok Rivers: "The sandbars down below our shore also have names. From the mouth of Qukaqliq River, they call the one that is farthest back Qikertarpak. Then the one beyond it is Kangicuaq, then Kangicualleq. Then there is Cuirneq, then Iretkuk, Qavlunaqvak, Aangaguk, and Marasvak." Frank noted that some sandbars are covered by high tides while others are not: "Even the very high tide doesn't cover the [sandbars] at the mouth of Qukaqliq, the river of Kipnuk. And gulls lay their eggs on them; they nest over on Qikertarpak. And there is a lot of wood there because the tide doesn't go over it. But the very high tide covers those [sandbars to the south]. Then from Iretkuk, they become lower. The tide covers the [sandbars] below our shore. It covers Aangaguk, Qavlunaqvak, and Marasvak."[1]

John Eric (December 2007:231) described a similar system of named sandbars below Chefornak: "There are many sandbars down below our village, and they are difficult to learn. The channels are hard to distinguish, but when the current is flowing down on the ocean, the *tevanqut* [deep areas] are obvious." Simeon Agnus (January 2007:90) described three large sandbars between the mouth of Qalvinraaq River and Toksook Bay: "There is one down below [the channel called] Kangiryuar, one down below Kangirpak [Channel], and one in front of Taqukatuli [Channel], a sandbar that stretches out toward the ocean."

John Eric (December 2007:244) described five deep-water north-south channels between the mainland and Nunivak Island: "There is a [channel] that barges travel through on the [east] side of Nunivak Island that is ninety feet deep for seven miles. There are actually four [channels], with the one located closest to land as the fifth." John (March 2008:114) joked that one of these channels, Kuiguyurpak (lit., "big channel"), goes north to south, "probably reaching Seattle."

These channels have strong north-running currents, often filled with debris, including everything from seaweed and fishing gear to dead seals and even whales. While halibut fishing in deep water, John Eric (March 2008:117) has snagged anchors and fishing lines on the rocky bottom of Akuluraq (Etolin Strait): "Down along the bottom [of the ocean] reaching Nunivak Island, there are many large volcanic rocks. There aren't any [near the mainland], but starting from the middle [of Akuluraq], they extend down toward the ocean. When you drop anchors, they make a banging noise as they hit rocks along the ocean bottom."

Although invisible at high tide during calm weather, John Eric (December 2007:231) described how sandbars were discernible through waves and foam: "There are constant breakers in those shallow areas, but deep places have only waves created by wind. We don't approach [areas covered by breakers], because we know they are shallow." Water color is another indicator, according to John: "The water varies along the ocean, getting dark and then light. When it changes to light-colored, green water, it is a sign that it's getting deep, but it isn't deep if it's brownish or grayish." John Eric (March 2008:125) emphasized that travelers must stay in deep-water channels: "We cannot travel in the ocean down below our village using just any route during fall or we will get stuck in shallow water. Since the water isn't all one depth, we tried to pay close attention when we traveled and not leave deep waters."

John Eric (January 2007:101) described traveling through sloughs during an incoming tide: "Some small sloughs next to sandbars are referred to as *imangyaratulit* [ones that fill with water early]. They said that if I traveled through [a channel] that fills up quickly with water, it would be the first place to get deep." In general, it was best to travel in the deeper channels during an incoming tide: "During summer when the ocean is ice free, one mustn't veer from the waterways between the sandbars. If a boat travels in the channels, even though it's windy or the water is low, one won't run into difficulty. The waterways between the sandbars are actually good to travel through" (John Eric, March 2007:288).

Getting stuck on sandbars in shallow water is a constant hazard, even for knowledgeable travelers. Peter Matthew (March 2008:139) exclaimed: "The ocean is also daunting. Although the weather is calm sometimes, a person can encounter danger down on the ocean. If a person happens to get stuck in shallow water, and if the tide no longer comes up high enough to reach his boat, that person will be in a situation similar to being lost in bad weather." Peter then described his experience as a young man getting stuck with other men and boys in an open boat during fall when entering a river mouth on an outgoing tide. They were there for eight days, and provisions ran out: "Eventually, there was nothing left on the bones of the fish we had eaten, and they turned white. . . . And the tide never came up high enough to reach our boat. The water would almost reach us, then recede." Finally, early in the morning of the eighth day, they floated their boat and returned home, having survived on the meat of a beached seal.

John Eric (March 2008:122) also was concerned about the dangers of getting stuck on sandbars, especially those with steep edges:

A boat that comes upon sandbars near the old Qalvinraaq [Channel] in the dark will get stuck there in shallow water. If a boat comes from the south or southwest, it won't go anywhere, but waves will fill the boat.

The sandbars on this [east] side of Qalvinralleq have sharp edges, since this area is a river down to that island. The west side [of that sandbar] gradually deepens. We call that *civlirtellria*, an area that has a gradual downward slope. And they call the deep-water sides of sandbars *isqut*. Those are dangerous.

[The sandbar named] Urrsukvaallrem Marayartaa is also like that. Its northwest side abruptly ends, since it is where a river exits. From Qalvinraaq River, from Issurituli [a sandbar] toward the northwest, that's what [the sandbars] are like.

Nelson Islanders noted that the north end of the sandbar Cingigyaq was particularly dangerous. Simeon Agnus (December 2007:244) explained:

There is nothing dangerous in the area down below Kangirrluar [Toksook Bay], but Cingigyaq is the only place that arouses concern for young men who don't know what it's like, since the west wind produces high waves there. Its point has lengthened because it's gotten shallow. There is a channel that cuts through, a *tevyaraq* [route one travels through, portage trail]. The [channel that flows through Cingigyaq] doesn't have large waves, but the other side where it exits is the only place where it's shallow [and has waves]. It's distressing to try to travel down below that area.

John Eric quipped that "its zip code is 911 [because it's so dangerous]."

Maria Eric (March 2007:296) shared a hair-raising account of traveling up the coast from Chefornak past Cingigyaq.

We also went to Umkumiut to harvest fish during summer one time. There were three boats. Amaqigciq towed us, and we were in David's boat. Then Qiurtarralek and his family were at the very back. We went across from Aqumgallermiut. It was calm and windless all day as we traveled slowly.

We traveled and were okay. Then we evidently got too close to Cingigyaq, which they mentioned forms large waves. The area ahead started to turn white. They were breaking waves.

When we got to that area, Amaqigciq's boat would go up on top of a wave and then go down. Then our boat would go down like that also. Then those in front of us would disappear on the other side of the wave. We also went through the path that he had taken.

Then the people who were at the very back, Qiurtarralek and his family, when they reached a wave, after going down, when it went up, the rope that towed the people behind us made noises from getting taut.

Then David shouted to Amaqigciq, who was towing us, that if we moved a little toward shore, the waves would be much calmer.

We continued on through those waves. Those of us inside the boat were going through really rough water. Then Umkumiut came into view over there. The waves got much calmer when we moved in toward shore.

We went across, and when we were about to get to the mouth of the Toksook River, when we turned toward shore, it seemed to me that we arrived in calm waters. It was like we slid down, and all the waves behind us were gone. Evidently, down below Umkumiut, they watched us as we came. There were no waves the rest of the way.

The people who were in the very last [boat] cried along with their mothers. Kayungiar said they cried out of fear, including their father. And Tacuuq was extremely terrified since he had never gone through large waves.

To this day, some people in the Canineq region attribute awareness to sandbars, which are said to recognize those who are abstaining and should not touch them. Sandbars, they say, are dangerous when angered. Such a view may reflect the very real dangers these invisible features pose to ocean travelers.

Elders note that sandbars are forming in new places. Joseph Patrick (January 2007:82) said: "These days, more and more sandbars are forming. Places where there weren't any sandbars in the past have now formed sandbars." Channels are also changing, new ones forming while others lose current. Simeon Agnus (January 2007:88) described the condition of the channels known as Kangirpak and Taqukatuli: "Kangirpak is where Taqukatuli goes out [to the ocean], and there is an island right below it. That was once an ocean channel. Back when my cross-cousin Kangrilnguq [Paul John] and I first started to hunt, boaters and kayakers used to hunt for adult bearded seals up and down the channel. But nowadays, since its outlet has filled with sediment, Kangirpak no longer has a strong current, and there are no longer seals in it."

Qairet Anuqa-llu | Waves and Wind

"It is joyful to be on the ocean when the weather is completely calm," recalled

John Eric (March 2007:289). Wind and waves, however, were more often present than not, and men who hunted on the ocean had great respect for their power. Paul Tunuchuk (March 2008:307) spoke from long experience:

> There are three types of waves. *Qairvaak* [long-period ocean swells] don't stop. Although the weather is calm and windless, those waves continue to flow.
>
> Then there are *qairet* [waves] formed by the wind. Those waves exist because of the wind. When the weather is calm and windless, the [ocean water] glistens. And when a light wind starts to blow, those called *qairuayaaret* [small waves] form.

As mentioned, *qairvaak* can be used as directional indicators, as ocean swells originate in deep water far out in the Bering Sea and always flow toward land. John Walter (March 2007:1363) recalled: "When we became disoriented, we would stop and look at the water, and it looked like it was breathing as it flowed toward land."

Paul John (March 2008:570) explained how ocean swells create breaking waves when they reach shore:

> *Qairvaak* are waves that head toward land from the deep water. When they reach a shallow area, they break the shore ice. They refer to that as *qairvaarluni*.
>
> *Qairvaak* break when they reach a shallow area after not breaking in the deep area. As they break, they sometimes create waves that meet and slap against each other in shallow areas. When those waves are large, they refer to them as *qaicurrluut*. Those are bad [waves] when traveling and will make the boat wobble.

Swells are present in summer and fall but usually not in spring. Paul John (March 2008:572) explained: "There are no *qairvaak* during spring because the ice holds them back; the *qairvaak* will extend a great distance from shore in an area that doesn't freeze and has no ice." Frank Andrew (October 2003:66) noted that normally *qairvaak* arrived when the walrus started to be seen in late April and early May.

Paul John (March 2008:572) described how Nunivakers used swells to predict coming wind: "When the people of Nunivak Island only paddled [by kayak] from the village, they evidently kept an eye on the small pieces of ice down on the ocean. They say when the ice out there went out of view and appeared again [above the waves], that was a sign of coming wind, and they would turn back. They say the waves hit first. Although the weather was calm, they'd head to safety to escape danger when those small pieces of ice started to go out of view." John Phillip (October 2003:61) also used swells to predict coming wind on the lower coast: "When the wind is going to blow from the north, the swells get high and deep."

While swells are not dangerous to travel on in deep, ice-free water, Paul Tunuchuk (March 2008:307) noted the warning not to stay in the ice in spring if swells arrive: "When *qairvaat* are around, we are admonished not to [be on the ocean] when there is a lot of ice. The [sheets or chunks of ice] suddenly move toward one another, and we are told not to go between them. They are dangerous when they collide with one another. And when there is a lot of ice, the swells obviously advance toward shore."

John Phillip (October 2005:282) compared the effects of shore-fast ice and seal oil on water:

> They say that the *tuaq* [shore-fast ice] is like seal oil down there. You know when you spill seal oil in a bowl, it's smooth. They liken the smoothness [of ice] to oil when it's calm. When it's windy and [seal] oil is spilled [in the water], there will no longer be waves. It is smooth, although the waves are small. The waves will no longer break.
>
> When you catch an animal in the ocean when it's windy, when its blood and oil flow out, [the water] becomes visibly smooth. And if [a sea mammal] that was shot in the wrong spot swims away and is diving and bleeding, although it is deep underwater, oil will rise in the direction it's heading and help us locate its whereabouts. That bloody area will be calm because it has a trace of oil. When there is a slight breeze, [the animal's] oil will surface. They call it *uquarluni* [oil surfacing on water from a wounded animal].

John (October 2005:283) added that the wind also can carry the scent of seal oil, scaring other sea mammals away: "When it's windy and we put [dead seals] up on the ice, the oil scatters as well. Then if a sea mammal inadvertently surfaces around the oil, if it smells it, it will quickly dive. Even though that seal oil is their oil, [the seal] does what they call *teplilluni* [smelling an odor that makes it flee]. It's true that seal oil has a strong smell, and the wind carries it."

Paul John (March 2008:571) noted that the wind can create smaller waves on the surface of swells: "When it's windy down on the ocean, those *qairvaak* have waves on top of them formed by the wind. The waves that the wind formed are obvious, but the ones underneath are waves from the deep ocean." Smaller waves—*qairuayaaret* and *qairuayagaat* (lit., "pretend small waves")—are caused by local conditions and, unlike ocean swells, cannot be relied on for navigation. Small waves and ripples, however, can be used as weather indicators. Camilius Tulik (March 2007:568) noted: "Down on the ocean when one sees ripples while it is calm and they look like fish wakes, they say that indicates coming wind. The [ripples] dash away on top of the water, looking like fish, and the wind then builds up." Paul John (October 2003:61) observed: "They say that the waves hit first before the wind arrives onshore. When that happens, they say that it is going to get windy and not good for traveling." Camilius Tulik (March 2007:568) noted that the opposite is also true: "When the top of the water is a little bit shiny, they say that the weather will become calm."

The relationship between wind and waves was important for every ocean traveler to understand. John Phillip (September 2009:109) explained: "The waves created by the north wind are closer together than waves made by the south wind. In fall the waves get shorter when winter gets nearer and it's windy. They warn of the danger [of waves that are close together] when the north wind blows, and we have to be very careful." Breakers indicating sandbars and shallow water also vary seasonally. John Eric (December 2007:218) explained: "Shallow areas and sandbars are often covered by breakers. During spring, the breakers are much calmer, but in late September, although the weather is calm, there are always breakers."

Areas where wind and currents meet can develop huge waves. Simeon Agnus (December 2007:260) observed: "If the current flows one way and the wind is blowing in the other direction, there can be waves as large as mountains. Since the current gets strong around Ugcirraq [Cape Vancouver], where it curves toward the ocean, that area usually has large waves." Many noted that the waves in Etolin Strait and around Cape Vancouver and Nelson Island generally were much larger and stronger than those encountered on the lower Kuskokwim coast. John Eric (March 2007:289) related his personal experience: "One day, it suddenly became windy while we were in the area below Up'nerkillermiut. The wind began to blow from the northwest around thirty knots, and large waves started to form. I thought I had experienced traveling through large waves in Bristol Bay. Their waves aren't actually rough, but when the wind is blowing from the northwest inside Akuluraq [Etolin Strait], it's really rough. My intestines were moving up and down in my body while traveling in huge waves because the current is strong down there."

In general, waves created by the south wind are less daunting. According to John Eric (December 2007:265): "I heard in the past that waves generated by the south wind are good to travel on because they are deep. And they say that the waves generated by the west wind aren't good to travel through, because they are close together." John (March 2007:295) added that since the current flows north in Etolin Strait, waves created by the south wind are much calmer, whereas the west wind creates bad conditions.

Paul Tunuchuk (March 2007:307) was warned about nearshore waves when the south wind was blowing:

That person named Qaivigaq arrived here [in Chefornak] one day when it was windy from the south. I told him, "My, how fearless you are."

Then Qiavigaq said to me that the only waves that are dangerous are the ones along the shore. He said waves aren't dangerous far from shore in deep water when the wind is blowing from the south because they are easy to avoid. But he said they are dangerous along the shore where it is shallow because they are difficult to avoid.

It is true that when the south wind is blowing and the tide is coming in, although it's windy down on the ocean and there are large waves, they are easy to avoid when traveling. That's what Qiavigaq said.

Stanley Anthony (December 2007:266) described the waves he encountered in a place where currents meet: "When I traveled to Kipnuk with our wide wooden boat, the wind was blowing from the south directly against the shore, and I found the waves down below Ingriik unusual. Those waves were colliding and did not appear like real waves. When the waves hit the boat, it would move from side to side." Simeon Agnus explained: "In places where *carvanret* [currents] meet, waves are like that."

Elders' conversations about traveling in wavy, windy conditions stand out for their emotional power. Simeon Agnus (December 2007:239) described his feelings when traveling in the dark through large waves:

> When a wave covered us, I suddenly panicked at that time, and I thought, "This is probably the only time we will be alive." Although my *ilurapak* [cross-cousin] was constantly bailing, the waves that splashed from along the front of our boat filled it with water. Then after a while, these waves got a little smaller. . . .
>
> When we docked at Toksook Bay, my body and mind felt heavy. There was no dry spot on my entire body, and here seal-gut rain garments are actually very airtight. I wonder how water got inside? Indeed, it's not good when you suddenly panic.

Paul John (December 2007:247) encountered an equally desperate situation traveling from Bethel to Nightmute with his brother-in-law, Jobe Abraham of Chefornak. They had good conditions past the mouth of the Qalvinraaq River but hit rough water heading east into Toksook Bay: "That is the most dangerous place. When the west wind is blowing, there are large waves there that head to shore from the ocean, especially in fall. It's not easy traveling through there." Paul and Jobe had made it past the breakers and through the portage route flowing through Cingigyaq when a wave covered their boat, broke the steering device, and killed the motor. The men then lowered their anchor line and let the boat face the waves while they restarted the motor. After making a new steering device from a five-gallon bucket and nailing it to the back of the boat, they continued on:

> We removed our anchor and at a slow speed, constantly dipping a stick in the water [to check the depth], we followed the edge of the waves and headed up slowly. Just as Qurrlurta appeared, the surrounding area started to get deep. We made it to Kangirrluar.
>
> Jobe expressed his thought when we reached the area down below Umkumiut that was sheltered from wind, "I thought that we would die over there."

Many noted that waves along the lower Kuskokwim coast are not like those around Nelson Island. John Eric (December 2007:263) explained:

> The breakers below Chefornak are smaller than the waves along Toksook Bay. [The waves] move slowly for a long time out there when it gets windy down below our village. Although a breaker catches up to [that boat] and hits it, it passes by. . . .
>
> I really enjoy it when the [waves] down there get deeper. When the boat is constantly going up and down, the waves generated by the wind are calm sometimes; when the wind is blowing fifteen knots from the southeast, they are just right, since [the waves] are deep although there are large waves.

Carvaneq Ula-llu | Current and Tide

They say the ocean is always changing and doesn't stay the same. That's why they say the ocean cannot be learned. After the tide comes up, when the tide goes out, it changes.

— Paul Tunuchuk, Chefornak

"There is a saying that there is nothing stronger than ocean water," John Eric (January 2007:98) recalled. This adage captures sea conditions around Nelson Island and the lower Kuskokwim coast. Although tidal variation is moderate, ranging between six and nine feet, the coastline's low elevation (less than a meter rise over 7.5 kilometers in some places) can translate into extensive mudflats during low tides.[2] Alternately, fall storm surges can push water and ice inland up to thirty miles, depositing driftwood and ocean debris high on the floodplain.[3] Harvesting activities were carefully planned, both to take advantage of the areas that high tides made accessible by boat and to avoid the dangers of getting stuck when minus tides drained river mouths and channels.

Frank Andrew (June 2005:3) recalled the names for incoming and outgoing tides on the lower coast: "When the tide is going out, they say *aterliluni* [from *aterte-*, "to drift with the current"]. And when the tide is coming in, during high tide, they say *tagliluteng* [from *tage-*, "to go up from a body of water or up any gradual incline]." Simeon Agnus (December 2007:97) said that on Nelson Island they referred to it as *tagcarneq* (incoming tide) when the current ran toward land, whereas they said *ketmurneq* (from *kete-*, "area away from shore") or *ancarneq* (outgoing tide, from *ane-*, "to go out") when the current ran out to sea.

Simeon Agnus (December 2007:56) shared what every coastal hunter knows well—animals are readily available only during the morning outgoing tide: "One must go down to the ocean in the early morning. When I used to hunt, I learned that surfacing animals were readily available some days during the outgoing tide.

Storm surge along the lower Kuskokwim coast, with water driven up to the doorsteps of homes in Kwigillingok, December 7, 1931. *Augustus Martin, Martin Family Collection, Anchorage Museum B07.5.4.518*

And when the current started to flow north, it was as if animals were no longer around."

Simeon Agnus (July 2007:683) also explained coastal currents, which run north during the day and west only during the early-morning outgoing tide: "They call the current that flows out there, directly toward the ocean and away from the shore, *ancarneq*. The tide goes out toward the ocean in the early morning, sometimes before it gets light. It goes toward the west. Then when the tide comes in, the current flows north all day. And although the tide goes out, the current continues flowing north. In our village it turns toward the ocean a little, but the current continues to flow north." Others agreed that although the strong current in Etolin Strait weakens slightly during slack water, it continues flowing north all day.

Simeon Agnus (March 2007:551) noted the strength of the morning outgoing tide: "In the morning when the tide goes out, the ice in front of the mountains goes out fast. The outgoing tide is strong, and it is dangerous." Hunters were cautioned about the strong currents in Etolin Strait. According to Simeon Agnus (December 2007:56): "Akuluraq isn't like other places; since it is narrow, its northern current is strong." James James (January 2007:96) of Tununak agreed: "That deep area has a strong current. Starting from Akuluraq, when it flows north, it's powerful." The same is true for the Ningliq River, which drains Baird Inlet on the north side of Nelson Island, especially during incoming or outgoing tides: "When we reach the Ningliq River, the current is strong because it is narrow" (John Walter, July 2007:376).

Coastal channels could harbor hazardous currents. Camilius Tulik (March 2007:530) recalled: "Our fathers said that the current in Kuiguyuq [Channel] used to be so strong coming [north] from Akiliit. It was dangerous when there was ice, and one never dared to go through that when they went by paddling [kayaks]." Simeon Agnus told of watching a walrus fighting the current at the end of Kuiguyuq:

> A young walrus was going east, and how fast it was! As it got to the ice, it turned west. It dove and surfaced, not having moved much, and it was bellowing! It seemed to be in the same spot, and it bellowed and turned around again.
>
> So we watched a young walrus when it had a hard time swimming against the current. Even seals probably panic a bit when they do not have a place to surface.

Frank Andrew (October 2003:66) noted that channels were most dangerous during an incoming tide: "The edges of the deep channels appear when the tide goes down. [The current] is strong when the tide is just rising. When they're full [of water], the current gets weaker." Currents in some channels have weakened, but many remain forces to be reckoned with for humans and animals alike.

Qanruyutait Ellarrlirqan Aarnarqaqan-Ilu |
Instructions for Traveling in Bad Weather and Dangerous Conditions

Calm days could quickly turn treacherous, and hunters recalled time-tested adages that covered such dangerous situations. First and foremost, hunters were admonished to curb free expression of blame or concern, especially around young people. Roland Phillip (November 2005:94) stated:

> There is the following *qanruyun*. During times of distress, when the ice or wind may give us difficulty, they told us not to say, "Because this person has done this, this person has caused us to get into a hazardous situation."
>
> They also say that if one of us wasn't as confident as we were, during times of distress down on the ocean, we elders should try to watch our conduct. They say that we should try to keep those who aren't as confident as we are optimistic. One should try not to tell them fearful things or to inform them about what might occur but appear cheerful, as though nothing was happening, although their situation was grave.
>
> More than once, I was with a man who talked foolishly when we were in that situation. One time while Albert Beaver was with us, although we were in that situation, he didn't seem worried. He was in good spirits. But the other man tended to say whatever he pleased, and I understood that what he did was wrong. Then someone in our group who was older than him told that person, "When we have young people

with us, it is an admonishment not to express one's thoughts and feelings but to try to keep those less-confident ones optimistic." Those are *qanruyutet* of the ocean.

Paul Andrew (November 2005:96) added: "Being around someone who is talking foolishly out of fear can actually be dangerous out on the ocean. It makes you feel desperate and lose confidence when you're with someone who expresses fear during times when waves are very large. I experienced that. I almost gave up at that time."

John Phillip (October 2003:239) noted practical advice for coping with fear: "They told us that sometimes if we were very scared on the ocean, we should eat until we were full. Once we had no trail and no way to go up to land. Since [my partner] was getting upset, I told him to cook a long-tailed duck, even though I wasn't hungry. I didn't feel hopeless. He cooked, and we ate until we were full. He really changed. He even looked better physically and got braver. It really is true." John Eric (March 2008:373) noted that "when traveling on the ocean, the only time you'll be afraid is if you're hungry. If you have a full stomach, you won't be afraid at all."

Hunters were also admonished not to value their quarry more than their own lives: "Even though the animal was easy to catch, we were told not to trade our lives for it, that it belongs to the ocean. If it was dangerous, we were told not to go after it at all" (John Phillip, October 2005:20). As a result, fewer men caught large bearded seals. Roland Phillip (October 2006:105) explained:

> All the men of the lower Kuskokwim coast up to Tununak have become *nukalpiat* [great hunters] these days. When I was young, I heard of only one man in Tununak who caught bearded seals. We never caught any bearded seals, but we began to catch them after we got [motor] boats.
>
> Back when I used kayaks, I caught a bearded seal twice. I couldn't hunt bearded seals although I saw them while I was alone, afraid that the animal that I caught might capsize me.

Paul John (October 2003:81) recalled the admonition to act like wood— that is, not fight the currents or winds—when traveling on the ocean: "While some travel it gets windy. They don't know if they will get home. When that happens, they were instructed not to give up. When they instructed us, they used the small pieces of wood that float on the river as examples. They told us to compare ourselves to them, pretending to be pieces of wood. If we can get through it, we would get out of it like the wood, but we can drive our boat the wrong way or tip over if we panic. If we act like floating wood, it can help us stay alive." Paul John (December 2007:151) also talked about the importance of not panicking:

There is also an instruction that a person should make an effort not to panic down on the ocean although he encounters peril. They say if he doesn't suddenly panic, he will recall the things that those who were constantly speaking had talked about. "Oh yes, this is what they said we should do." They say that person will find a way to save himself.

But the following type of person will suddenly panic if he encounters peril. Overcome by fear, he will be like one who isn't cognizant. It's as though he is only aware of what he sees in front of him. They say *qamelria* [one who panics] is more likely to get into a mishap. Although he is fully capable, one who panics won't be able to function, submitting to his fear.

Peter John (March 2007:1187) had been taught not to panic in difficult situations lest he lose strength:

Although I was afraid, they said that I shouldn't try to make myself feel fearful. They say when a person panics, even if he is strong, he will be unable to handle that situation.

One time, I thought Tom and those who were with him had panicked. When pieces of ice went up [toward shore], since he had gone on the other side of the ice, they were stuck down there. The three of us saw them. I was wondering, "Since it isn't heavy, why don't they have the strength to haul their boat?" I thought they had panicked.

Since they could hear my voice, I yelled to them that although they were in a hurry, they should stop and rest for a while. They rested down there. Then I told them to head up [to shore]. My, they easily hauled up the boat that was too heavy before. I thought that they had panicked at the time since I recalled how people will lose strength if they suddenly panic.

Men also recalled many practical admonishments on what to do in windy conditions. Paul Tunuchuk (March 2007:90) said that hunters were told to immediately secure their seal-gut rain gear around the openings of their kayaks: "That's why they made the bottoms of seal-gut rain garments wide, so that they could easily fit over the kayak cockpit coaming. Although the water [splashed on them], they could get on their knees and let water trickle down into the ocean." Paul noted that men were also admonished to face the wind: "They'd use those small long-tailed ducks as an example when they spoke about windy conditions. [Those ducks] would face the wind with their younglings behind them; although they didn't move forward, they were told to always face the wind and they would make it through."

John Eric (January 2007:112) shared practical advice, including the importance of using a heavy boat in rough seas. John (March 2007:292) also noted the importance of traveling through large waves at an angle: "When large waves started to form while a boat was traveling, they told me not to face my destination directly and take a straight path. They said that when waves are close together, I should face them at an angle. Since Anii's father evidently experienced these things when traveling with a kayak, he passed on that lesson to us."

Peter John (March 2007:1186) described factors determining his paddle choice: "My father said that there is a paddle with a large blade that I could use when my kayak was carrying a heavy load. And he said those thin-bladed paddles are good to use in windy weather. They don't make a person tired right away. He also said that if the weather suddenly got bad and conditions were perilous, I shouldn't paddle using all my strength but at a comfortable speed, not trying to exhaust myself." Peter was also instructed to use his kayak sled as a rudder in windy conditions: "They said that if I was unable to handle the conditions and the wind was blowing from the side, to use my kayak sled. I should use a skin line from a harpoon to tie it along the center and place it along the side where the wind was blowing. As the kayak went up against the wind, [the sled] would keep the kayak [on course]. He said that although the wind was blowing me, if I continued on my way, I would gradually head in that direction."

Coastal hunters might also raft their kayaks together in rough weather. According to Frank Andrew (June 2003:292): "Sometimes they all tied themselves together quickly when it suddenly became windy. And they put their kayak sleds beside them, tying them with sealskin line and throwing them in the water, towing them to prevent [the waves] from overturning the kayak. Even though the waves hit them, they wouldn't hurl the kayaks because they were weighted down by their kayak sleds. They tied them on to the kayak deck stiffeners. Kayaks that were tied together side by side were solid. They could walk on top of them, and they weren't tippy."

Coastal hunters routinely carried *qerruinat* (sealskin floats) for seal hunting as well as in case of emergency. Frank Andrew (June 2003:292) continued:

> They placed floats on the outside in front, putting harnesses between the kayaks on the bottom, placing floats on the ends, tightening the spaces between the kayaks, using them to keep afloat. They said the waves wouldn't overturn kayaks that were tied together. Floats cannot go underwater, even though waves hit them. They say they really use them when the ocean is dangerous.
>
> That's how those of us from the coast who used kayaks were instructed. And one could use the float if he capsized while alone. They said they always carried them inflated. The [float's] neck area had a large skin loop that could be placed on [the foot]. And they tied them on the end of the sealskin line. He could use it if he capsized, getting out [of his kayak] and standing in the water on top of the float. Then he would turn his kayak upright, take the water out, and get in.

Landing a kayak in windy conditions required skill. Frank Andrew (June 2003:300) explained how hunters balanced their crafts in preparation for landing:

> When it suddenly becomes windy on the ocean, it is impossible to dock and exit the kayak in the usual way. When the waves break, water flows on top of the shore ice, and it's dangerous. They taught us about exiting [the kayak]. And they put ice chunks in the backs of kayaks that didn't have a load to weigh them down. We would have the stern lower when we were going to land. Even though waves hit us, the bow piece wouldn't go into the water. Because the stern was heavier, it was farther underwater.
>
> Some used their catches for weight. And those who had too much would let others who were with them use them for weight, distributing them.

Hunters might also abandon their catch in windy weather. Simeon Agnus (July 2007:361) remembered: "Their wives made their husbands seal-gut rain garments with very wide neck openings. When [the men] had no way of paddling, they could remove their load by taking it out through their neck opening, discarding the load they were carrying. It is said that's what they did when there was no ice around and it suddenly became windy and dangerous."

According to Frank Andrew (June 2003:301), when everyone was ready, hunters rode the waves to land, one at a time:

> When we approached the shore, one person who was going to let the wave carry him as it broke [on the ice] would face land, as he was perched on top of one steep wave. He would go at a good speed, riding on top of a steep wave, and head toward shore, and he would go on top of the ice without touching it. And when the wave broke, when he finally touched the shore ice, he would quickly exit his kayak and pull it forward. Then another person would go on top of the next wave, and the next person would follow. They would exit from the water one by one. The experienced ones would instruct their young people. They all did that. They would let the wave break underneath them. That's how we got out. I joined those who went through that twice.

Stanley Anthony (December 2007:261) reiterated the importance of following elders' instructions.

> One must follow the admonishments and instructions of elders. Evidently, a person must not try to follow his own judgment. Although an elder admonished me, when I followed my own judgment, my boat tipped to one side and almost capsized while I had my two younger siblings with me.
>
> We had a small wooden boat. Although people hunting seals in open water warned me that there were large waves, I wanted to go across [Toksook Bay]. I had

my two younger siblings in my boat. We went through large, breaking waves. Then before we reached Qikertaugaq, our boat tipped to one side.

I told the other [sibling] to give me her scarf, just so that I could wipe my eyes. Then a breaking wave caught up to us and hit us, and the boat moved. When the gas tank slid [and was about to fall out], I just placed my foot on it to try to hold it, and it stayed down there, and I was holding the edge [of the boat]. And my two younger siblings stopped along the other side where [the boat curved]. When [the wave] passed, the boat suddenly became upright again.

That's the time I almost panicked when my boat almost overturned. We got to Toksook Bay and went inside their river. When I went up to land, my hip boots were full of water. My poor companions. When we docked and they stood up, it looked like someone was spilling water over them—the water immediately cascaded down. [*laughing*]

Elders like Stanley Anthony were frank in speaking of their own mistakes, using them as opportunities to teach others. Roland Phillip (October 2006:108) shared the adage that a person becomes cautious only after facing adversity: "They said a person cannot learn a lesson without encountering danger from time to time. That's why sometimes [when I hear that] someone has encountered danger but hasn't died, I say, 'It's okay. When he encounters that again, he will recall that. He will learn from it.'" Roland noted another memorable saying: "In the past they also mentioned that a person doesn't go somewhere by just pointing his index finger. They do not travel without ever encountering peril. They encounter danger once in a while. Sometimes, death is close. That's why they mentioned in the past that a person doesn't point his index finger and travel to a place. That's an example they used."

Despite precautions, accidents did occur. David Martin (January 2002:86) explained: "When they used only kayaks, many men had accidents. They call it *cayukaulluni* [weather suddenly turning bad while people are traveling] when it happens on the ocean and not the land." John Phillip (October 2003:239) noted the term for surviving an accident: *anamyiiqerrluni* (from *anag-*, "to escape, to get away"). Frank Andrew (October 2003) described what happens after a fatal accident occurs: "No one stays on the ocean when someone is taken away out there. I experienced that once. When they were crying, it felt strange. It makes one unable to respond. Everyone who was down on the ocean was sent home when one had an accident. They put the [dead one] in the kayak and brought him home. They all left their kayaks down there, just moving them farther up."

As a dramatic example of how the ocean cannot be learned, John Phillip (October 2006:85) shared the story of a fatal accident that occurred at the mouth of the Kuskokwim in which many men perished when caught on the ice in strong winds. This event took place when John's own father was young, and elders still recall it as a warning. A group of men from Qinaq were seal hunting in spring and camped on a *nacaraq* (ice pile) attached to the sandbar Marasvak. Two men from Quinhagak joined the Qinaq men, and Roland Phillip (October 2006:86) recalled what one of the survivors told him:

> After sitting in silence, he said that there were two of them who had come from Quinhagak. While paddling, they arrived on this side [of the Kuskokwim] and joined those [other hunters].
>
> After saying that, he said that his grandfather would tell him the following, "Now, if conditions down on the ocean become unsafe, if you are among people from Canineq, stay with them. Those people from Canineq know about ocean faring."
>
> Then he mentioned that Alaqteryaq and his group, including Asngualleq and my father and Inaqaq, Piiyualria, had warned them about the coming wind, but [the men from Qinaq] stayed there [on the ice]. He said that a short while after dark, they started to hear a rumbling noise. Then after a while, the wind began to blow steadily, and the area above the ocean became dark.
>
> That wind arrived, and the tide was extremely high, as it usually gets during spring. He said that at daybreak, when the water got high, even though [the *nacaraq*] was large, [the waves] began to break it.
>
> The *nacaraq* began to break apart, and the pieces of ice began to float away. He said that some people went down into the water and stayed afloat, getting in [their kayaks] before it became too difficult to do so. And he said that [he and his partner] got in [the water] also. He said that some people stayed on the piled ice, but it became too steep to go down into the water. When those people attempted to go down into the water, their kayaks would float away. And he said that although they watched them, they couldn't help them in any way. They couldn't get their kayaks for them, as they were trying to keep themselves safe.
>
> And some kayaks would capsize once in a while. He said some people would climb on the ice without a kayak, but after a wave splashed over it, the person would be gone. The waves floated the person away. He said they couldn't help them, although they watched them.
>
> He said that eventually only a few of them were left. And the piled ice where they had stayed broke to pieces. He said their only elder told them that there was nothing they could do and that they should try to go back to shore. So they listened to that person and traveled back to shore with the wind blowing along their backs. Before they left, [their elder] told them not to stay too close to one another, and to avoid breaking waves.
>
> After some time, the ice up [along the shore] appeared. He said the windward side of the ice was covered with spray, probably from the waves splashing against the ice.

The Quinhagak survivor then described how the men reached the shore ice and, following their elder, successfully rode the waves one at a time to shore, just as Frank Andrew had experienced. The men stayed on the ice until the weather became calm, then left for home. Roland Phillip concluded: "Those people, Alaqteryaq and his group [were safe]; Alaqteryaq had tried to warn them when he came upon those people in calm weather, that they should find a safe place, as the weather out there wasn't going to stay calm any longer." George Billy (October 2006:94) immediately responded: "We heard the story correctly just now. They followed their elder's instructions and ended up surviving. I heard in the past that if an elder who had smelled the world first admonishes me, I should listen. The story we heard is a testament to the fact that one should not go against one's elders. That person evidently led them when they were in a desperate situation." John Phillip agreed, underlining the cause of the disaster: "Although those people were instructed, they didn't follow their instructions, and they died as a result of being defiant. Those people had experience going out to the ocean, but because they followed their own minds, they died."

Throughout our meetings, elders expressed their desire that—like the hunters who survived the Qinaq disaster—contemporary young people listen to what they were saying so that they, too, would have the tools they needed to avoid accidents and travel safely. Paul Tunuchuk (March 2008:7) expressed his concern: "The ocean is something that we especially should know about, but these days there's no one to teach about it." Camilius Tulik (March 2007:569) was more optimistic: "It would be good if we bring up these instructions that are still practiced when we talk about the ocean. If some recognize what they hear, they will further reveal those things to their relatives."

Theresa Abraham (March 2008:235) observed that the ocean is still daunting: "Although things are easier these days, the wilderness and the ocean haven't changed and still require suffering and hard work. For that reason, I really want our young people to understand the *qanruyutet* and to be aware of their Yup'ik identity." Moreover, many still view the ocean as responsive to human action. David Martin (January 2002:86) interpreted the Qinaq disaster as an intentional act on the part of *ella*: "After the weather was good, when the ocean felt annoyed, it got angry." John Phillip (October 2005:95) recalled the ocean's reaction: "They said that after that [storm], the weather was extremely calm and joyous. They refer to that as *cirimciluni*. It is said that [the ocean] was satisfied with the people whom it had killed, since *ella* can also have the desire to kill and hunt [for people]." Frank Andrew (October 2003:239) agreed: "They say the morning after the disaster, it was so very calm. There was no breeze on the ocean. It was so proud of its catch." John Phillip (October 2003:65) concluded: "I have heard the saying that only people will change to be like *kass'at* [white people], but the ocean and weather will not become *kass'at*."

ancarneq. Outgoing tide (from *ane-*, "to go out"). See also *ketmurneq.*

aterte-. To drift with the current.

carvaneq/carvanret. Current/s.

civlirtellria. One (sandbar edge or shore) with a gradual downward slope.

en'aq. Sandbar.

etgalquq/etgalquut. Shallow area/s.

iginim engelii/iginiq/iginit. Edge/s of deep water.

ilacarneq/ilacanret. Place/s where two currents meet.

imangyaratulit. Ones that quickly fill with water during an incoming tide.

imarpiim puyua. Saltwater condensation (lit., "the ocean's smoke").

imarpik. Ocean (from *imaq*, "contents").

isquq/isqut. Deep-water side/s of sandbar/s.

kangiq/kangit. Bay/s, open water bordered by ice or land, ice bay/s.

ketmurneq. Outgoing tide (from *kete-*, "area away from shore"). See also *ancarneq.*

kuignayuq/kuignayuut. Channel/s in mudflats filled with water during low tide.

kuiguyuq/kuiguyuut. Ocean or river channel/s.

kuineq/kuinret. Deep part of channel/s.

makuat. Particles/ice crystals in water around floating ice that are visible in reflected sunlight; also referred to as the ocean's eyes.

marayaq/marayat. Sandbar/s.

negetmurneq. Northerly current (from *negeq*, "north").

qaiq/qairet. Wave/s.

qairuayaaret. Small *qairet* (waves) caused by local conditions. See also *qairuayagaat.*

qairuayagaat. Small waves caused by local conditions (lit., "pretend small waves"). See also *qairuayaaret.*

qairvaak/qairvaat. Ocean swells; long, high-amplitude waves originating in deep water far from shore (lit., "big *qairet* (waves)").

qikertaq/qikertat. Island/s.

qukaritet. Sandbars in the middle of channels.

tag'aq. Surf.

tagcarneq. Incoming tide (from *tage-*, "to go up from a body of water or up any gradual incline").

tevanquq/tevanqut. Deep area/s in water, valley/s.

tevyaraq. Route one travels through, portage trail.

tungyuq. Incoming tide.

ula. High tide.

ulcuar. Small swell.

ulerpak. Large swell, flood.

Ciku

ICE

There are many different names [for ice formations] down on the ocean during winter, spring, summer, and fall. Their names are always changing. It would be good if they are written in books at this time, and how great it will be if our young people read about them before they get into a perilous situation.

—John Eric, Chefornak

ELDERS REPEATEDLY SHARED THE OBSERVATION THAT UNLIKE the land, the ocean was always changing. In this they touch upon a defining characteristic of the Bering Sea coast relative to both Bering Strait and the Arctic coast of north Alaska. There tidal variation is modest (at Barrow less than six inches), and the shore-fast ice typically is much thicker during most of the season, making it less likely to break and deform. On the Bering Sea coast, however, tides can vary by as much as nine feet. Moreover, the coastline's low elevation can translate into extensive mudflats during low tides, and fall storm surges can push water and ice inland up to thirty miles. Finally, Nelson Island and lower Kuskokwim coastal hunters can rely on neither thick, multiyear ice nor a well-defined separation between freeze-up, winter, spring, and breakup regimes that, through the 1990s, helped impart more predictable conditions in most high Arctic regions.[1]

Reviewing elders' statements, glaciologist Hajo Eicken noted that while it is sometimes assumed that Bering Strait hunters on Diomede, St. Lawrence Island, and King Island hunt in the most diverse and demanding ice conditions in Alaska, conditions on the lower Bering Sea coast are equally if not more challenging due to the complex interplay between tides, currents, and wind. An added element of complexity is the important role played by deformation of the thinner (less than three feet), weaker, and hence more dynamic ice of the Bering Sea coast.[2] The result, as Yup'ik hunters know well, is a remarkably rough ice environment, where there is really no safe place, as all types of ice have some propensity to break up or deform or behave in potentially dangerous ways. Elders are indeed correct when they say the Bering Sea is always changing. Off Barrow, the rule of thumb is that older, first-year ice and multiyear ice thicker than three feet are usually safe from deformation and offer a somewhat-safe haven even in a dynamic ice environment.[3] Along the Bering Sea coast, however, calm growth conditions are seldom achieved, and thus the ice rarely reaches that thickness and even level ice is likely to deform in some fashion. As a result, hunting in and around the ice requires a wealth of knowledge regarding its formation, physical characteristics, and behavior—and its dangers. John Eric (March 2008:293) noted that although a person could learn the normal seasonal cycle, no year was ever the same: "When an entire year has passed, the [ice formations] cannot be exactly the same as the year before. A person cannot learn it, but we can talk about it based on our observations of what it looked like in the past."

Before ever venturing out on the ocean, young men in the past were taught what to look for and what to do there. Such instruction is no longer universal. John Eric (December 2007:83) remarked:

> When someone encountered danger before GPS devices were around, I asked him, "What type of ice is there where you are located? Are there *kaulinret* [broken pieces of ice] in your area; are there *etgalqitat* [ice piles beached in shallow water] that are large?" He told me, "I don't know what those are because I've never heard about them."
>
> Indeed, although we ask these people who have no one to instruct them about what the wilderness is like, they won't know. It would be good to tell these young men the names of things on the ocean.

Imarpiim Cikuyaraa Uksuarmi | Fall Formation of Sea Ice

Along the Bering Sea coast sea ice forms anew every year, usually in November or early December, whereas close to 30 percent of Arctic ice remains at summer's end. John Eric (December 2007:65) briefly summarized the annual cycle in southwest Alaska: "You know how water starts to freeze during fall. Then [ice] gradually forms, and eventually after the snow starts to fall, it gets thick. The ice forms up until winter, and then it starts to break to pieces in spring."

The snow-covered mountains and frozen bay on the north shore of Nelson Island, March 2007.
Ann Fienup-Riordan

Paul John (December 2007:66) described how fall rains contribute to rapid ocean freeze-up: "Toward fall, when it starts to rain a lot, fresh water accumulates on top of the saltwater. They say that leads to the ice freezing at a faster rate during fall." Fresh water accumulating along the ocean leads to the formation of *cikullaq* (newly frozen ice, frozen floodwater on the ocean), also known as *nutaqerrun* (new ice) or, in English, frazil ice or grease ice, referring to its random crystal structure, as opposed to the long, regular crystals of older ice.

Paul John (December 2007:66) recalled the dangers associated with traveling in fall when *cikullaq* began to accumulate along tidelands:

> They said that when the tide comes up [during high tide], the water that covers the mud also freezes. Back when they used kayaks, they warned a person not to travel anywhere when it started to get cold and *cikullaq* started to form because it would tear the kayak [skin coverings].
>
> And when *cikullaq* started to form during fall, when the [water] on top of sandbars froze, when the tide brought *cikullaq* toward shore, it was impossible to walk on top of the ice without falling through.

Paul Jenkins (March 2007:254) confirmed that traveling in *cikullaq* was dangerous: "A kayak could not travel anywhere when *cikullaq* was around. The weather might be calm and windless, but a person wouldn't be able to [hunt]. But when the sun started to get warm and the *cikullaq* started to soften, we would travel in a group down on the ocean."

Hunters could also find themselves stuck when rapidly freezing *cikullaq* prevented them from returning to shore. Paul John (December 2007:67) described how hunters dealt with this situation: "If there were a number of people, they would connect their kayaks together with ropes. They lined up behind one another. One of the men would place his wooden plank seating [along the front of his kayak] to block [the ice] and would paddle through that ice. That was the instruction they gave back when *cikullaq* formed quickly during fall."

Paul John (December 2007:67) noted other essential equipment during early freeze-up:

> Also, they told them to start bringing their snowshoes with them when *cikullaq* was forming. If *cikullaq* happened to obstruct his path to shore, those snowshoes would prevent him from falling through when he put them on and towed his kayak on top of the *cikullaq*.
>
> And since they were never without sealskin floats, they used their floats to help them when falling through [*cikullaq*]. There was a small loop along the front [of the float]. One would insert his foot through [the loop] and use his float if he fell through the ice.

Following the onset of cold weather, *cikullaq* steadily built up along the shore. Peter Dull (March 2007:574) explained: "When it is cold and the tide comes in, the ice comes up on the mudflats and stacks up. They get thick by stacking on top of each other. Even though they are thin, they multiply. When it is freezing, one must not pretend to be fearless. Hunting for seals toward winter is more daunting than hunting in spring." Peter John (March 2007:1208) learned this teaching through experience: "In fall when the water tended to freeze, it was an admonishment not to continue to venture from shore since ice floats up to the surface along muddy areas and gets thicker. One day we went down toward the water, and when the tide started to come in, we went toward shore. As we went, the ice got thicker and thicker. We started to travel by swaying our boat from side to side. Indeed, I understood how true that teaching is." *Cikullaq* does not pose the same risks to contemporary hunters using aluminum skiffs. Paul Jenkins (March 2007:254) concluded: "These days, there are no obstacles for those younger than us. They are starting to travel through *cikullaq*."

Paul John (December 2007:68) explained how *cikullaq* led to the formation of *tuaq* (shore-fast ice) in fall: "*Cikullaq* freezes when it's cold out, and the ice gets thick, and starting along the shore, the *tuaq* gradually forms and extends out toward the ocean. When [the shore ice] reached the area where it usually ended, it stopped extending. It seemed that [shore ice] couldn't extend past the area where the ocean gets deep, what they call the *iginiq* [edge of deep water]." John Eric (March 2007:251) said: "People mentioned that the ice that froze during fall was good [solid] ice, and they called it *tuaq* [shore ice]. *Tuaq* will stay until the ocean swells arrive in spring."

Mark Tom (March 2007:1208) observed that "when *cikullaq* continually layers and piles, that ice gets thick right away." John Eric (December 2007:77) called this layering process *qasmegulluni*. *Cikullaq* also thickens when it breaks apart and refreezes. John Eric (March 2008:348) described *cikullallret* (lit., "former *cikullaq*"): "These [pieces of ice] that the wind broke to pieces are *cikullallret*. And if the *cikullallret* freeze again, they will become solid."

Elliqaun | Newly Frozen Ice along Shore-Fast Ice

As the shore-fast ice continued to extend, *elliqaun* (thin, newly frozen sheets of ice, lit., "something quickly put in place"; also *cikunerraq* or nilas in Western sea ice typologies) began to form along its edges. John Eric (March 2007:300, 251) explained:

> *Elliqaun* is smooth, new ice that freezes at night. It is attached to the shore ice and gradually becomes thin as it extends out over the water. But the *tuaq* that is behind it is covered with snow. . . .
>
> When it got cold after it had been warm, they called the ice that formed *elliqaun*. When *elliqaun* froze, the surface was always moist. And when snow covered it, the ice underneath didn't freeze but only became solid when it was extremely cold out. But they say that the *tuapiaq* [genuine shore ice] that froze during winter was always dry.

Paul Tunuchuk (March 2008:361) commented on the viscous quality of freezing water: "In fall when [ice] starts to solidify, it freezes fast. The *elliqaun* is like that. Because it's like Jell-O [and solidifies fast], although it's thin, a person can walk on top of it." John Alirkar (March 2008:545) also recalled walking on *elliqaun*: "When one strikes it [with an ice pick] and it doesn't make a hole [through the ice], one can walk on it. But one cannot walk on it if a hole forms [when struck]. Freshwater ice breaks easily, but when stepping on saltwater ice, one doesn't fall through."

John Phillip (December 2009:292) noted that the process of *elliqaun* adhering to shore ice, referred to as *elliqauciqerluku*, can be interrupted by currents as well as warming temperatures: "Although it doesn't melt, when the current is strong when the tide starts to come in, it breaks the *elliqaun* to pieces and removes it." *Elliqaun* can form in spring as well as fall when weather stays cold, and John had seen it extend over two miles out.

Paul John (December 2007:70) described the formation of *elliqaun*: "Back when it used to be cold, along shallow areas were *etgalqitat*, small ice sheets beached in shallow water. And when it gets cold, the newly frozen ice gets stuck along those *etgalqitat*; that [ice that forms] is called *elliqaun*. The [ice] that the *etgalqitat* prevented from moving is *elliqaun*." Unlike shore-fast ice, *elliqaun* is unstable and therefore dangerous. Paul continued: "*Elliqaun* is ice that formed during the night or over several days and sticks to the shore ice. *Elliqaun* is not as solid as real ice. And it can float away when the weather is warm and there are wet conditions caused by melting."

Although not as safe as shore ice, *elliqaun* is denser than freshwater ice. Paul John (December 2007:130) explained:

> Our elders mentioned that since river ice is fresh water, it breaks easily. But they said that since the ocean is saltwater, its ice is much denser. Even though it shifts when a person walks on it, some areas [along the ice] don't collapse right away.
>
> Once I became terrified. Thinking that it had been cold enough, I reached *elliqaun* and traveled on, back when we started to use Ski-doos [snowmobiles]. Then I noticed that the ice was moving as I was traveling, so I increased my speed and reached shore. I was amazed that I hadn't fallen through, having traveled with the ice moving beneath me. If it had been river ice, I would have fallen through and drowned. Since it was saltwater [ice], I didn't fall through.

John Eric (March 2008:301) distinguished between *elliqaun*, which is attached to the shore ice, and *cikullara'ar*, thin ice forming in open water—a distinction that Western typologies do not make, referring to both as nilas:

> Sometimes when it is cold and calm, there is a large [ice sheet] that is thin, and some areas have water. We call that *cikullara'ar*. Although it's windy, large waves cannot form. When it's cold, when it's five above or zero degrees, ice like that forms all night down on the ocean.
>
> Although they are similar, *cikullara'ar* is more extensive than *elliqaun*. *Elliqaun* is always attached to the shore ice, but *cikullara'ar* is located down below the shore ice and extends toward the ocean.

Manialkuut Manigat-llu | Rough Ice and Smooth Ice

The surface of the shore-fast ice could be more or less rough or smooth. John Alirkar (March 2008:548) explained: "The current breaks up pieces of ice, causing them to be rough and jagged. They become packed together, and then they freeze in place. [Ice] that the current gathers around the channels is like that." Jagged pieces of ice pushed on shore by the high tide formed *manialkuut* (rough ice). Paul John (March 2008:550) noted that the location of jagged ice depends on conditions during freeze-up: "It's different all the time. When the wind freezes [ice] after it's windy, the ice isn't smooth. The broken pieces of ice gather together [and freeze]. Waves made by the wind cause that to happen." Although rough ice has always been present, it is becoming more prevalent, as fall storms have increased.

John Alirkar (March 2008:550) noted the importance of smooth ice for traveling: "When traveling, they follow *manigat* [smooth ice] although they are constantly turning, traveling all over the place." When the trail was rough, men worked to smooth it. Paul John (March 2008:550) explained: "We make a trail, smoothing areas, chopping the worst areas with ice picks. Sometimes when they hunted by kayak, when they went down they'd say that the ice was too jagged after it had been cold. Those who had gone seal hunting would return home, only going down after fixing a trail." John Eric (March 2008:330) recalled: "Sometimes the water down there is completely calm, but when the ice is rough, although we want to go down to the water, when we can't travel through the trail, we don't go."

Travel is impossible over rough ice until snow fills in the low areas. John Eric (December 2007:74) noted: "Eventually, in December and January, the snow fills the rough areas. . . . Boats couldn't go down when [the ice] was too rough, so they'd wait for snowfall. Only the snow smoothes [the rough ice]." Conversely, Paul Tunuchuk (March 2008:331) warned that water-soaked snow can make a smooth trail over shore-fast ice impassable: "Sometimes really smooth ice cannot be traveled on. The snow is soaked with water. They call that *mecqiitaq,* and that sometimes ruins the trail." John Eric (March 2008:332) agreed, adding that *mecqiitaq* is also hazardous when covering *elliqaun:*

> The surface of *elliqaun* is usually very smooth. And since that *elliqaun* isn't like shore ice that froze during winter, when it is covered with snow, water fills [the *elliqaun*] from underneath for a great distance. Although the surface is good, the bottom is bad.
>
> When it's below zero, [the *elliqaun*] is good and solid, but when it's warm, it immediately melts [the ice] underneath the snow. It's not good to walk on. One's shoes get heavy as the snow tends to stick to them like mud. That's how I've always viewed *mecqiitaq.*

John Eric (March 2008:331) noted that travel was also difficult if the trail was too slick: "Before there is snow around, the ice is extremely smooth. When there is no snow, a snowmobile cannot go fast, as its [tracks] have nothing to grip. But when there's snow, it's good for traveling. That's why we wait for the snow more than anything else [to go down to the ocean]."

Nepucuqiq | Rough Edge of the Shore-Fast Ice

John Eric (December 2007:73) described the gradual expansion of the shore-fast ice: "When ice forms and becomes thick, open water gradually moves down [farther from shore]. When the tide comes up in cold weather, the ice packs and sticks [to existing ice]. [The ice] forms rough and smooth, but the ice is mainly rough inside the channel since the tide comes in for six hours. It packs up like ice that accumulates in one place, and ice is placed haphazardly and then freezes together inside the channel. Then [ice] stays there when it gets cold although the tide goes out." Wind and tides, however, could break up the edge of the shore-fast ice, creating a rough, impassable ice edge known as *nepucuqiq* (from *nepute-*, "to stick on to something").

David Jimmie (March 2008:297) described the formation of *nepucuqiq*, which did not occur every year: "They say [*nepucuqiq*] was placed there by *qacaqneq* [a south (onshore) wind]. The wind turns and then blows toward shore. [Ice] that forms in that way is solid." John Eric (December 2007:74, 76) explained:

> Sometimes when the wind is blowing from the south directly against the shore, when there is a lot of ice, the large ice floes head toward shore and pack up; I used to hear them called *nepucuqit*. When [ice blown against shore] becomes thick, although [ice floes] are about three to four feet thick, the south wind breaks them to pieces and packs them.
>
> Evidently, *nepucuqit* were created by wind. They were extremely rough and jagged, and a person moved on top of them like a dog [on all fours] and couldn't walk on them. When the tide comes in down on the ocean and it's windy at the same time, it packs up a lot of ice, extending a mile in any direction. And the shore ice that is usually smooth, although it is thick, [the incoming tide and wind] can break it to pieces. . . .
>
> One time the shore ice was around three feet thick. Just as the tide started coming in, the wind direction changed and it started to blow. As the water level went up, the ice that had seemed impossible to move started folding toward the south and north, breaking to pieces. And here we thought that nothing could [break the shore ice].

John Phillip (December 2009:307) remembered using *nepucuqiq* as cover when seal hunting along the ice edge. John Eric (December 2007:75) noted that *nepucuqiq* could be so steep it could not be broken by ocean swells: "Back when Aassanaaq was alive, he said that *nepucuqiq* formed that was very steep. And he said that before it melted, [people] came up [to the Kuskokwim River area] in June. Steep ice had packed up [against the shore] at that time. And he said that ocean swells headed toward shore couldn't break up that *nepucuqiq*."

Tuam Cimillra | Changes in Shore-Fast Ice

Freeze-up today occurs later than in the past, and the shore-fast ice does not form as far out. Nelson Island elders provided detail on what coastal residents generally observe. Phillip Moses (January 2007:494) noted: "Right now it takes so long to freeze up, and [the shore-fast ice] stops right there by Qikertaugaq. In the past, it always froze nice and smooth up to Ulurruk. The ice extended when it kept freezing, and when it got to Ulurruk, it stayed there for quite a while." Paul John (December 2007:252) agreed: "When I came to observe it, [the edge of the shore ice] was never too far from Ulurruk. The ice reached that area and was solid like lake ice." One year, when men still hunted using kayaks, shore ice extended all the way to Up'nerkillermiut, and all of Toksook Bay was frozen. Camilius Tulik (March 2007:540) recalled: "That was when the ice really extended out. Now it has come way into the bay."

Michael John (March 2007:1038) observed the same loss of shore ice north of Nelson Island: "Sometimes the shore ice isn't very extensive. And it hardly forms extensively below the village of Tununak because there are more sandbars nowadays." Peter John (March 2007:1158) recalled: "When [the shore ice] froze far from shore, past these [ice piles named Qikertarraat (lit., "few or small islands")], it was pretty far down. I went seal hunting once when [the ice edge] extended that far out. After that it gradually receded toward shore."

Many associate lack of shore ice with warming temperature. According to Paul John (December 2007:72): "Since it no longer gets extremely cold today, the shore ice that forms is no longer extensive; it has changed. Back when it used to be [extremely] cold, there was ice [on the ocean] and those *evunret* [piled ice] formed." Paul (December 2007:68) noted that shore ice was thicker in the past: "Back when it was cold, the shore ice was solid and stable; it formed thick ice.

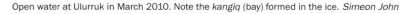

Open water at Ulurruk in March 2010. Note the *kangiq* (bay) formed in the ice. *Simeon John*

And the ice sheets that floated [along the ocean] were thick." Peter John (March 2007:1158) added:

> In those days [in the 1950s], shore ice froze about three to four feet thick. When we were on top of [the shore ice], when the tide was coming in and the current caused a large piece of ice to bump into [the shore ice], the [piece of ice] would break to pieces although it was thick. When it hit, the sheet of ice would move under the shore ice although it was large and thick.
>
> These days, these areas have become shallow, now that the current isn't as strong. The shore ice has gotten thinner, and sometimes it's about one foot thick. This year it's thin, and the water is close to shore since the weather was constantly warm last fall.

Shore ice also does not stay as long in spring. Edward Hooper (March 2007:1306) remembered: "There used to be shore ice down there along the edge of the channel, almost three miles [long], when we went seal hunting. These days, one can't go and spend a night [on the ice] thinking that the ice may detach and float away while they're on it. We used to spend nights down below the shore, right behind the water." Susie Angaiak (March 2007:1306) shared her experience:

> If you happen to spend a night, you will end up calling a helicopter. [*laughter*] When I became aware of life, the shore ice extended far out to the ocean and stayed for a long time. They'd tow [their kayaks] for a long time to go seal hunting [to reach the ice edge]. And how pitiful these days, the shore ice will be extensive, and then the next day one will check and see that it is completely dark [from open water].
>
> Once when that person was fishing down there, I went to him and saw that he had caught lots. I suddenly had a yearning to go fishing. So I made a fishing hole, planning to fish the next day, and I was eager. Evening came. When I went to check the ocean the next day, it was very dark [from open water]. My poor fishing hole had already drifted far away. [*laughter*]

Qanisqineq | Snow in Water

Both fall and spring hunters encountered *qanisqineq* (from *qanuk*, "snow"), snow that gathers on the ocean, also referred to as slushy snow or slush ice. John Eric (March 2008:302) explained: "Sometimes, before the shore ice goes away, a large amount of snow accumulates down on the ocean, when it continually snows and cannot melt. Although it appears thin, the snow is thick, as the current [gathers] and piles it up. When the tide goes out, it takes [the snow] away but doesn't let it scatter. It accumulates deep in the water, and most of the area is covered by *qanisqineq*." John Phillip (December 2009:298) once saw a seal surface in thick *qanisqineq* where it hung without sinking after it was shot.

Qanisqineq, or snow that gathers on the ocean as a viscous mass, obstructing one's path and potentially dangerous, January 2009. *Nick Therchik Jr.*

Qanisqineq can obstruct a hunter's path and is potentially dangerous. Men were warned not to travel during an incoming tide when *qanisqineq* was present, as the tide would gather it and block their path to shore. Travel through *qanisqineq* gathered along current lines was also difficult. Simeon Agnus (December 2009:301) explained: "*Qanisqineq* becomes like old *elliqaun* when two currents gather it in one spot; it quickly turns white and thickens when two currents meet. Some are extremely thick. Some boats cannot go through them." Simeon Agnus (December 2007:123) also recalled the following admonishment: "They say when *qanisqineq* suddenly surfaces on top of sandbars during spring, some is thick. When *qanisqineq* accumulates, a small outboard motor cannot travel through it, as it won't be able to take in water. [*Qanisqineq*] is dangerous when it is thick and white, and it's better if a person doesn't try to travel through it." John Eric (March 2007:263) described what to do if caught in it: "When *qanisqinaq* accumulated in one place, among them were other pieces of ice. If a boat traveled fast, its motor would get caught. A person driving would sway from side to side as they moved forward, since it's difficult to see [ice chunks] when *qanisqineq* covers them."

Qanisqineq could also be helpful. Paul John (December 2007:131) recalled how in the past kayak hunters used it as a source of fresh water: "When they were [traveling] when no ice was around to melt for drinking water, they were told to gather that *qanisqineq* and mash it with their hands to squeeze out the water and then place it on top of the kayak sled. They say that the salt leaches out, and after melting it, it is drinkable, although it tastes salty."

Pugteqrutet | Those That Suddenly Surface

Other sea ice forms important to recognize were *pugteqrutet*, pieces of ice stuck to sandbars that suddenly surface, also known as *tumarneret* or *tumarngalriit* (lit., "ones that are assembled"). John Eric (March 2008:345) explained: "I only hear about those [pieces of ice] that suddenly came to the surface in the fall; they are pieces of ice that froze on top of the mud, and when the tide comes up, they are then called *pugteqrutet*. Since the surface of the mud is mixed with water, it freezes it on the mud. But [the ice] located inside the river floats down toward the ocean, and then the area on top of the sandbar freezes. Then when the tide comes up, the tide keeps it afloat. They were afraid of that when they used kayaks, and they'd tell them not to go near them with kayaks." John (December 2007:73) concluded: "Since there is a lot of mud around, and it formed smooth ice, that *pugteqrun* that formed during fall was dangerous. They say when the tide goes out, that *pugteqrun* drifts away from shore and it can float a boat out to the ocean." Michael John (June 2008:192) shared a sobering example: "Mark Tom's older brother docked his boat along a sheet of ice, and he was carrying his snowmobile inside it. Right after he got out, a large, round piece of ice surfaced along the edge of the ice and hit his boat. It capsized his snowmobile and his boat, and they sunk. Still today, very large pieces of ice suddenly surface from underwater."

Uksumi Ciku: Evunret | Ice in Winter: Piled Ice

Huge piles of ice began to develop in fall in shallow areas along the coast, including channel edges and sandbars. Paul John (December 2007:64) explained: "*Evunret* only grow in shallow areas. In cold weather, when the current and wind moves the ice, when it lands along shallow water and isn't able to move, the ice breaks to pieces and piles, and from that time on, they are referred to as *evunret* [lit., "those that are piled"]."

John Alirkar (March 2008:534) described their formation: "These [*evunret*] form on top of sandbars, along the edge of deep water. *Evunret* form when the tide is extremely low and thick *qanisqineq* floats, and then the [*qanisqineq*] stops and gets stuck there. And when the area underneath [the *qanisqineq*] freezes, [ice] from the surrounding area continually gathers there, and eventually they become large." John Eric (March 2008:171) noted: "The ocean forms *evunret* by piling pieces of ice on top of one another and breaking them to pieces, and then the cold weather welds them and freezes them in place. The water forms [*evunret*] during winter. That's why there is an oral teaching that no one should underestimate the ocean at all." Paul Tunuchuk (March 2008:172) emphasized the ocean's power: "When the ice moves, it's like a bulldozer, as the wind and current move it. That's how ice piles down on the ocean, making some areas extremely steep."

John Eric (December 2009:306) added: "Huge *evunret* can form in twenty minutes or less, since the ocean is so strong. A very large ice floe can break apart and form an *evuneq* that becomes high right away."

Once frozen in place, *evunret* were stable. According to Peter John (March 2007:1160), "Those large *evunret* were strong; they didn't move and didn't break although other sheets of ice came upon them." Phillip Moses (January 2007:487) described an ice floe colliding with *evunret*: "I was one of those who was fleeing by kayak when [an ice floe] was approaching. When it got to the *evunret*, [pieces of ice] went up and broke off, even though they were thick. And the ice floe just kept going uninterrupted and never stopped. Oh, how strong the current is absolutely! And those *evunret* never break off, but more weight will be added to them." Paul Tunuchuk (March 2008:174) noted that *evunret* formed in cold weather are safe: "They had an admonishment about those steep [*evunret*]. They said the ones formed during winter are stable and aren't cause for worry, as the cold weather welds them together in place. But they said that if that [*evunret*] formed after the weather got warmer, a person should not stop along its sheltered side when in a desperate situation. They say those easily break to pieces."

Paul John (December 2007:64) described dark-colored *evunret* mixed with sand: "Those *evunret* that have sand mixed in were called *asvailnguut* [lit., "ones that are solid and immovable"]." Sediment-laden piled ice was also known as *marayilugneret* (from *marayaq*, "mud") and *tungussiqatiit* (lit., "ones that are dark"). Paul (March 2008:537) noted that sandy *evunret* are sturdier than those formed in deeper water: "Since the area around those is shallow, when the wind blew the waves and sand, the splashing caused them to become sandy and dark and they became *marayilugneret*. They considered the ones with sand in them stable and strong, and the sand prevented them from breaking to pieces right away; [waves] splashing on them in cold weather cause them to stay solid." Paul (December 2009:204) later added: "Although the tide comes in, those *tungussiqatiit evunret* cannot float to the surface since their bottoms are frozen in place. Since they cannot float, the current's murky water starts to cover them. Those layered with sediment are really welded in place and frozen to the bottom." Not only are *marayilugneret* immobile; they melt in place and are among the last ice formations to disappear.

John Alirkar (January 2007:487) noted that *evunret* that are formed at the edge of deep water are light in color: "Because the ice floes that come from down [in the ocean] get on top of them, those *evunret* are always white." Paul John added that *qatqitiit* (white-colored *evunret*) are merely piled ice and are more dangerous and break more easily than sandy *evunret*. John Phillip (December 2009:200) explained: "They said those [*evunret*] that piled much later, after winter came, are white. They said those don't stick [to the bottom]; those aren't solid and sturdy. But those [sediment-laden] *evunret* grow first, and they hold our shore-fast ice in place. *Evunret* are on all the sandbars along the channels down below our shore."

John Eric (December 2007:82) noted that *evunret* could be huge: "Once we came upon very large *evunret* during June. We climbed up, and when we looked down, our boat was small. I think we climbed up about fifty feet. Since the ocean is large, since the current is strong, it creates [ice formations] that cannot be broken to pieces."

Large *evunret* grew in predictable locations year after year. John Alirkar (March 2008:535) affirmed: "[*Evunret*] always formed along the *iginiq* [edge of deep water], the area where the ocean bottom suddenly gets deep. They always formed in their usual places." Times have changed. According to Simeon Agnus (March 2007:539): "In the past [*evunret*] formed in their usual places, but now they have moved. They are going toward the points, and those points are extending farther out in the ocean, and they are scattering." Reasons for this change include warming weather as well as sandbars forming in shallow areas at the mouths of bays and rivers.

Evunret that grew in predicable locations every year were given names, often in the plural, as they were not considered singular monoliths but groups of ice pieces. John Alirkar (January 2007:487) named *evunret* south of Cape Vancouver: "Those that were visible from [Toksook Bay] were called Ullagciigalnguut. And the one farther down was called Kangirpiim Isqurra." Tununak and Newtok elders also named several, including Nuakiliit and Qikertarraat (lit., "few or small islands"). Large *evunret* called Kavialget (lit., "ones with foxes") form to the east of Nunivak. John Eric (March 2008:338) explained that many names were based on use and appearance: "In the past, when they came upon those *evunret*, if a person had seen an animal around or if a lot of ice surrounded it, they gave them those names. If the ice was packed there and there was no water around, they refer to those *evunret* as Ullagciigalnguut [lit., "ones that can't be reached"]."

Visible from far away, *evunret* are navigation aids. David Jimmie (March 2008:327) explained: "They were all different. Some were called *qatellriit* [white ones], others *tungussiqatiit* [dark ones]. Back when they didn't have compasses, they used them to determine where they were. They would tell us to observe the *evunret* closely and try to distinguish their features." Paul Tunuchuk (March 2007:304) concluded: "In the past, *evunret* served as markers for us. Although they traveled beyond them, they knew where they were located by means of these *evunret*."

Evunret were also used as lookout points. According to Paul John (March 2008:536): "They also climbed on top of *evunret* to scan their surroundings. Since crevices are covered by snow, because they may fall into the crevice by accident, they were told to always hold the *negcik* [gaff] when they climbed *evunret*." David Jimmie (March 2008:327) described how hunters used *evunret* to search for a trail to open water: "They would climb up and look at their surroundings, searching for a route to take." John Eric added that this is still true: "Those *evunret*, when we go down to the ocean and search for open water, since those steep areas are located along the shore ice, we climb on top of them to search for a trail. A snowmobile cannot just take a straight path, but they have to scan the surrounding area. That's what we do nowadays when we go down." Peter Dull (March 2007:561) said simply: "When people used to seal hunt, they went on top of *evunret* and looked around to see open water."

Evunret were also places where hunters sought safety. Paul John (March 2008:535) explained: "*Evunret* were used for safety when ice floes suddenly prevented them from traveling. When there was a lot of ice around, they'd go along their side sheltered from the current to wait until they had a way to go. And when it suddenly became windy and there was no ice around, they would go along their leeward side to wait for the wind to calm down, when it was impossible to paddle." Paul Tunuchuk (March 2007:280) added a caution: "Although they gave us that advice, they admonished us not to go on top of newly formed *evunret*. They said that although they look safe, they can break apart. However, they said *evunret* formed by winter weather are safe."

Nacarat | Piled Ice on Sandbars

Piled ice on sandbars surrounded by shore-fast ice was known as *nacarat* (lit., "places to look around from a high vantage point"). John Alirkar (March 2008:535) explained: "If there is ice surrounding [piled ice] beached along a shallow area, they call that a *nacaraq*. *Qayivigluteng* [They are places where a kayak can dock], as they say, and they have low places where one can climb on to the ice. If there isn't [ice around them], one cannot climb on top of them since they are steep. They are extremely large." Jobe Abraham (March 2008:175) noted: "These *nacarat* are different from *evunret*. *Evunret* are piled ice surrounded by water. But the ones they call *nacarat* have shore ice between them, and they are situated along shallow areas." John Eric (March 2008:173, 175) described their formation:

> Down there, the place that usually forms a *nacaraq* always has [a *nacaraq*] since it is shallow. If a large piece of ice beaches in shallow water, and *cikullaq* [newly frozen ice] gets stuck there, it grows, and eventually it gets very large. Then it comes to be called a *nacaraq*. It is on top of a sandbar. *Cikullaq* continues to stick to it in the cold weather, and it expands.
>
> And those *nacarat* form shore ice during this time now [March], as smooth ice forms in the area around them. And when we come upon them, we recognize them.

Because *nacarat* form along sandbars in predictable places year after year, like *evunret* some are named. For example, the Akiliit are *nacarat* that grow along the north side of the mouth the Qalvinraaq River. Some report that they are larger today: "At this time, those Akiliit are growing larger, as well as these other *nacarat*. In the past during March, shore-fast ice was around the *nacarat*. When the weather stays cold during winter, since there are many sandbars, many *nacarat* form" (John Eric, March 2007:303). John (January 2007:99) described climbing a large *nacaraq* to view his surroundings: "Once when we came from over there by boat, we started to see a mountain down near Cingigyaq, seemingly beyond it. When we came upon it, it was a very large ice pile. We just brought up our anchor and climbed up, using all fours, since we couldn't walk up. We started to look down at Nelson Island because it was so steep."

Etgalqitat | Ice Beached in Shallow Areas

Smaller piles of ice beached in shallow areas close to shore were known as *etgalqitat* (lit., "ones that reached a shallow spot"). They differed from *evunret* and *nacarat* in size, location, and stability. David Jimmie (March 2008:174) explained: "We call ones that are closest to shore *etgalqitat*, and the ones down below them are *nacarat* or *evunret*." Paul Tunuchuk (March 2008:339) added: "There are *etgalqitat* and *etgalqitayagaat* [lit., "small *etgalqitat*"]. When the tide comes in, they will drift, and when the tide goes out, they will beach along shallow areas. They are not large, but some are extremely steep."

John Eric (March 2008:339) noted: "These *etgalqitat* are located right beyond the shore ice. Some are floating, and some aren't. They are [ice piles] located around sandbars behind the *nacarat*. They were once part of the shore ice, but when their surrounding ice drifts away, they start to be called *etgalqitat*." Paul John (March 2008:552) described conditions that create *etgalqitat*:

> Some *etgalqitat* are [ice pieces] that have beached in shallow water. And when the tide goes out while the thick [ice chunks] are in shallow water, the cold weather [freezes them] and keeps them in place. But the surrounding ice rises when the tide goes up, and when the tide goes out, [the surrounding ice] goes down. The type of [ice] that is beached along shallow water and doesn't float to the surface is called *etgalqitat*.

John Eric (December 2009:233) attested to the safety of *etgalqitat* formed in cold weather: "They call those *etgalqitat uksullaat* [ice formed during winter that is beached in shallow water]. They are down in our river, sometimes on top of a sandbar. They say during cold weather, ice continually accumulates there and adds to them; they call those *pengegnailnguut* [lit., "ones that are not worrisome"]. Although ice collides with them, they stay and don't break."

John Phillip (December 2009:313) recalled the instruction to avoid the side of *etgalqitat* facing the current: "When [ice] starts melting, the current eats away the sides of *etgalqitat*. They say that sometimes that area suddenly cracks, and they fall toward the current. That's why when we used to paddle, they admonished us not to get close to those *etgalqitat* in spring."

Like *evunret*, *etgalqitat* formed in their usual places year after year. Unlike *evunret*, Paul John (March 2008:535) said that they are not named. Paul added that *etgalqitat* are more plentiful nowadays and are also forming in new places. According to Phillip Moses (January 2007:487): "Across from us is a straight line of *etgalqitat*. They never used to appear inside Kangirrluar [Toksook Bay] but now they do because it has become shallow. Even when one goes quite a ways from shore at low tide, one can encounter *etgalqitat* at the present time."

Culugcinret Ircaquruat-llu | Erect, Pointed Ice and Heart-Shaped Ice

Ice frozen in different shapes had distinctive names. A *culugcineq* was a pointed piece of ice, named for its resemblance to a *culugaq,* or fish's dorsal fin. Paul John (March 2008:538) explained: "When ice piles up, when a piece of ice is erect [lit., "has an erection"] and doesn't break and freezes in place on top of other pieces of ice, they call it *culugcineq*." John Eric (March 2008:320) added: "That type of ice looks distinctive at its tip. Those *culugcinret* are on top of an *angenqaq* [large ice floe]. They are not large, and although they are upright they are slanted. They are attached to other ice around them and stick up in the air, and they won't break off until they melt."

John Eric (March 2008:340) also described heart-shaped pieces of ice known as *ircaquruat* or *ircaquruarpiit* (lit., "large, pretend *ircaquq* [heart]"): "Sometimes these ice formations that water placed somewhere appear different. I used to hear that there is an *ircaquruaq* over there. You know how hearts are shaped nicely; if a piece of ice was sticking upright and looked similar to [a heart], then a person can call it that, based on what he saw."

Qulinret Aayuqat-llu | Crevices and Cracks

Elders admonished hunters to always carry a *negcik* (gaff) to prevent them from falling into *qulinret* (cracks or crevices in the ice, from *qulig-*, "to crack). These *qulinret* could also be helpful. David Jimmie (March 2008:310) recalled that, like *evunret*, *qulinret* formed in predictable locations: "In the past, we were told that the area below Ingriik tends to crack, probably when the tide comes in or when it goes out. We were instructed when we first went seal hunting that the ice below Ingriik tends to split and that leads open that we could travel through."

Large crevices in the ice, extending a long distance, were known as *aayuqat*. Nick Therchik Jr. (March 2008:557) of Toksook Bay explained: "*Aayuqat* are very large crevices that extend a great distance. There used to be some *aayuqat* here and there down the coast from our village, inside Kangirrluar. There was an *aayuqaq* down below Umkumiut last year." John Alirkar warned: "It's dangerous to suddenly come upon those. The snow conceals them. Although [the ice] is thick, it has a crack." Women also knew their whereabouts. Lizzie Chimiugak (January 2007:427) recalled: "I know what *aayuqat* are since they revealed their location in the past in case people encountered mishaps along them."

John Eric (December 2009:304) described how a rough ice edge promoted the creation of *aayuqat* in the past: "Our elders used to say that only the *nepucuqiq* [rough edge of the shore-fast ice] forms *aayuqat*. *Nepucuqiq* is rough ice that the wind piled [along the shore ice], mixed with snow. They said that only when *nepucuqiq* is present, those *aayuqat* open. But they said that the ice from the winter doesn't have *aayuqat*. Indeed, when I started to go down after *nepucuqiq* had formed, there would be *aayuqat*, since they weren't frozen properly, as the pieces of ice were situated haphazardly."

Qiugaar | Reflection of Open Water

Before setting out on the ice, coastal hunters searched for *qiugaar* (*qiu-*, "to become blue"), the reflection of open water known as "water sky," seen in the sky as a line of darkened clouds. Although ice-free water might be miles away, its reflection could guide hunters toward it. The opposite was also true. Paul Tunuchuk (March 2008:322) said that a white line in the sky indicated distant ice in the open ocean, what sea ice scientists call an "ice blink": "Sometimes [when going down to the ocean] there is no ice around at all. Then you will see a white mark up in the sky in the distance. It will be obvious out there and will be small and white, or sometimes it will be long. Then you will travel toward it and see that it is ice. When you search for ice when the sky is blue, you can see that ice."

Simeon Agnus (July 2007:287) shared a well-known story of two hunters from enemy villages who met when searching for the reflection of open water: "When they were experiencing starvation, that person who searched the sky for the blue reflection of open water evidently came from Hooper Bay. Somewhere down near Kavialget was the only [open water]. The entire ocean froze, but that reflection of open water was probably located down near Kavialget. That place had a blue reflection in the sky all winter." As the two men approached from opposite directions, they saw a bearded seal perched on the ice by open water. Together they harpooned the seal and butchered it. The man from Hooper Bay then helped the man from the north regain his strength, after which they divided their catch equally and returned home.

They also told us that after winter, when spring comes, the ocean doesn't stay in its usual state. For that reason, although someone says that they have learned the ocean, they will not learn to predict its conditions.

— Paul Tunuchuk, Chefornak

Peter John (March 2007:1162) soberly recalled that young men were carefully instructed as spring approached: "They would have started to talk about the ocean at this time [March], giving instructions about what to do when they began to travel during spring. Teachings for fall and spring weren't the same." As sunlight returned and days lengthened to over twelve hours by mid-March, snow melted and conditions on the sea ice rapidly changed. John Eric (March 2007:265) recalled: "During the time when these bearded-seal and spotted-seal pups are born, the names [of ice formations] change. And some become rounded with a high periphery and deep center. I call them *avatait qertulriit* [ones with a high periphery]. The water gathers pieces of ice in a circle, and they freeze together. If a boat suddenly goes up on one of those, it will [hang over the edge]."

With returning sunlight, warm weather begins to melt once-solid shore ice. John Eric (March 2008:181) explained how in April the places where the channels are located will form *kangiqutat* (small bays): "When the sun gets closer, it heats the shore ice, and it rots and becomes thin. Then [river mouths] become [concave]. The current also melts the ice down below my village during spring. When the sun gets warm, the [ice] inside the channel rots." Paul Tunuchuk (March 2008:182) warned against entering river mouths and *kangiqiugneret* (coves in the ice) during spring: "If there is a bay in the ice ahead and you unknowingly con-

Spring ice: breakup on the shores of Toksook Bay, May 2008. *Nick Therchik Jr.*

tinue to travel and hunt along the ice edge and enter far into that bay, when you look back you'll see that your way out has already become obstructed, and the ice has already enclosed you, *cikum qumikeqerluten* as they say. You will stay in the middle of ice, and you won't be able to go anywhere. People are told to pay attention while [hunting] inside bays. They say they quickly close in."

Paul John (March 2008:552) also recalled the admonishment to pay attention to the *qapuut* (foam) that forms along ice floes in spring: "*Qapuut* start to appear through small holes. When it melts and holes appear, *qapuut* become visible." John Eric (December 2007:110) also noted that *qapuut* are a sign of melting ice: "When *qapuut* start to be seen, although [the ice] is white, it becomes thin and dangerous as the current melts the ice from underneath. When foam starts to form, the shore ice rots quickly." Indeed, along the Bering Sea coast where ocean currents can bring in warm water, rapid ice melt often occurs from the bottom rather than from the top, aided by snow cover acting as a thermal insulator.[4] John (March 2008:333) recalled the admonishment to avoid cracks where foam is visible:

> It seems that when the sun starts to get warm, *qulinret* [cracks] start to form [on the ice]. It is said they cause the shore ice to melt faster. Although the area where the crack formed is small, the area around it melts. The foam melts the ice underneath. That's why my grandfather told me not to step on these *qulinret*, that my leg would fall through, but to step over them. When the amount of foam increases, when the sun gets closer, they continue to melt and don't stop. The current underneath the ice probably builds up the *qapuut*. *Qapuut* are seen along cracks on the surface of the ice.

Paul Tunuchuk (March 2008:334) reiterated John's warning:

> If you see foam, that area is dangerous to walk on. Ice along the middle of rivers forms holes. But they tell us to travel only along the *iginit* [edges of deep water].
> When it's not dangerous, they can travel inside the river channel. But when holes start to form, some rivers become dangerous to travel on. Starting from foam, holes form [on the ice]. If you see foam and happen to travel through it, you will fall through.

John Eric (March 2008:320, 333) agreed with Paul that traveling on the sides of rivers in spring was safer than following the channel:

> *Iginit* [edges of deep water] are covered by reddish ice along the shores of rivers. [*Iginit*] melt later than the deep river channel.
> When the tide comes in, the sides are mud but its channel goes down. The [ice] on top of the mud is rougher since it's shallow, but the deep area that fills with water is less rough. . . .

Also, they said that we should always take the trail along the side of the river, along the *iginiq*, to go down [to the ocean]. They say that the side of the river is a good trail, even when the weather has gotten warm. Although [the channel] has water, the [*iginiq*] has no water since it's slanted. We always take this route when our [winter] trail is no longer suitable for travel. We go only along the side of the river. [The channel] is *mecqiitaq* [water-soaked snow], but its *iginiq* is good.

John Phillip (December 2009:264) noted that the same warning applied to bays in the Canineq area: "[The ice] along the bay in the area around my village is smooth. But we go down to the bay along the edge [of the smooth ice]. We are admonished not to go down the center [of the channel] when conditions become dangerous; the current thins [the ice] at the channel's outlet far up toward shore. And foam appears, revealing a dangerous area, and holes form around them."

John Eric (December 2007:113) noted that snow might hide dangerous ice: "Dangerous areas [along ice] are hard to distinguish because it's white. But when some areas are a little dark, you can tell [that it's dangerous]." John (March 2008:343) then described falling through thin shore ice covered by snow:

Once I was alone on top of shore ice. That ice was actually old ice that had formed during winter. When I wanted to have tea, I went up to the shore ice, and the area in front of the boat looked [solid]. I took the kettle and used its lid to scrape a small amount of snow so that I could boil water, and when I stood on top of the shore ice, I fell through up to my waist. The ice was evidently thin over an air bubble that had formed [in the ice].

I'm lucky that it wasn't extremely cold out. I talk about that experience. Although [the ice] was extremely smooth, since that type of ice is dense, they say the area where an air bubble formed couldn't form thick ice, and water was inside it. Those are seen down on the ocean sometimes and also on rivers.

Paul Tunuchuk (March 2007:281) described how rapid melting could take place behind the *nepucuqiq,* the thick, rough edge of the shore-fast ice.

Some time ago, when the shore formed *nepucuqiq,* when I was bringing [a seal] to shore in the evening, I started to see pools of water that had melted just as I was about to reach the boats.

The next morning when I went outside, I saw that along the area that I had walked through there was open water. What I had thought were pools of melted water were apparently holes. When there's *nepucuqiq* along the shore, the area behind evidently melts rapidly first.

John Eric (March 2008:298) also described the effects of *nepucuqiq* on the melting process: "They said long ago that huge *nepucuqiq* was steep. And they said that although the rivers around this area broke up and were free of ice, that [*nepucuqiq*] was still there. And they said that they eventually went to the Kuskokwim to harvest salmon at summer fish camp [while the ice was still there]." The opposite can also occur. John continued: "They say that sometimes the ocean [ice] melts before the river ice. [The ocean ice] will drift away as one large ice floe by detaching from the shore."

The ocean could be a peaceful place during calm weather in spring. David Jimmie (March 2008:338) recalled: "I would get sleepy when James used to bring me with him, since he tended to get up and leave before sunrise. But down on the ocean, when it started to get warm, I'd get up on the ice beyond him and sleep and not hunt at all. [*laughs*]" David Jimmie was not alone, as Paul Tunuchuk (March 2007:258) recalled:

> They said those who lost sleep [when hunting] back when it didn't get windy all the time would go down to the ocean to sleep on top of floating ice. How nice it would be to sleep and wake up and immediately hunt.
>
> During those two years [when I hunted with kayaks], it was hard work. It is especially difficult to go without sleep. Although I tried not to sleep, when I'd go up on top of the ice and get warm, I'd fall asleep there. I'm not lying when I say that I used to sleep, but not for long, as that old man would tell people to hurry and get up.

John Phillip (December 2009:340) described finding freshwater ice on *evunret* in spring: "When we hunted by kayak, I was in awe. They would go and get ice from the top of *evunret* and after melting it, they'd have tea and it wasn't salty. When *evunret* sit for a while, the saltwater evidently seeps down and comes to be situated underneath [fresh water]." John Eric (March 2008:360) noted that ocean water was saltier in spring, as the saltwater ice—which freezes at a lower temperature—melts first: "Also, in spring, the water down on the coast isn't like it is during fall. Saltwater in spring is stronger than in the fall. Freshwater [ice] accumulates along the ocean, and there is more saltwater during spring, because the fresh water is still frozen; they say fresh water freezes first [and melts last]."

Qairvaat | Ocean Swells

Qairvaat can break [ice] to pieces from way out there toward the north for half a mile to one mile toward shore. They say qairvaat are what break the shore ice to pieces.

—John Eric, Chefornak

The occurrence of *qairvaat* (ocean swells; lit., "big *qairet* [waves]") marks the transition from relatively stable winter conditions along shore-fast ice to the myriad sea ice forms of spring. As described above, these long, high-amplitude waves originating in deep water far from shore can break the ice pack hundreds of miles away from the open ocean. Swells travel relatively fast and present a direct link to processes far away.[5] John Eric (December 2009:235; 2007:113) commented:

> Since there are many sandbars along our shore, *qairvaat* that reach shallow areas are constantly breaking. They don't break in deep areas. Only *qairvaat* that reach shallow areas are dangerous. . . .
>
> When *qairvaat* first arrive, conditions are bad, as they are powerful [when they hit]. When that starts to occur, the sun starts to heat the near-shore ice. Once in a great while the *qairvaat* break the ice and take it out to sea. [The shore ice] gradually recedes, and eventually open water gets closer to shore. That's how the ocean down below my village is during spring.

John (March 2008:298) wisely observed that current and wind are as important as heat in sea ice breakup.

> The ocean breaks the ice to pieces only during the time *qairvaat* are around. Then after [the ocean] breaks them into pieces, the sun melts the areas around the rivers, it rots the ice in those areas and forms holes there, and eventually the ice starts to recede toward shore. Bays are formed.
>
> The current also melts the ice from underneath. Although the ice is white, it will be extremely thin. That's why they tell us that a trail that we took in the morning can melt by the afternoon and be impossible to travel on.
>
> When *qairvaat* arrive and begin to flow, every incoming tide will break the ice to pieces, advancing toward the land. Eventually it has [no ice] to break. But the river [ice] also melts on its own and recedes, and some pieces [of river ice] drift away.

Ocean swells are a force to be reckoned with. John Eric (March 2008:296) emphasized their power:

> *Qairvaat* are large, and a sheet of ice that is a distance away disappears from sight when [the swell rises]. They say they come up toward land from the ocean and can rise over three feet. Because they are large and wide, no matter how thick an ice sheet is, [*qairvaat*] can break it. . . .
>
> They lift *nacarat*, and when they land too violently, they break to pieces. They also said that [*qairvaat*] are large when they first arrive, but the next day or day after, they are calmer.

Michael John (June 2008:209) contrasted waves formed by the wind and ocean swells, which make one feel dizzy: "These pieces of ice make an *eng* sound [when *qairvaat* are advancing toward land], and one will start to feel strange." John Eric (December 2007:75) noted that ocean swells come from far out to sea and can be hard to distinguish: "When these *qairvaat* are present down on the ocean, they go down slowly, and then they peak slowly. They flow very slowly. . . . [*Qairvaat*] aren't apparent. They say that *qairvaat* come to shore from somewhere in Japan. [*laughter*] *Qairvaat* are astonishing."

According to John Eric (March 2007:262), *qairvaat* can also smooth the shore-fast ice: "Back in those days when it was cold, the ice submerged in the water after ocean swells. They said that [the ocean swells] were forming ice for us when we went down to the ocean the following day. Some areas would be extremely smooth a great distance away, and it would be light in color when it formed a great distance from shore."

Simeon Agnus (December 2007:105) commented on the arrival of ocean swells: "They appear during spring when the shore ice starts to disappear. *Qairvaat* start to be around when fledglings are first seen down on the ocean." Paul John (December 2007:106) noted that ocean swells arrive when the ice that held them back begins to recede: "When ice that is floating on the ocean starts to melt, waves that are present far from the [shore] ice no longer have ice to hold them back, and they start to reach shore." John Eric (March 2008:176) associated the arrival of *qairvaat* with the birth of seal pups: "We also heard that when baby spotted seals and baby bearded seals are born, the ocean swells arrive at that time. . . . They say the motion of the swells puts them to sleep. They are like a babysitter."

Sometimes *qairvaat* arrive before their usual time. Many residents of lower Kuskokwim communities, including Chefornak, attribute this to the failure of hunters and their families to follow *eyagyarat* (abstinence practices) surrounding the birth and death of family members. John Eric (March 2008:295) explained: "If his wife had a miscarriage, her husband had to wait for the red-necked grebes to arrive [to go down to the ocean]. Or if a person's [immediate family member] died, he had to wait for the arrival of adult bearded seals and young spotted seals [before going to the ocean]." As described above, both the grebes' feces and seals' blood are said to blind the ocean's eyes, rendering those going through *eyagyarat* invisible and thus enabling them to approach the ocean without negative consequences. A person ignoring *eyagyarat*, however, could cause the shore ice to crack and detach along his or her tracks.

Ocean swells are, in fact, a greater hazard in the Canineq area than around Nelson Island, which is protected by nearby Nunivak. Paul John (December 2007:106) explained: "Although the weather is calm sometimes, [the waves] are deep in shallow areas. Akuluraq [Etolin Strait] has small [waves] like that since Nunivak Island blocks the *qairvaat* that head toward shore from the ocean. That's why there aren't large *qairvaat* around Nelson Island." Simeon Agnus (July 2007:264) shared his experience:

> I am afraid of Canineq, the way that the [shore] ice breaks to pieces when there are large ocean swells.
>
> In spring camps [on Nelson Island] our means of transportation aren't cause for much worry. Since [shore ice] here isn't extensive, it doesn't detach and float away. But when the wind is blowing directly against the shore, when the ice gets thin, the area where [Canineq-area people] go down to the ocean collapses.

John Eric (December 2007:111) described in detail the dramatic effects of *qairvaat* south of Nelson Island:

> Around 1969, seven snowmobiles sank as the *qairvaaq* broke the ice where they were situated. Only two snowmobiles didn't sink.
>
> The two of us brought a number of boats to shore at that time. Since there were many [hunters], they'd quickly lift a boat and place it inside the sled. Then we moved [the boats] back toward shore and returned home. People were downcast, as they had no snowmobiles.
>
> The *qairvaaq* broke the shore ice almost one mile back toward shore into fairly large pieces. Since it's hard to tell [when ocean swells arrive], at the time we were at Qalvinraaq when the tide came in. Angutekayak said to me, "I think there are *qairvaat* around here."
>
> When I put my head down alongside the boat and looked toward shore, I saw that there were *qairvaat*, and our snowmobiles were over there. As he traveled [back] along the [ice] edge, and we'd come upon snow in water, he never decreased his speed. We saw that people in our hunting party had already arrived down below the snowmobiles. Water would splash in the air when two pieces of ice collided. It was like that all the way toward shore.
>
> Evidently [an old man's snowmobile] would sink. [The ice] broke to pieces, and they had actually placed [the snowmobiles] on top of *etgalqitat*. [The ice] was completely broken to pieces.
>
> And if a boat attempted to travel in the area around the river channel, [the ice] would damage it. We stayed for one hour, waiting for the tide to slacken. Then although pieces of ice were around, when they stopped colliding, we headed toward shore.

Michael John (July 2007:265) added that although *qairvaat* break the shore ice, smaller pieces of ice ride out the waves unbroken: "The ocean swells easily break thick [ice]. But these thin [pieces of ice], even the ones that are lifted by the water, aren't easily broken by the ocean swells. And the *qairvaat* cannot break the ice in those bays, but when they reach the ice on top of the sandbars and move them, they will ultimately overtake them." John Phillip (December 2009:262) explained: "Since thin ice follows [the movement of the swells], it cannot break right away. But *qairvaak* quickly break thick ice to pieces. That's why we call them *mulutuugek* [hammers]."

John Eric (December 2009:235) noted that ocean swells were strongest when they first arrived: "The ocean swell evidently first arrives during the incoming tide, and it's powerful at that time. Although ice sheets are thick, they collide when the water suddenly rises, since water is so strong. After that it is much calmer, although swells still arrive." John Phillip (December 2009:262) added a warning for the Canineq area:

> When the incoming tide is about to end, the current is strong. Before the end of the incoming tide, they tell us to go to [our snowmobiles], since the ocean swells suddenly become deep at the end of the incoming tide. And when the tide is starting to come in, the ocean swells suddenly get large also.
>
> [Ocean swells] aren't obvious inside the channel, in a deep area. But when they get to a shallow area, they become obvious. Even though the weather is calm and windless, when those ocean swells arrive, they break our shore ice to pieces.

Cikuq Up'nerkami Nallini | Ice in Spring

John Eric (March 2008:341) clearly articulated the guiding principle of ice typology: "When we are asked about the ocean, we describe [ice formations] that we come across based on how they look. Since there are different types of ice, we give them names. It looks similar to something, like a barge or a boat, or we say that it looks like a heart. Or if a piece of ice is completely white, we call it *qatellria* [lit., "one that is white"]. Or if it's jagged, we call it *manialnguq* [lit., "one with rough edges"]. These names increase during spring breakup. John continued: "People in the past mentioned that there are few names for [ice] down on the ocean during winter but that the names increase after the occurrence of ocean swells. Starting in March, there are actually many names [for ice formations] down on the ocean."

Angenqat wall'u Tualleq | Large, Drifting Ice Floes

As the ocean swells break the ice, they create free-floating ice formations of various sizes. Some are particularly dangerous, and all are important to recognize for safe traveling. *Angenqat* (lit., "biggest ones of a group"), also called *allenret*, are large, moving ice floes that break away from the shore ice and drift in the ocean after ocean swells. Paul John (March 2008:539) recalled: "During spring there are *angenqat* that the south wind or north wind brings to this area." John Eric (December 2007:81) noted their size: "Before *qairvaat* broke them apart, these *angenqat* could be half a mile in diameter and about three to four feet thick." John Eric (March 2008:316) also noted that *angenqat* were detached pieces of shore ice: "*Angenqat* cannot get thicker. And their surface cannot form snow, since blowing snow cannot form along the ocean. The only time that [ice] becomes covered with snow is when it is still attached to the shore. And when [ice] drifts away, it is then called *angenqaq*. Since it was once part of the shore ice, it is also called *tualleq* [old shore ice]."

John Eric (March 2008:317) described seals resting on and around *angenqat* during spring: ·

> An adult bearded seal will be grateful for it, so that it can go on top of it and sleep. And a young bearded seal or small spotted seal will be happy for the bed that it will have. When the sun starts to get warm, it will bring its mother up [on the ice] and nurse and sleep as much as it wants. To [seals], there is nothing frightening around.
>
> Before the sun got [higher] and birds arrived, it was colder around these *angenqat* than the area near shore. And *cikullaq* formed in the area around an *angenqaq*.
>
> In the past, when we'd go down [to the ocean] with dogs, an *angenqaq* would drift north, and an adult bearded seal would immediately appear [in the water].

John Eric (December 2007:76) declared: "They say those *angenqat* that are about half a mile long are powerful and don't stop moving right away." Paul John (December 2007:71) advised against staying on *angenqat*:

> If their trail to shore was obstructed—*atreskaki* [if they were drifting], as they say— they were supposed to stay on a sheet of ice that was thick but not too large. They say that if they go on a large sheet of ice, since large ice floes tend to travel farther, when it starts to move, it parts the smaller ice around it and enters closely packed ice. They also told them not to stay on top of an *angenqaq*.
>
> They told them to stay only on ice floes that are thick but not large because those smaller ice floes actually stop when they collide against other ice, and they don't drift inside closely packed ice. That's what they said when they told them about dangers.

John Eric (March 2008:326) gave an example of men finding safety on thick ice:

> [Hunters] evidently experienced that once not far from the shore ice. There was a lot of ice around. The tide was going out, and they couldn't paddle among the ice. Since there probably weren't any *evunret* around, someone in their group searched and

found a place for them and told them to try to come toward him. There were a number of kayaks, and they all fit [on that ice].

Then that sheet of ice he found touched bottom, along the mud. When it [touched bottom], the drifting ice floes continually passed them and never piled on top of [that ice]. [Ice] would drift alongside [the ice they were on], and eventually the open water got close and their bay came into view. Then they went up to shore. Those are things that we need to tell our young men who go down to the ocean.

Paul John (March 2008:541) added that they were told not to go between *angenqat* that are advancing close to one another in case they collide: "If two large ice floes collide, they break each other to pieces. The [broken pieces of] ice pile on top of one another along the edges. *Evuluteng* [they break and pile]." David Jimmie (March 2008:313) described his experience caught between two floes: "That happened to me, Yang'aq, and Otto John. We had a small boat at the time. The ice that was advancing toward the shore caught up to us, and I thought that [the ice] was going to split our boat when it started to make popping sounds. The ice came upon the back of our boat. I was afraid, but we got through it. These *angenqat* are evidently powerful when they hit." Paul Tunuchuk (March 2008:314) also recalled being caught between two large, drifting ice floes: "I was driving the boat while it was foggy. It was actually two large ice floes. I thought I was driving along the edge of the shore ice. Then we reached a dead end and saw that our trail was obstructed, and we couldn't travel through. I immediately turned around, brought the boat [close to the ice], and increased the motor's speed. The ice was advancing toward us and arrived, and it pushed our boat along. There is evidently nothing we can do when those *angenqat* do that."

Peter John (March 2007:1167) noted the admonishment not to stay in the path of large ice floes:

> If [an *angenqaq*] came toward us, they told me not to stay in its path but to immediately move out of its way. If that ice hit [the ice I was on], although I escaped, the ice would break and fold.
>
> And if we saw an ice floe, maybe three to four feet thick, although it was calm, they told us to immediately go up to shore. If that thick [ice floe] arrived, it would break to pieces. They told us to watch out for those.

Simeon Agnus (December 2007:119) recalled how a man's kayak was caught by piling *angenqat* in the Canineq area: "Large ice floes pile on top of one another [near *evunret*]. They are dangerous, [as the ice] rises when the current is strong. Kayungiar evidently experienced that, when *cikullaq* was going underneath [a piece of ice] and another [piece of ice] was moving on top of it. [The ice] evidently got him from behind. The *cikullaq* hit his kayak under the stern and pushed him toward shore."

John Eric (March 2007:262) described what happened when *angenqat* encountered *nacarat*:

> They were admonished about the area those *angenqat* were approaching when they were moving north on an incoming tide. They said an *angenqaq* drifts, and although it reaches [the *nacaraq*] and stops at first, it moves again. It breaks to pieces and makes [the *nacaraq*] higher. Ice that is very smooth piles on top and goes around [the *nacaraq*].
>
> Since these *angenqat* are large, those people feared them. But when it's foggy, it's difficult to tell what direction that large ice floe is heading.

Tommy Hooper (June 2008:20) described land animals walking onto the ice in search of salt and sometimes floating away on *angenqat*:

> When the ocean had shore ice, [tundra hares] would lick salt from there, salty snow. They like the taste. Once when they all went down there, when the wind was blowing from the south, the ice detached and floated them out [to the ocean].
>
> And when reindeer were in this area, they also went down and licked salt there in the past, but [people] had them return before the ice detached and floated them away.

Angenqat might also detach from the shore ice and drift away with people on top. Paul John (December 2007:117) recalled: "They refer to it as *angenqiurluni* when a large piece of ice detaches [when someone is on it], sometimes along a crack over a mile or two from shore, and drifts out to the ocean." Paul Tunuchuk (March 2008:315) noted that this was more common in the Canineq area. John Eric (March 2008:316) explained: "Since the area down below Kwigillingok is a point, Nuqarrluk said that when they finally reached the water, the land disappeared far in the distance. They'd take a compass and place it right beyond them. The compass would show them when that ice detached up near shore [and was drifting]. They'd take a long time to go back to shore. He said that's what they used to experience down below Kwigillingok, where large ice floes tended to detach and drift away." John Phillip (September 2009:215, 220) recalled the sudden crackling noise and movement he experienced, as well as the change in current flow along the ice edge, when hunting with his father on ice that detached in the Canineq area when he was young: "When one is on top of the ice when the tide is going out, the ice edge is obvious, as [the current] always [moves along it]. Then when the ice detaches somewhere, the current suddenly stops flowing when the ice starts to drift away."

Paul John (December 2007:118) noted the admonishment to head to shore when ice detaches: "When visibility is good, they search the area behind them and see water toward shore, and a person can seek safety by heading to the other side of the ice. When the ice drifts away, when he sees the place where the ice detached, he tries to get out of that situation by heading toward shore." John Phillip (December 2009:215) added: "When a large piece of shore ice detaches and breaks to pieces, they told us to travel through the areas with thick [ice], since thin ice breaks and piles." Simeon Agnus (December 2007:118) described his experience when ice he was on detached at night in calm weather:

> One time we slept over at Kangiryuar with a number of boats down below us. I woke at night, and it was very calm out. I saw that the ice had detached while Mancuaq and I, accompanied by some other people, were on it, and it was floating us down [the coast]. I just ignored him and didn't wake him and went to sleep. I was thinking that we could immediately hunt when we woke up. [*laughs*]
>
> When my cross-cousin became aware of what happened, I told him, "It's okay if the ice detached and floated us. Just make some tea." [*laughs*]
>
> I wasn't afraid at all, since there was a lot of open water, and there wasn't a lot of ice.

Like Simeon Agnus, Joseph Patrick (March 2007:1188) noted that ice detaching in calm weather was not necessarily scary: "Once ice detached while we were on it during extremely calm weather, and we had kayaks. While we were just sitting there, the current gradually stopped flowing where we were. Then I looked toward shore and saw water up there. Our hunting partners were within shouting distance, and so we yelled to them that the ice had detached and was drifting away. I think the poor thing panicked at that time, but here the weather was extremely calm."

Paul John (March 2008:542) noted that there were many accounts of people floating away on detached ice: "Some people floated away when the ice detached when it was too windy down on the ocean. They would drift [on the ice] when conditions weren't good for paddling. But when conditions got better, they paddled and went up on the ice that detached and headed along the ice toward shore." Michael John (March 2007:1086) shared his experience when the shore ice detached:

> One time the ice detached and almost took us out. Angayiq and that person from Cuukvagtuli, we all went seal hunting together.
>
> We spent two nights along the edge of those large *evunret*. After the second night, we woke and saw that the wind had started to blow from the south. We immediately got ready, and when it got light we headed to shore.
>
> When we came upon that [crack] that had not been dangerous before, it had

gaped open. We arrived before the others in our group. The water was churning [from the current].

We walked along the ice to check it, and when the gap was a certain width, the late Angayiq stopped and tossed his lead dog to the other side. Although [the dog] didn't reach the other side, when he pushed the other [dogs into the water], [the lead dog] went up, and they were jumping up and down.

When my lead dog jumped over [the gap], although it didn't reach the other side, the other followed, and [Angayiq] took it by the neck and lifted it up [onto the ice]. They jumped one by one, and I crossed, and my dogs pulled the sled.

If we had delayed going back, the ice would have detached from the shore with us and our dogs still on it. When the weather got calm, Nakrialnguq and I stayed there and went seal hunting.

Ice continues to detach today. Nick Therchik Jr. (March 2008:543) said: "Ag'apaq's Ski-doo sank near Umkumiut last year. When the ice detached, he said that his snowmobile slanted and sank, since it was along the edge."

Akangluaryuut | Ones That Roll

Among the most important ice formations to recognize were *akangluaryuut* (lit., "ones that roll"), rounded sheets of floating ice that could tip over. Some were small, but others were quite large. Paul Tunuchuk (March 2008:311) explained: "Back when they used kayaks, they were admonished about ice called *akangluaryuut*. If they happened to [be paddling] close to their edge when [*akangluaryuut*] suddenly rolled, it would immediately capsize that kayak. They told them not to get close to them. Boaters and not just kayakers are admonished not to travel near *akangluaryuut*."

Akangluaryuut occur in spring as ocean water melts the submerged part of the ice, making it lighter. Paul John (March 2008:544) explained: "When [the ice] was frozen, the bottom was heavier. And when the water began to melt the bottom [of the ice] in spring, the top becomes heavier, and what was once the top becomes the bottom. Since the bottom down there becomes lighter, it suddenly rolls." *Akangluaryuut* are also hazardous when stuck in shallow water. Edward Hooper (March 2007:1086) recalled: "This side of Nuakiliit is a shallow area. A piece of ice that is about eight feet down in the water will get stuck in shallow water there. When the tide starts to go out, they watch in case it might roll this way."

Marayilugneret | Piled Ice Mixed with Mud

Elders also shared valuable observations about sediment-laden ice—a common

Upper: *Iqalluquat* created by the north wind and used as guides along the Bering Sea coast.
Ann Fienup-Riordan

Lower: Driftwood pushed miles inland by a storm tide along the southern coast of Nelson Island.
The line of wood marks the *ingigun*, the rise on to the permafrost plateau where marshland and
tundra meet. *Ann Fienup-Riordan*

Upper: A Kwigillingok hunter using his kayak sled to pull his kayak over the *tuaq* (shore-fast ice), late 1940s.
Warren Petersen, Petersen Family Collection

Lower: The edge of the shore-fast ice as seen from Ulurruk, on the north shore of Toksook Bay, February 2008. Note the formation of *elliqaun* at the ice edge and the *cikulleq* forming in distant open water.
Nick Therchik Jr.

Upper: *Manialkuut* (rough ice), with Toksook Bay in the distance, February 2008. *Nick Therchik Jr.*

Lower: *Cikullallret*, or pancake ice, circular pieces of ice with raised rims due to the pieces striking against one another. A common process of sea ice development is when these pancakes raft together to increase in thickness, eventually freezing together to form larger floes. January 2008. *Nick Therchik Jr.*

Upper: A hunter atop *evunret* (piled ice), using the pressure ridge as a lookout over a field of *manialkuut* (rubble ice) gathered and packed by the current. Note the *negcik* (gaff) he carries for safe traveling. *Leuman W. Waugh, 1935, National Museum of the American Indian, L2236*

Lower: *Tungussiqatiit* (piled ice mixed with sand). Unlike many coastal residents in the Arctic, Nelson Islanders have given proper names to *evunret* forming in the same places year after year. May 2009. *Mark John*

Upper: *Etgalqitat*
(ice beached in shallow
areas) along the mouth of
Kangiryuar Channel, March
2010. *Etgalqitat* were not
found here in the past
but are common today
as the bay has become
increasingly shallow.
Simeon John

Lower: *Culugcineq*,
a pointed piece of ice
resembling a *culugaq*
(fish's dorsal fin), March
2008. *Ann Fienup-Riordan*

Upper: Broken ice along the north shore of Toksook Bay, May 2008. *Nick Therchik Jr.*

Lower: Ice mixed with mud on the north shore of Toksook Bay, May 2008. Paul John noted that when the wind blows ice to a shallow area, the waves continually cover them and their surfaces get muddy. This is one of many ways in which ice interacts with the coast. *Nick Therchik Jr.*

Upper: Large pieces of floating ice known as *qerturaggluut* (lit., "those that are high"), May 2009. Climbing on these is dangerous, as they can suddenly flip, and one must be cautious traveling close by, as they extend deep underwater. *Simeon John*

Lower: Small, rounded pieces of *cikurrluut* (rotten ice, lit., "ice that has departed from its original state") that gathers along current lines during spring, May 2009. *Mark John*

Upper: *Qilungayiit* or packed ice floating in the current line, May 2009. *Mark John*

Lower: A Kwigillingok hunter using his canvas-covered kayak in ice-free water in the late 1940s. *Warren Petersen*

Scammon Bay hunters standing on *marayilugneret* (piled ice mixed with mud) pushed inland by a fall storm, November 2009. *George Smith*

characteristic of the shallow, muddy coastal environment, where clear ice is the exception rather than the rule. Hajo Eicken notes that whereas coastal erosion is often attributed to a lack of sea ice that allows fall storms to eat away the shoreline, in fact sea ice is the most effective mover of sediments in Arctic and subarctic waters with seasonal ice cover.[6]

Simeon Agnus (December 2007:145) explained how *marayilugneret* (piled ice mixed with *marayaq* [mud]) form along sandbars in fall and become dangerous in spring when they suddenly break loose:

> They caution people about those *marayilugneret* that surface from underwater and suddenly wobble, rising tilted to one side. A person must watch out for those; when the shore ice has gone, there are many [*marayilugneret*] along sandbars.
>
> If a *marayilugneq* happens to surface and hit a boat, it can capsize. And when it suddenly surfaces, some are extremely steep and high. They don't [surface] flat but vertically. They also caution about those *marayilugneret* that froze in place underwater along sandbars.
>
> When the ice starts to melt, they suddenly surface once in a while. If that type of ice happens to [surface], it could injure the people inside a boat. And if it hit the center of a kayak, it could break it or break the bow.

Paul Tunuchuk (March 2008:321) noted that *marayilugneret* are dangerous because they are hard to see: "They say when small parts of those dark ice sheets poke out of the water a little, they are dangerous. They are muddy and hard to see. They anchored in the mud and could suddenly surface as they detach."

Tamarqellriit Quyungqalriit-llu | Scattered Ice and Packed Ice

Stanley Anthony (December 2007:92) described *tamarqellriit,* or scattered ice: "They call those [ice] pieces that are large and easy to travel through *tamarqell-*

riit. When the tide comes up, when the ice scatters after being packed together, it is possible to travel by boat through that area." John Eric (December 2007:80) added: "A boat can travel through and around *tamarqellriit.* They are pieces of ice drifting down on the ocean that aren't in close contact."

Conversely, Paul John (March 2008:543) noted that *quyungqalriit* (packed ice) prevented hunting: "When the ice pieces are packed together like one solid piece, it wasn't possible to paddle and there wasn't open water. They cannot go down to the ocean when they don't have a path to travel through. Then they always hunted for seals along the ice edge. They call it *kangiliyarluni* [going to the bay or edge of the shore ice]. They stay at the bay." John Phillip (January 2011:71) added: "They call those gathered by the current *qec'ulriit* [(ice) that tends to gather]."

Pugtalriit Kaulinret-llu | Floating Ice and Ice Pieces

John Eric (December 2007:79) described *pugtalriit* (floating ice): "Some are about the size of this table. Although a boat goes on top of it, it won't break. And seals climb on top of them and sleep as much as they want. We call those *pugtalriit.*"

Specific forms of floating ice in spring are small, rounded pieces of broken ice known as *kaulinret* or *kaimlinret* (from *kaimlleret*, "crumbs"), which ice scientists call brash ice or ice pebbles. Paul John (March 2008:548) noted: "*Kaulinret* are small pieces of ice that broke when they constantly bumped into one another. Those can also be called *ciamneret* [broken pieces (of ice); lit., "those that are crumbled"]." John Eric (March 2008:344) described their formation: "Toward summer, when they begin to melt but haven't really melted, some *kaulinret* are small and round. Although some are a little scattered, some drift together as a pack. Some are nice pieces of small ice, and some are small pieces of ice mixed with sand. They say that those are left over from ice that melted." In the past *kaulinret* could be difficult to travel through: "One cannot go through *kaulinret* with a kayak but only with an outboard motor. If you travel through an area that is covered with small pieces of ice, you will see *kaulinret*" (Paul Tunuchuk, March 2008:309).

Kaulinret were also used as toys by young bearded seals. According to John Eric (January 2007:104): "They say young bearded seals played with those *kaulinret.* Some are as large as Pilot Bread crackers, and some are small. They grasped those [*kaulinret*] and played with them, probably wanting to play basketball. Young bearded seals played sometimes when they were staying in one place for a while, since they were young, the adult bearded seals' offspring that were born in March. They say that some kayakers saw [seals] doing that with *kaulinret* long ago." Here John may be describing the ball-like lumps of ice known as shuga ice in English.

Kaulinret can beach on the incoming tide. Paul John (December 2007:89) said: "[That ice condition] is like *elliqaun*. When [ice] sticks to the [shore ice], they refer to it as *mauniirluku*."

Qilungayiit Kigumaat-llu | Packed Ice Floating on Current Lines

Packed ice floating in a long line was referred to as *qilungayiit* (from *qilunguaq*, "pipe, tube"; lit., "imitation intestine"). John Eric (March 2008:369) provided a good description of these extensive ice belts, sometimes seen in the channel between Nelson and Nunivak Islands:

> *Qilungayiit* are floating ice pieces, lined up for a great distance following the current. Since the current continually flows north in [Etolin Strait], the ice pieces on top of [the tide rip] are called *qilungayiit*. *Qilungayiit* are pieces of broken ice that are gathered by the tide.
>
> *Qilungayiit* are long, starting from down below our village and extending a great distance north along the ocean channels. Some places are wide and some are narrow.

John (December 2007:105) later added: "When there's a large [linear] accumulation and the current is unable to scatter them, we call them *qilungayiit*. Although they are twisted, we call the packed ice *qilungayiit*. You know how along the road, the snow that a tractor plowed through stays [packed]. It looks like that." John

Tamarqellriit (scattered ice), March 2008. *Nick Therchik Jr.*

Eric (January 2007:105) described traveling around *qilungayiit*: "You can see what's on the other side of some *qilungayiit*, and some are wide. A boat will have a difficult time coming out on the other side since they are tightly packed, but it can travel along the edge."

Paul John (December 2007:128) described similar belts of floating ice known as *kigumaat* gathered in current lines in both ocean channels and river mouths: "When there isn't a lot of ice [on the ocean], when the ice extends toward the north or downcoast along places where the currents meet, they call it *kigumaaq*. It is long, even though it isn't very wide, with vast open water on each side. It's actually broken pieces of ice, but the two currents that meet cause it to gather in one spot. It is floating [ice], and the current is constantly moving it. When the tide goes out, it brings the ice down, and when [the current] starts to flow north, it brings it north." Peter John (March 2007:1165) noted that packed *kigumaat* preceded wind: "All the mouths of rivers have [*kigumaat*] in a line on the edges of their channels. Although those areas have a lot of ice and the wind is blowing, if it is going to get calmer, they will scatter and one can travel in their midst. And although the weather is calm, if the wind is going to start to blow, those *kigumaat* suddenly pack together. They knew [that it would get windy] through those *kigumaat*."

Unlike *qilungayiit*, hunters could travel through *kigumaat*. Peter continued:

> If there isn't a lot of ice, they'd wait for the current to slacken and the packed ice to scatter. Then by navigating around them and constantly turning, we'd head toward our destination.
>
> And when we began using boats, sometimes when we got stuck, if it wasn't too packed and the other side wasn't far we'd bump the ice. When our motor started to go, we'd continue to bump [that ice] and move forward until [the boat] went out into water.
>
> When the current starts flowing, [that ice] gathers and goes into the river, along the channel. The deep channels where the current is flowing have those *kigumaat*.
>
> When the tide is about to come in, they begin to scatter, and then there is open water in their midst. And when the current starts to flow again, they begin packing together.

Icinret | Overhanging Ice Edges

As warming continued in spring, ice sheets began to develop *icinret*, or thin, melting, overhanging edges: "The water melts the ice underneath [the overhang]. When a wave slaps [against the ice], it goes underneath and eventually [the ice edge] becomes an [overhang]" (Paul Tunuchuk, March 2008:312). Paul John (March 2008:553) added: "When the edge of the shore ice doesn't detach, it can form *icinret*. And the edges of floating ice can also develop *icinret*. Melting causes the bottom to start to hover [above the water]. Since the ice in the water melts faster, the bottom of the ice forms an *icineq* [overhang]."

Paul Tunuchuk (March 2008:312) noted that *icinret* were dangerous: "If you step on top of that [overhang], it will break and you will slide [into the water]. One has to continually use a *negcik* to check for those *icinret*." Paul John (March 2008:553) added that humans were not the only ones who fell in: "Since these walrus that perch on the ice have a tendency to sleep a long time, when they shoot a walrus and kill it, or even an adult bearded seal, if the *icineq* happens to break when it dies, it can roll into the water. They mentioned that that occurred many times; they would say that the animal they caught got away when the *icineq* broke and the animal fell into the water."

Paul John (March 2008:554) described how seals were said to perceive the small waves constantly slapping against the ice edge:

The boy whom the bladders took with them [to live under the ocean] evidently told stories about the [waves lapping]. When [the seals] were about to leave [their under-water *qasgi*], those who instructed them said, "Now when you travel, try not to sleep too much." Some [seals] would say, "The two that drum, when they start to drum constantly, make us want to sleep." Evidently, when they were perched on the ice, they would refer to the slapping noise made by waves on the *icineq* as one who was constantly drumming. They would get sleepy listening to that.

John Alirkar added: "They say those [seals] knew the hunters who would catch them. When [the hunter] was hunting [a seal] and was about to reach it, although [the seal] tried to stay awake, eventually it fell asleep when [the hunter] hunted it. The [slapping] caused [seals] to feel sleepy."

John Eric (March 2008:317; March 2007:265) described ice forms transformed into *kinguvkutiit* (ice pieces with large overhangs) in spring:

Sometimes when we're searching for ice, somewhere we will see *etgalqitat* or *nacarat* that are broken to pieces. Then when we come upon them when ice is no longer abundant, we see that they don't look good. The water [melts them], and they form large overhanging ice along the edge. They call those *kinguvkutiit* ["those that come after, the descendants (of previous ice forms)"]. There is snow on top, and this [lower] section is in the water. The waves had melted it. And you can't get on top of some of them. . . .

When it starts to get warm, the name for those [ice forms] changes to *kinguvku-tiit*. They say a seal won't climb up on top of them.

Cikupiat Cikurpiit wall'u Cikulugpiat | Thick, Clear Ice from the North

Clear, blue-green floebergs (massive pieces of sea ice) coming from the north are known as *cikupiat* (real ice), *cikurpiit* (big ice), or *cikulugpiat* (original or authentic ice). *Cikupiat* can appear thin but are actually quite thick. Paul Tunuchuk (March 2008:358) explained: "Sometimes they mention *cikulugpiat* down below our village, *cikupiat* from the north. Only a small part is visible, but it is submerged deep in the water. That clear ice doesn't form in our area but comes from up north when the wind is constantly blowing toward shore." Paul John (March 2008:340) noted: "When it is floating, it isn't steep. But when it beaches in shallow water and the tide goes out and it becomes visible, we'd see that it's actually thick. . . . They call them *essmiarlriit* [ones floating out of the water a little]." John Eric (March 2008:359) added: "They mentioned that adult bearded seals cannot go under them. They stick out of the water a little but are extremely thick and submerged a great distance underwater."

John Alirkar (March 2008:339) noted the clarity of northern ice as a defining feature: "Ice from the north is very clean and clear since it probably formed in a deep area. Those they call *qagkumiutaat* [ones from the north] are bright and transparent and absolutely clean. And their snow is completely clean without sand in it." According to John Eric (March 2007:262): "Sometimes, when the wind continuously blew from the west or northwest, those from the north arrived that were extremely clear and a green color. They said that the wind brought them to our area."

Cikupiat could be so large that men could camp on them. John Alirkar (January 2007:499) recalled: "One time one big ice floe came from out there to Etolin Strait. We spent a night on it a couple of times with others. *Cikurpall'er cikuq* [A huge piece of ice]. It was thick and greenish all the way to the bottom. When ice from here collided with it, [local ice] would break since it wasn't as sturdy. [Local ice] kept crumbling when it hit because it was thinner. And it couldn't come into this Kangirrluar [five meters]; it was too shallow for it. It would hit the bottom because it was so thick."

Imarrlainaq | Ice-Free Ocean

In spring hunters begin to search for open water in earnest. John Eric (January 2007:106) explained: "If there's water down below [the shore ice], we call that area *imarrlainaq* [open ocean, lit., "nothing but ocean"]. There's no ice. And we refer to a sheet of ice down in the open ocean as *imarrlainarmi ciku*." John (December 2007:82) described good spring scanning conditions:

> In March or April, in the morning, a layer of *cikullara'ar* [thin ice in open water] forms during the night because it's cold.

Although it's constantly windy, if there is *cikullara'ar*, it creates ideal conditions for scanning one's surroundings. When one comes upon that [ice], one would call over [the VHF radio], "This area is good for searching surroundings because there is *cikullara'ar* around."

But when [*cikullaq*] is [layered], the water will not be able to break it to pieces. Only the area where the wind first reached is rough.

Paul John (December 2007:107) described how the northeast wind created open water around Nelson Island:

When the wind is constantly blowing from the northeast, the ice no longer touches the shore. It takes the ice down toward the ocean, and the water opens. Animals that surface also start to become readily available along the shore ice as they head up to shore. These seals know what the ice conditions will be like by the wind's direction, since it is their environment.

When the weather is good, one isn't hesitant to go down to the bay, since seals start to appear more frequently when an offshore wind is constantly blowing along the edge of the shore ice. That wind direction is better than other wind directions.

Conversely, an onshore wind and incoming tide create poor hunting conditions. Paul John (December 2007:108) explained: "When the wind is constantly blowing from the ocean, when there is a lot of ice around, the resulting conditions prevent hunting. And when the tide comes in, the ice begins to fill the bays. They no longer have open water, when the wind pushes the ice up against the shore, *patgulluku unavet cenamun cikuq* as they say. When that occurs, they cannot hunt as they desire since ice has blocked the edge of the shore ice." Although hunting is limited, seals still surface along the edges of *etgalqitat* (ice beached in shallow water). Paul continued: "When that [onshore wind] occurs, [animals] aren't completely gone but surface once in a while. Since they have surfacing holes along the edges of *etgalqitat*, although there is a lot of ice, sometimes [seals] are seen in the bay. Since the ice constantly rises [with the tide], and since the edges of *etgalqitat* frozen in place aren't completely sealed, [seals] sometimes surface in those places."

Before guns were widely used, men set nets under the ice for seals as well as hunting seals near their surfacing holes. Paul John (December 2007:109) commented:

They say that when [seals went underwater] along the edge of the bay and swam without air, they set nets [to catch them]. They didn't [set nets] in areas where they couldn't chop the ice, as ice entering [the bay] would ruin [the net]. But they'd estimate when [the seals] would suffocate from lack of air after diving underwater from the water's edge, and from that they'd determine where to set their nets.

There were no stories of people hunting [seals] at their breathing holes. But sometimes, when there was no open water for hunting, when they saw [a seal] perched on the ice by a surfacing hole, they'd hunt it. They used their small kayak sleds to obscure themselves and approach it and shoot it when it was at a good distance. Back when they hunted on foot, they'd just lie on their kayak sleds and kick [to glide the sleds along the ice] with their feet, and when it was a good distance they'd shoot it. They said that's how they hunted *ugtat* [seals perched along the shore ice].

Nanviuqerrneret | Ice-Free Areas within the Ice Pack

Elders also described *nanviuqerrneret*, lake-sized areas of open water within floating ice. According to Paul John (March 2008:562): "They call those ice-free areas within floating ice *nanviuqerrneret*, since they look like *nanvat* [lakes]." Hunters were tempted to check these areas, as animals swam inside. Simeon Agnus (December 2007:93) noted: "There are bearded seals that make their mating calls inside them, or adult bearded seals are perched on the ice along the back."

Paul Tunuchuk (March 2008:305) spoke of the importance of *nanviuqerrneret* in enabling paddlers to proceed through packed ice:

A kayak cannot travel through *qanisqineq* and *cikullara'ar* but only through open water. When the current gathers the ice, you will not be able to go through with a kayak. And while you are among the ice, you will see *nanviuqerrneret*, and you can continue on by paddling through them. But if you [try to cut across the ice], you won't be able to paddle through.

While you're paddling you will see that you no longer have a path to travel through. Then you will continue on down the coast and see that there are more and more *nanviuqerrneret*, and you will travel through them when it is possible.

In some places currents combine to open some areas and close others. Paul continued: "If you head in one direction, you'll see that there is nowhere to go, but there are many places to travel to and hunt in if you go in another direction when two currents suddenly combine."

Like bays, *nanviuqerrneret* can trap hunters when ice closes in. Stanley Anthony (December 2007:93) noted: "A person must not go inside a large *nanviuqerrneq* without paying attention, since the area where he entered can immediately close and trap him. He should go to that *nanviuqerrneq* with constant caution. That was an instruction they gave, that if I started to travel with a boat, I should watch the place where I entered the *nanviuqerrneq* although that place was serene, that it closes when there is a lot of ice [when the current changes]." Simeon Agnus (December 2007:97) said that *nanviuqerrneret* in some areas are particularly dangerous: "When the ice is situated far out in the ocean, when coming upon a *nanviuqerrneq*, no one should try to go inside although it looks calm. That [type of *nanviuqerrneq*] can close up; the *nanviuqerrneret* that are situated toward Nunivak Island are dangerous. The northern current is strong there. If he tries to exit, he will have a difficult time when [the ice] closes in on him and obstructs his path."

Cikum Akuliikun Ayagalleq | Traveling through Ice

As noted above, significant tidal variation along a shallow coastline can translate into miles of mudflats or "ice flats" at low tide. In addition, Eicken notes, there are more steady currents driven by freshwater inflow into the ocean and other factors that can be strong and seasonally variable.[7] As elders' statements make clear, the Bering Sea has strong and changeable winds. These different currents, tides, and winds can drive the ice at different speeds or in different directions. Generally, the rougher, more deformed the ice, the more effective is the transfer of momentum from the air or water to the ice. Moreover, the mass of the ice and the size of individual floes determine how ice responds to changes in forces such as tidal oscillations that drive ice motion.[8] That means that under the same general current or wind conditions, different types of ice and different sizes of floes can move at different speeds and even in different directions. Elders reveal a deep and subtle understanding of these complex forces in what they say about the ice and how to travel through it.

Traveling through sea ice was fraught with peril. Men paid close attention

A *nanviuqerrneq*, or ice-free area, within the ice pack, January 2008. *Nick Therchik Jr.*

to instructions regarding specific conditions. Paul John (December 2007:69) recalled: "They were instructed to watch for the possibility of wind and current drifting the ice [out to open water] and causing it to collide and pile [onto other ice]. If [the ice he was on] happened to hit another piece of ice and break to pieces, he was told to try to avoid breaking and piling ice from reaching him."

When traveling in and out of open water, hunters were always on the lookout for the onset of dangerous onshore winds. David Jimmie (March 2008:306) explained:

> We are also afraid of a *qacaqneq* [onshore wind] when a lot of ice is around. One time we were unable to return home when the ice closed in on us. The ice blocked the trail our boat had taken down at Kangirpak, and we stayed there for three days. We were lucky that one of us caught a spotted seal. When it was time to eat, we'd eat that spotted seal.
>
> We couldn't return home, and although the tide went out, it wouldn't take that ice out since the wind was blowing against the shore.
>
> Since there were lots of us, we towed our boats on top of the rough ice after smoothing it, clearing a trail. When we finally reached the real shore ice, two people in our group went to get snowmobiles from the place where we'd gone down to the water in front of the Urrsukvaaq [Chefornak] River.

Some areas along the ice were impossible to travel through when ice gathered there. John Eric (March 2008:339) noted: "They call a long sheet of ice surrounded by packed ice *qesngakii cikuq* [from *qet'e-*, "to embrace, hug"]. If we reached the water below it, we might tell our companion, 'The other side of the one down there is visible, but we can't reach it. Let's just go boating behind it.'"

As noted, hunters were warned against entering bays along the ice, as changing currents could quickly pack the ice and close their exit. Paul John (December 2007:124) recalled: "After being scattered, when [the current or tide] takes [the ice] to a place where two currents meet, they refer to it as *qet'aa* [it packs] when the ice gathers. They mentioned that [condition] earlier, not to go inside a *kangiqutaq* [small bay (along the ice)]. When [a bay] is located along the edge of where two currents meet, it closes in."

Areas of open water appeared and disappeared with the changing tide. Paul John (March 2008:563) explained:

> When there is a lot of ice, one cannot immediately travel down on the ocean when the ice is touching against the edge of the shore ice, *patgusngaaqan* as they say. Bays are also filled with [ice] that the current has brought in. They waited for the ice to open up, *kiitnercirluku* as they say. When the tide goes out and [the ice] begins to drift away from the edge of the shore ice, they say *kiituq* [it has opened].
>
> Although it seems that there won't be a place for animals to surface, when it

opens and a lead forms, a ringed seal or a spotted seal surfaces in the water. But here, when the tide was coming in, there was no water at all.

Simeon Agnus (March 2007:554) commented on the mutable character of the ocean's surface: "When there is a lot of ice, it changes every single day; there is no way that anyone can learn it. When it is absolutely calm, one has no reason to fear anything about the ocean. But when there is a lot of ice, it is dangerous to be on the outer side of the ice in case the wind picks up."

Hunters paid close attention to places where currents came together, as these tended to fill with ice. Paul John (December 2007:125) explained:

> When the current is flowing down the coast, and another current is flowing from the land out toward the ocean, the place where they meet is called *ilacarneq*. In such places, bay openings tend to fill with ice and close in.
>
> I also experienced that. When [my son] Mark was a boy, he was the only one to accompany me. When I went to check [inside the cove], the trail that we took closed in [from ice] like that. Since he was a boy, I tried not to cause him to panic, and after bringing our boat up [on the ice], I told him to eat. I was careful and watched what I said, thinking that he might panic.
>
> There are *evunret* called Qikertaat below Kangiryuar that extend down the coast; when we got close to their sheltered side, the surrounding ice started to disperse. We pushed our boat down [in the water] behind [the ice] and went out [of the ice].

An outgoing tide in Kangirrluar (Toksook Bay) could be particularly dangerous. Simeon Agnus (March 2007:557) said: "When the tide is going out of Kangirrluar, if the ice is trying to split, it gets really tightly jammed. And the northbound current helps it. And the ice in Kangirrluar pushes it out toward the ocean. It is extremely dangerous." Simeon (March 2007:557) recalled how boats stuck in the pack ice finally exited by going with the current:

> A lot of boats got stuck once when there was lots of ice. By talking on a radio we got our boats together in one spot when we were trying to go toward land, and no more solitary boats were around.
>
> We were about to go into the midst of the ice pack. As we proceeded, there was a current, and the open water of Kangirrluar was far off.
>
> When the tide came in, it was sideways. When some boats could make no progress, we waited for them.
>
> So I yelled to him when he stopped, "Go with the current! If we keep going [against it], it will be in vain!"
>
> We went with the current in the middle of the ice floes, and we got to where it wasn't as strong.

Since they were not going to be successful [coming out of the ice], I woke them up [brought them to their senses] when they got into the pack ice. [*chuckles*] These young men really listen when there is a lot of ice.

Tommy Hooper (March 2007:1355) spoke of comparable dangers associated with an incoming tide on the north side of Nelson Island: "When we are seal hunting in that area in open water, when the tide is coming in, a lot of ice flows near the shore from Cape Vancouver. It is packed and one tends to get stuck. But before [the ice pack] gets too close to shore, if one goes ahead of it, one doesn't get stuck in it." Hunters must also be wary of floating ice colliding with shore-fast ice during an incoming tide. John Eric (December 2009:234) admonished: "Although pieces of ice were large, when they hit the shore ice, they could flip over and enter [under the ice]. That was something one had to watch for when the tide was coming in."

Phillip Moses (January 2007:292) recalled how hunters sometimes had to leave their catch behind when ice closing in threatened their path: "When they went by kayak and the ice floe was about to close in on them, they alerted their companions. After he pulled the seal he had caught onto the ice, one of the hunt-ers left it [and went toward land] because he might not have a way to get out of the ice. They are very careful in those places." In the same way, John Eric (Decem-ber 2009:236) reminded his listeners that the only irreplaceable thing is a human life: "They also told us that when we didn't have a way up to shore, the snowmo-biles up [along the ice] will be replaced although they sink. They said we have to watch the people while we are down there."

Stanley Anthony (December 2007:93) described the need for vigilance in areas where currents meet during an outgoing tide, packing ice together and obstructing one's path: "Also, since currents flow in different directions in our area, when [the ice comes down] when the tide goes out [away from land] and the current is flowing north at the same time, since [the two currents] meet, they told me not to travel through there but to flee toward open water. They said not to flee toward that area where [ice] tends to meet but to go toward the ocean. They said when the tide comes in again, the area to the north forms a trail; the ice starts to scatter and separate." Camilius Tulik (March 2007:535) was also admonished to head toward open water if trapped in the ice: "They told us not to panic if the weather is fine but to flee out to the ocean. One does not experience hardship in Akuluraq." Stanley Anthony warned: "Also, farther down, when there is an outgoing tide, one has to be alert because the outgoing tide is strong and heads toward Nunivak Island. Although a boat is situated on top of floating ice, it will move at a rapid speed toward the western part of Nunivak."

Simeon Agnus (December 2007:98) recalled the instruction not to struggle if caught in ice in Kangirpak but to follow the current north. He then described the time when packed ice closed in, preventing a number of hunters from returning to shore:

> One time, all the men of Up'nerkillermiut evidently went missing; *cikullaq* obstructed their trail [to shore] back when they paddled [with kayaks]. Many men in kayaks drifted out in the ocean.
>
> During that time, since the weather had been cold, I think it was beyond Nunivak Island. They say that [after testing the ice with their gaffs], they traveled with their kayak sleds on top of the *cikullaq*. When it was no longer safe, they'd get inside their kayaks and continue on. They probably reached shore near the village of Chefornak.
>
> Those people were out in the ocean, but everyone evidently reached Up'nerkillermiut. If it had been extremely windy with bad weather, many men probably would have died at that time.

Newtok hunters were also instructed not to go to shore but to head north, into deep water, if ice obstructed their path. Peter John (March 2007:1160) recalled: "They told us that if there was too much ice, that we should escape to that area [north of us]. They said although there was a lot of ice there, the ice wouldn't be as packed and there would be a way to get [to shore]. . . . They told us not to try [to go to shore] if there was *cikullaq* around but to let the wind drift us."

Michael John (July 2007:269) described how he led a line of boats out of the ice, one behind the other:

> If the ice obstructs your path, if the *kigumaaq* [belt of floating ice] is broken to pieces and you have no way up to shore, if the area behind you is open water, no matter how far [the shore] is and a number of boats are grouped together, if you bump into the ice and face [the shore] facing the current, you will go through the ice and continue on. It is possible to travel through it.
>
> I experienced that twice down below our village. I faced the current and went on top of ice that wasn't large with the motor running. It didn't start moving right away, and my motor just made noise. Then when it started to move, the ice didn't stop but continued to move.
>
> I went out into open water. And I left that ice and finally went toward help without any hesitation.

Some attributed more than human power to those who could lead their fellows out of the ice. Paul John (October 2003:47) described *qupurruyulilget* (lit., "ones with the ability to make them split or crack"):

> Some people have that ability. They call them *qupurruyulilget*.

There was that person who had that ability passed down to him, and when he used it, his companion watched him. Since they couldn't do anything and the ice blocked their trail to shore, he said, "I am going to try. If we happen to move, don't bump me whatsoever while you're behind me." When they began to advance, the ice on the trail in front of them kept splitting. The one who was continuously paddling in front never reached [the place where the ice parted]. He saw what looked like human hair underneath them and hands that were splitting the ice outward.

The long hair was under their kayaks. The other person was trying to touch the hair with his paddle once he saw that they were going to make it out of the water. When he did that, he bumped the one who had told him not to touch him whatsoever. [*laughs*] Gee, then the first one stopped. He looked back and said to him, "I told you not to touch me. Let's use our kayak sleds and go the rest of the way." They finally pulled up their kayaks and went with their kayak sleds.

They say that one was *qupurruyulilek* [one with the ability to part the ice]. He used his ability to split the ice only when he was desperate for help.

Paul Jenkins (March 2007:255) reminded his listeners that packed ice could prevent hunting, even when animals were present: "Sometimes when [the ice] drifted [sea mammals] toward the north in the pack ice, we couldn't do anything but watch them pass." Simeon Agnus (July 2007:276) noted that even when hunting in the ice was possible, men could not take a straight path: "The ice along Etolin Strait that goes back and forth [from the ocean toward the shore] is extremely packed and dangerous during spring. Although a person knows the location of his destination, if ice obstructs his path, he will be unable to go there even if he wants to."

Michael John (June 2008:207, 222) was advised to seek safety on rough ice in windy weather: "He said that if the wind became too strong for me, I should go on top of a large ice sheet with a rough surface and stay there. He said that [rough ice sheets] go deep in the water. And when the wind blows them, they move slower than others. Although the waves are large, *manialnguut* [those (ice sheets) with rough surfaces] don't break easily." Indeed, the depth of floating ice was a crucial factor. Michael continued: "They say that when the wind blows thin ice like *cikullaq* away from shore, its speed increases. But [ice] that extends deep in the water moves slower. When [thin pieces of ice] come upon it, they quickly pass by."

Finally, Dick Lincoln (April 2009:75) noted the admonition not to stay out on extensive shore ice in spring: "Even at this time, there is a saying that if there is extensive shore ice toward spring not to stay there. They told us that if we had dogs, we were to leave them up there around the *etgalqitat* and go down [to the water] to hunt. We were also told never to return home from far away by traveling along the shore ice if shore ice extended far out. They said when shore ice was extensive, it could easily detach."

Aarnarqellria Ciku | Dangerous Ice

//

They were always vigilant, constantly watchful of the unsafe places along the ice when they were down on the ocean.

—Paul John, Toksook Bay

Elders shared abundant information on sea ice, in large part motivated by the danger it poses to uninstructed youth today. Camilius Tulik (March 2007:558) declared: "We should certainly talk about these things at this time, since these boys have come to be daredevils. [The ocean] is dangerous." Both men and women openly discussed their experiences—including errors in judgment—in hazardous situations. Some types of ice are inherently dangerous, like *elliqaun* (thin, new ice sheets) and *kaimlinret* (small pieces of broken ice). Simeon Agnus (December 2009:219) noted that around Nelson Island, the biggest danger was ice blocking one's trail: "In the area where we head down to the ocean the biggest concern is the possibility of losing one's trail [back to shore] when the tide is coming in. When there is lots of ice, Kangirrluar quickly fills when the tide starts coming in toward [shore]."

Simeon Agnus (July 2007:277) also recalled the warning not to venture far from shore when *cikullaq* (frozen floodwater) formed. Paul John (December 2007:69) agreed: "When there is an accumulation of ice, when *cikullaq* tends to form toward spring, a person has to be careful." Paul Tunuchuk (March 2007:260) added: "A kayak could go through *cikullaq* only when it thawed and got soft when the sun got warm. Also, they warned us not to travel when ice was piling. When ice piles, it thickens, and they would be unable to help that person. They said when the ocean is in the process of freezing, it prevents even a paddle from moving. These days, people travel wherever they want." Simeon Agnus (July 2007:280) shared his experience when caught in piling *cikullaq*:

> We went down toward the ocean. Then Cakiculiq called me and said that the *cikullaq* in the area where we came from was piling up while I was with my son, their eldest brother.
>
> We were pursuing an adult bearded seal, but since I was in a hurry, we turned back. As we were heading to shore, we came upon that ice that was piling and getting thick from the current. I put my motor on full speed, and then our boat went up [on the ice] and we went toward shore on top of the *cikullaq*. And the other boats were watching us.
>
> When the motor couldn't go through [the ice] any longer, we lifted the motor and towed our boat toward shore, stepping along the ice that piled and layered, towing our boat through an area that we were unable to walk on before. That's one time that I was afraid.

When *cikullaq* piles, it isn't easy to get out of. That's why when Cakiculiq called me, I immediately turned toward shore.

Qanisqineq (snow that gathers on the ocean) could be dangerous. Peter John (March 2007:1180) recalled: "When *qanisqineq* formed and became thick during spring, it was extremely abrasive to paddle through, and it kept one from traveling fast. We were cautioned not to travel through thick [*qanisqineq*] and to watch the current when carrying our catch. And they told us not to be confident in traveling through the lighter-colored [*qanisqineq*]. Even though [*qanisqineq*] is thin, it is extremely rough to travel through. And when traveling with a boat and going through thick [*qanisqineq*], the motor can't take in water [and will overheat]."

John Eric (March 2008:318) noted the dangers of dark-colored ice: "Some pieces of ice are hard to see. That's why we tell people that one must pay close attention while down on the ocean. If a person is driving the boat fast and comes upon that kind [of ice], his motor will suddenly split or fly off. They say they are smooth pieces of ice floating a little in the water and covered by sand. Even a small piece of ice can cause a motor to break down." John (December 2009:279) also described the danger posed by *marayilugneret* in spring: "In my village, two people had a mishap before the ice was all gone during spring. They evidently suddenly came upon a *marayilugneq* partially covered by the incoming tide, and when their motor got caught, it threw them from their boat and they died."

Some dangers are man-made. John Eric (December 2007:114) told the story of men stranded on floating ice:

> While we were boating, I saw a boat, and there were people standing on the ice pretty far away, and their boat was drifting in the water. They got on top of that ice, and when they all turned away from their boat and were searching for something, their boat slowly and quietly drifted away.
>
> Acivaq was among them. I watched [the boat] over there. Since Evon wasn't far from [the boat], although he would eventually go to them, he was joking around, watching Acivaq wave his arms, and eventually [Acivaq] removed his small coat and threw it in the air. [*laughs*] Although [Evon] would eventually go to the boat, he delayed going.

Simeon Agnus concluded with feeling: "My, when there's no one [inside the boat], when his hunting partners are walking [on the ice], one must not walk [leaving his boat]."

Simeon Agnus (December 2007:123) described Paul John's narrow escape when left without a boat on melting ice:

> This person [Paul] experienced danger that they were warned about. His hunt-

ing partner was butchering an adult bearded seal on top of a small piece of ice. When [another] adult bearded seal surfaced while they were there, [his partner] left [Paul]. Ice has a tendency to melt fast when it forms foam during spring. But before the ice broke to pieces while he was on it, a person from Nunivak Island came upon him. His partner was pursuing an adult bearded seal, and he had gotten pretty far away.

He experienced that [teaching] that one should not leave someone without a boat.

Negcik | Gaff

They say that negcik extends one's life.

—Michael John, Newtok

Paul John (December 2007:69) noted that the ocean was referred to as *ayaperviilnguq*, something that a person cannot lean on with his hands for support to avoid danger. Since the ocean itself could not be leaned on, it was essential that a hunter always carry a *negcik* (gaff) when traveling on and around sea ice. According to Peter John (March 2007:1183), "They told us not to walk around on shore ice with nothing in our arms." Paul John (December 2007:132) recalled:

> We were instructed to always hold the *negcik* when we left our kayaks and walked a short distance. If a person happened upon and fell through a crevice or thin ice, the ends of his *negcik* that are touching [the ice] won't let him fall [all the way] down.
>
> And when we climbed *evunret*, they say some have crevices covered by snow after blizzards. If we suddenly fell into a crevice, [the *negcik*] would prevent us from falling through.

Stopping a fall was only one of the gaff's many uses. Paul John (December 2007:132) explained: "Along the binding on this *negcik* was a dangling piece [of line]. If his kayak happened to slide down [from the ice] and was drifting away, and he was about to swim and follow it, he would pull out the end of that [binding] and bite it and take his *negcik* along like that when he swam [to retrieve his kayak]." The *negcik* could also be used to retrieve a person. Paul continued: "If someone fell and there was no way to retrieve him, and if his arms were constricted inside a crevice, they were told to hook him with the gaff [under his jawbone] and pull him. They say although there is a gash [under his jaw], a person doesn't die from it. If he is able to grab [the *negcik*] with his arms, he could grasp it and have [his partner] pull him up, but if he is unable to move his arms [to grab the *negcik*], he could [retrieve him] through that part of his body."

Paul John (December 2007:136) also spoke of the hunter's use of the *negcik* to announce his catch: "If he caught a bearded seal and walked to the village to announce his catch—*uurcaquni,* as they say—he'd use his *negcikcuar* [small *negcik*] as a walking stick. And if he was announcing that he had caught two bearded seals, he'd hold his large *negcik.*"

Simeon Agnus (December 2007:144) noted that the *negcik* was used to stabilize a kayak when entering and exiting: "When one is about to go up [to shore], they just jab that *icineq* [ice overhang] to break it, and then one can get out of the kayak [and onto the ice]. The *negcik* was a counterpart; they just placed it up [on the ice], and when it hooked in place, they grasped it holding the opening of the kayak [at the same time], and the kayak wouldn't wobble. The person in the kayak depended on that *negcik* for balance although he was alone."

The ice pick at the bottom of the *negcik* could also be used to chop the edge of the ice before bringing up a seal for butchering. Paul John (December 2007:144) explained: "When a person caught a bearded seal, he would chop the ice [edge] to make it slanted and not too perpendicular—*civlirrluku,* as they say—fixing the path it will take when he pulls it onto the ice. [Seals] are easier to lift onto the ice like that."

Paul Tunuchuk (March 2008:337) described how the *negcik* could be used as an emergency paddle: "If you lost your paddle, you could use that as a paddle. And there is an ice pick along the end of the *negcik.* You always check the thickness of the ice with that, to check whether it's dangerous or safe."

Concluding his discussion of this important tool, Peter John (March 2007:1183) regretted that young people today do not routinely use them: "These days they don't bring anything with them when they travel because they no longer hear instructions from us. During this time of year, we would have begun to give instructions to others concerning the ocean, about what they should do when they got into particular situations on the ice."

Murilkelluni Ayagallerkaq Cikumi | Safe Traveling on Ice

Men were taught about safe, as well as dangerous, areas in the sea ice. Paul Jenkins (March 2007:255) recalled: "The teachings of [the ocean] are numerous. A person who follows them and doesn't act impulsively gets himself out of danger." Hunters were admonished to seek safety on the sheltered side of *evunret.* Simeon Agnus (December 2007:119) explained:

> When there is a lot of ice, the ice breaks and separates along *evunret* that are solid and stable. And when a person didn't have a trail to take [after the ice passed], when the current started to flow again, he would go to the *qamaneq* [sheltered side] of [the *evunret*], seeking safety. Although ice advanced toward them, [ice] along the sheltered side was more dispersed.
>
> Also, *tungussiqatiit evunret* [dark, sediment-laden *evunret*] are safe. *Evunret* that

have sand in them, *marayilugneret evunret*, can save you. No matter how large an ice floe may be, it will never smash those blackened *evunret*. Those covered with mud are really sturdy and will not break.

Peter Dull (March 2007:560) confirmed the safety of *evunret*: "If one has no other resource to survive, they are admonished to take refuge on *evunret*. Even though the ice is packed, the lee side of *evunret* is always water. It cannot be pulverized. Those *evunret* keep dividing the oncoming ice floe. If we cannot travel by boat because of packed ice, they told us to flee to those *evunret*. And if the current changes, one moves to the other side and stays there. Those who hunt seals must be told that, if they are in dire straits, they must flee to the *evunret* if they don't want to be smashed on top of the ice pack."

Peter John (March 2007:1167) recalled his experience on *evunret*:

George Carl and I were together, and these other people had evidently gone up to shore before I did. When we started going back to shore, they were about a mile away and we suddenly couldn't go to them.

I went up since I had seen that [*evuneq*] closer to shore. When I went along its sheltered side, I told my cross-cousin, "Throw our anchor down." He was afraid, but since there is an oral teaching about that, I told him about it. After tying the boat down, I said to him, "Now let's have some tea and eat." After initially being afraid, he was no longer fearful. [*laughs*]

Simeon Agnus (December 2007:130) described seeking shelter on *evunret*: "This person [Paul John] also delayed me by shooting a ringed seal. Although I paddled hard and tried to keep up, I ended up behind my hunting partners. When I was about to [reach safety], the ice closed in. I was very afraid at that time. We stayed along *evunret*, where we sought safety, for four days. The ice obstructed our path to shore, and we stayed at Qikertaq, on top of *evunret*."

Ice not only provided shelter. Paul John (December 2007:101) noted how ice could also warn hunters of coming wind: "They said when floating ice starts to disappear [behind] waves, they would start heading toward the land for safety. They said when it was about to get windy, that was how floating ice appeared." John Eric added: "That's evidently how the east side of Nunivak Island is. Although it's completely calm and windless, those types [of waves] start to arrive."

Paul Jenkins (March 2007:253) recalled the admonishment to travel from ice floe to ice floe rather than across open water:

This was my most adamant admonishment. [Nurauq] told me, "When you are down on the ocean, paddle to all the *cikurraat* [small ice floes] that you see. These walrus tend to attack people who don't go to them."

When that poor person told a story about what happened, I understood that instruction. He said that he saw a sheet of ice below him in front of the Qalvinraaq River, and there was a walrus that was pretty far down. He paddled fast toward that [ice] and, when he got there, lifted his kayak. He said that just as he brought it up, [the walrus] appeared along the [ice] edge. It almost caught up with him.

I was afraid because walrus tend to attack kayaks, probably thinking that they are walrus. [*laughs*]

Women were also taught how to behave around dangerous ice. Pauline Jimmie (March 2007:285) of Chefornak recalled standing on the shore and beating on a bowl to guide dogs home over rapidly breaking ice:

The ocean is daunting, and we also experienced that. At that time, my mother and Panicuar were coming from the ocean [over the ice] to Qikertaar, and they said that this [shore-fast ice] broke to pieces when we were in Qikertaar during spring.

My grandmother told us to take a bowl and hold it up toward the [oncoming] dogs and beat on it. A number of us ran down to the shore and beat on the bowl. The two dogs over there dashed quickly, approaching along the breaking ice.

Another elderly Chefornak woman (March 2007:282) shared her experience when ice collapsed while she was traveling toward shore:

Back when we used to bring people down to the ocean, Paniliaraq's father constantly told me not to let go of the sled if the ocean ice broke to pieces before we got up to shore and we fell through.

One day I brought him down in the morning along with someone else. Then when we put the young bearded seal that Tuguyak had caught inside the sled, he said that if we happened to fall through the ice, we should push [the seal off the sled] without any regret and let it sink.

After we brought him to his kayak, after eating a small meal, we headed back to shore. While we were on our way, although the trail that we took looked safe, it began to shift and eventually crumbled and collapsed.

One of us held the lead dog, and when we'd come upon a large ice floe, we'd move onto it and make our way toward shore, but it was carrying us southward.

Then right after we passed the area below Kipnuk, we finally got onto ice that wasn't broken. And I wasn't afraid at all. Then we moved closer to shore and along solid ice, heading north, and when we got to the trail that we usually took, we went up, carrying that young bearded seal.

We tethered the dogs, and while we were eating they mentioned seeing a person down along the ocean. Back in those days, Ilanaq used to tell me that if I saw a person coming to shore, I should go and meet him with dogs.

Then he got close. He walked to shore very slowly. Since the dogs that I had taken probably came into view, he headed toward them as he came. Then his mother said, "What's wrong with your Qulicungaq?"

I went outside and went toward him. When the sled got a certain distance away, he looked up. When he saw me, he staggered toward the front of the sled and fell over. Thinking that he was ill, I ran to him and said, "Qulic', what's wrong?" He said that he suddenly faltered when he saw me.

He crawled over and when he sat in front of the sled, he said, "How did you all arrive?" Then I told him what we did, that we held the lead dog and when we'd reach a fairly large ice floe, we'd move onto it.

"But which direction did you head?"

"We went toward shore, toward the land."

"Who told you what to do?"

"No one advised us what to do."

He said that he had forgotten to mention the instruction that he was supposed to give me. He expressed his gratitude, thanking God. He thought that we had already died.

Then he said to me, "Did you carry the young bearded seal in the sled?" I told him that we carried it. Then he said to me, "I thought I told you to let it fall in the water." Then I told him that since we didn't fall into the water, we carried it and arrived home.

It's true that although the ocean looks safe, it quickly breaks to pieces and collapses. When we got to the sled and looked back toward the ocean, there was no ice at all. I experienced it, although I'm a woman.

Tommy Hooper (March 2007:1303) said that among the most important instructions was not to give up:

When a tidal wave occurred [after the 1964 earthquake], [the ice] broke to pieces while I was on it.

I was extremely desperate. But the one who raised us told me when I was small, pointing to my future, "If you are in a desperate situation when ice breaks while you are on it, don't suddenly lose hope." He spoke of the most important teaching down there.

[He said,] "If you don't give up, if you search for a way to get out of it, you might be able to get to safe ice up near shore." What he said was true. At that time when I felt desperate, I saw his face in front of me. When I recalled the instruction he gave, I no longer felt afraid, and [his face] disappeared.

Simeon Agnus (March 2007:562) observed the importance of not showing fear: "It is also a prohibition that when one has a younger companion, even though he is afraid, he must not talk in such a way to cause his young companion to panic."

As important as not showing or giving in to fear, one should not be fearless. Peter Dull (March 2007:561) said: "They are told not to be overconfident and climb up on [pieces of ice] right away. One must be cautious because some have holes and cracks that go all the way down." Young people, however, take risks. Camilius Tulik (March 2007:562) stated: "Nowadays, thinking that they have good gear to hunt with, we know that they take risks. They are fearless since they have not experienced a dangerous situation, and they probably are curious about doing some things in the ocean." Simeon Agnus recalled what his generation knows well: "The ocean is daunting. When one says that he has learned the ocean, he is gravely mistaken. Every single day, [the ocean's] condition changes when it has a lot of ice."

Cikuq | Ice

aayuqat. Large crevices in the ice, extending a long distance.

akangluaryuk/akangluaryuut. Rounded sheet/s of floating ice that can tip over, some large (lit., "ones that roll").

allenret/allenvallraat. Large, moving ice floes broken away from shore ice. See also *angengaq, tualleq.*

angenqaq/angenqat. Large, moving ice floes broken away from shore ice and drifting in the ocean after ocean swells (lit., "biggest one of a group"). See also *allenret, manigaq, tualleq.*

arumalria. Rotten ice. See also *cikurrluk.*

asvailnguut. Sediment-laden piled ice (lit., "ones that are solid and immovable"). See also *evunret, marayilugneq, tungussiqatiit.*

ciamneret. Small, broken pieces of ice (lit., "those that are crumbled"); brash ice. See also *kaimlinret, kaulineq.*

ciku/cikuq. Ice. See also *qenuq.*

cikullallerrluk. Old pieces of *cikullaq.*

cikullallret. Broken pieces of *cikullaq* (lit., "former *cikullaq*, what was once *cikullaq*"); pancake ice.

cikullaq/cikullat. Newly frozen ice, frozen floodwater on the ocean, ice that freezes along open water; grease ice; frazil ice.

cikullara'ar. Thin ice forming in open water; nilas.

cikulqaraq. Thin ice. See also *qenulraar(aq), yuulraaq.*

cikulraq/cikulraar. Glare ice.

cikulugpiaq/cikulugpiat. Thick, clear floeberg/s coming from the north (lit., "authentic ice"). See also *cikupiaq, cikurpak.*

cikunerraq. New, freshly formed ice; nilas. See also *elliqaun.*

cikupiaq/cikupiat. Thick, clear floeberg/s coming from the north (lit., "real ice"). See also *cikulugpiaq, cikurpak.*

cikuquq. Jagged ice.

cikurlak. Ice from wet weather, freezing rain.

cikurpak/cikurpiit. Thick, clear floeberg/s coming from the north (lit., "big ice"). See also *cikulugpiaq, cikupiaq.*

cikurraq/cikurraat. Ice chunk/s, small ice floes.

cikurluk/cirkurrluut. Rotten ice (lit., "ice that has departed from its original state"). See also *arumalria.*

cirmiute-. Ice in squirrel den.

culugcineq/culugcinret. Erect, pointed piece/s of ice, broken by ocean swells and resembling a *culugaq* (dorsal fin of a fish).

elliqaun/elliqautet. Thin, newly frozen ice that freezes along the shore-fast ice (from *elli-,* "to put, to place"; lit., "something quickly put in place"); nilas. See also *cikunerraq.*

etgalqitaq/etgalqitat. Ice beached in shallow areas, closer to shore than *nacarat* (from *etgate-,* "to be shallow"; lit., "ones that reached a shallow spot").

evuneq/evunret. Piled ice, once used as markers (from *evu-,* "to pile up"). See also *asvailnguut, marayilugneq, tungussiqatiit, uguut.*

icineq/icinret. Thin, melting, overhanging ice edge/s in spring.

imarrlainaq. Area of open water, open ocean with no ice (lit., "nothing but ocean").

ircaquruaq/ircaquruat/ircaquruarpiit. Heart-shaped ice formation/s (lit., "pretend *ircaquq* [heart]").

kaimlinret. Floating ice broken up and pushed together in spring (from *kaime-,* "to make or drop crumbs"); brash ice. See also *ciamneret, kaulineq.*

kanevcir-. To have tiny ice crystals in the air.

kangiq/kangit. Open water bordered by ice or land, ice bay/s.

kangiqiugneq/kangiqiugneret. Cove/s in the ice that hunters should avoid because ice might close in (from *kangiqutaq,* "small bay, cove").

kaulineq/kaulinret. Floating ice broken up and pushed together in spring, used as balls by bearded seals; shuga ice. See also *ciamneret, kaimlinret.*

kigumaaq/kigumaat. Belt/s of floating ice gathered in the current lines of ocean channels and river mouths; floating ice on the sheltered side of sandbars in an outgoing tide.

kinguvkutiit. Ice pieces with large overhangs (lit., "those that come after, descendants [of previous ice]").

kucuknaq/kucukat. Icicle/s.

makuat. Particles/ice crystals in water around floating ice that are visible in reflected sunlight, also referred to as the ocean's eyes.

manialkuk/manialkuut. Jagged piece/s of ice pushed onshore by the high tide, rough ice, rubble ice.

manialnguq/manialnguut. Ice sheet/s with rough surfaces (lit., "ones with rough edges"), rough ice.

manigallret. Floating ice that has broken away near sandbars.

manigaq/manigat. Smooth ice, both shore-fast and floating.

marayilugneq/marayilugneret. Sediment-laden piled ice; piled ice mixed with *marayaq* (mud), stuck to sandbars. See also *asvailnguut, evunret, tungussiqatiit.*

mecqiitaq. Water-soaked snow, making travel over sea ice difficult.

nacaraq/nacarat. Piled ice on sandbars surrounded by shore ice, shore ice piling up after ocean swells (lit., "place to look around from a high vantage point").

nanvarnaq/nanviuqerrneq/nanviuqerrneret. Ice-free area/s within a larger area of floating ice (from *nanvaq,* "lake").

nepillinret. Ice that is stuck to the mud all winter and doesn't move, even though the tide comes up; becoming *tumarneret* when they break loose in summer and suddenly surface.

nepucuqiq/nepucuqit. Rough edge/s of the shore-fast ice formed when ice broken by wind and tide refreezes (from *nepute-,* "to stick on to something").

nutaqerrun. New ice in the fall, new snow on the ground (from *nutaraq,* "new one").

patuggluk. Ice fog.

pugtalriit. Floating ice (from *pugte-,* "to come to the surface").

pugteqrun/pugteqrutet. Ice piece/s stuck to the bottom that suddenly surface in spring and can be dangerous (lit., "those that come to the surface"). See also *tumarneq.*

qairvaaq/qairvaak/qairvaat. Ocean swell/s capable of breaking shore ice (lit., "big *qairet* [waves]").

qanisqineq. Snow in water; snow that gathers on the ocean as a viscous mass, obstructing one's path and potentially dangerous (from *qanuk,* "snow"); also called slush, slush ice.

qapuut. Foam along ice floes indicating melting ice.

qas'urneq/qas'urneret. Place/s where ice forms on the edge of sandbars and fills with deep water; overflow on shore ice in spring, as well as on river and lake banks.

qatellria/qatellriit/qatqitiit. White-colored *evunret,* formed later in winter (lit., "white one/s").

qec'ulria/qec'ulriit. Ice that tends to gather.

qenulraar(aq). Thin ice. See also *cikulqaraq, yuulraaq.*

qenuq. Ice, broken or slush ice (from *qenu-,* "to freeze"). See also *ciku.*

qerturaggluut. Large pieces of floating ice (lit., "those that are high").

qilungayiit. Packed ice floating in a line, linear accumulations of sea ice (from *qilnguaq,* "pipe, tube"; lit., "imitation intestine").

qiugaar. Reflection of open water known as "water sky," seen in the sky as a line of darkened clouds (from *qiu-*, "to become blue").

qulineq/qulinret. Crack/s in the ice, crevice/s (from *qulig-*, "to crack").

quyungqalriit. Packed ice (lit., "ones close together").

tamarqellria/tamarqellriit. Scattered ice.

tualleq/tuallret. Large, drifting ice sheet/s (lit., "old *tuaq* [shore-fast ice]"). See also *allenret, angenqaq.*

tuapiaq. Genuine shore ice, freezing in winter and dry (lit., "real *tuaq*").

tuaq. Shore-fast ice.

tumarneq/tumarneret/tumarngalriit. Ice piece/s stuck to the bottom that suddenly surface and can be dangerous (lit., "ones that are assembled"). See also *pugteqrun.*

tungussiqatek/tungussiqatiit. Dark, piled ice formations; sediment-laden *evunret* (lit., "ones that are dark"). See also *asvailnguut, evunret, marayilugneq.*

tuuta. Ice bridge between floes

uguut. Piled ice. See also *evunret.*

yuulraaq. Thin ice. See also *cikulqaraq, qenulraar(aq).*

Yun'i Maliggluki Ella

Ayuqucimitun Ayuqenrirtuq

THE WORLD IS CHANGING

FOLLOWING ITS PEOPLE

The weather has changed before my eyes. It isn't like it was back when I first became aware of my surroundings. In those days they would mention that the weather was worsening and wasn't like in the past. When I think about it, the weather was actually good back then. These days the weather is changing before our very eyes.

—Lizzie Chimiugak, Toksook Bay

Ella Cimissiiyaagtuq | The Weather Has Changed So Much

AN UNDERCURRENT OF CONCERN RAN THROUGHOUT OUR DISCUS-sions of the land and sea. Everything—rivers, lakes, wind, waves, snow, ice, plants, animals, and weather in every season—seems to be changing in southwest Alaska, as in other parts of the Arctic.[1] All agreed that the weather is becoming warmer. Paul John (December 2007:324) stated what many know well:

> We happened to catch the time when it used to get [severely] cold. And frost formed, even inside homes, since wood [for heat] wasn't readily available to us down along the marshland.

When snow covered [their homes], only their smokestacks would be visible, and they made windbreaks for their windows and doorways. One entered the porch and took stairs down to the doorway because there was so much snow. The fact that it's changed has indeed become apparent.

Paul Tunuchuk (March 2007:215) noted that in the past, when men urinated outdoors in winter, the place where their urine hit the ground would freeze into an icicle. Lizzie Chimiugak (January 2007:303) remembered standing on containers of seal oil to press it out, as the cold had made it so thick it could not be poured.

Paul Kiunya (January 2007:5) declared: "It doesn't get really cold nowadays. The weather isn't at all like it was when I was a boy. When it gets hot during summer, it now reaches eighty degrees, which it never did in the past." Lizzie Chimiugak (January 2007:303) added: "At the present time, the sun's heat spoils some food. In the past, when they got herring, they left them alone for a while [before hanging them to dry]. Nowadays their intestines start to rot right away. And many fish hanging to dry are ruined when people don't turn them and cover them."

John Eric (March 2007:224) shared a number of vivid markers of cold weather:

The sound of people walking on the [dry] snow could be heard from inside the house, even when that person was far away. And sled runners used to make noise on top of the snow as they approached. I no longer hear that noise these days.

And dogs that had been bred were also wrapped with cloth as a precaution so their nipples wouldn't get frostbitten. And tarpaulins kept outdoors used to break, since cloth breaks to pieces when frozen.

John Eric (March 2007:228) noted that the tundra area was even colder than the coast: "Some people said that cold weather isn't the same along the ocean as in inland villages, where the cold is more severe. And the cold weather is also severe along the Kuskokwim River during winter." Times have changed. Golga Effemka (January 2006:30) spoke for upriver: "It used to get extremely cold when I was a child. It no longer gets cold these days, and snow has become scarce." Lizzie Chimiugak (January 2007:485) spoke for the coast: "When we were small, we really knew how to hunker down against the wind, but it does not get as cold now."

Some lacked food during the coldest months. According to Simeon Agnus (December 2007:328, 340):

Indeed, we haven't experienced the suffering and hardship that our parents endured. Their arms were their only tools, and ones who didn't have dogs, their legs. That's why when winter came, some of us experienced starvation. And it was difficult to travel during winter when it was cold. . . .

They say the shrinking of the stomach is extremely painful. My, I wonder if our ancestors who had many children felt such great compassion for their children when they were going through starvation. When a child asks for food and there isn't any, they are pitiful to see.

Paul John (December 2007:337) recalled starvation times: "I consider [those people] to be so strong when they talk about what they experienced. Since they were so thin, their anuses couldn't close any longer. When they became like that, they say those who went to subsist in the wilderness plugged their anuses so that cold air wouldn't enter."

Although cold, the weather was often calm. Frank Andrew (October 2003:123, 134) recalled:

They say when the world was in its original state and was cared for and respected, in January the wind would weaken, and the shore-fast ice would be far out there down near the edge of deep water. They say it was as if the people of the coast were at spring camp when it got warm in February, starting to catch sea mammals. . . .

When I first became aware, the weather wasn't constantly bad. And the wind stopped blowing during spring. The blizzards they had were very strong, but they never went beyond two days. When it was windy from the south, it got good before daybreak. But the [wind] from the north got stronger and the bad weather lasted longer.

Paul John (December 2007:336) also recalled calm, windless weather in spring:

Since I first became aware of life, I never forgot the time when the weather was calm for long periods. Once when a light wind started to blow, I saw that it hadn't been windy for fourteen days. And the sides [of trails] were steeper because snow blowing along the ground hadn't filled them.

Long ago, back when there were longer periods of windless weather, when a lot of frost accumulated, when it started to get windy they said that although they were standing, they couldn't see their feet because the blizzard was so severe.

Paul John (March 2008:596) noted: "Our elders used to talk about the fact that it has gotten windier nowadays. These days, it seems that wind has gotten more frequent and it gets windy easily. The wind appears right away when it wants to." Winds are also changing: "The south wind brings warm weather. And during fall, the west wind hardly ever brings cold weather. I also found the following to be strange: although it was fall, it tended to be cold from the west. Changes that are occurring are affecting the weather, since some of our ways as people are changing" (Paul John, March 2008:600).

Many commented on the increasing unpredictability of today's weather. According to Paul John (January 2007:11): "They said that wild celery became as tall as people back when cold weather formed frost without any erratic and stormy weather. Then they began to say that since weather conditions are now unpredictable and stormy, the wild celery no longer forms frost. Back during times when the weather wasn't erratic and stormy, I think our land always stayed cold. That's why they called it *ellarrliyuitellra* [a time when there wasn't any erratic and stormy weather]."

Camilius Tulik (March 2007:565) noted how today wind appears to be coming but does not arrive: "*Ella cimissiiyaagtuq* [The weather has changed so much]. And when it is about to do something, it doesn't come to be. When it seems that it is going to get windy, it gets completely calm. [*laughs*]" Paul John (March 2008:596) commented: "In the past, they used to say that the weather would improve during the evening. Sometimes their prediction would come true. These days, although evening comes, a severe blizzard occurs when it wants to."

Weather conditions used as predictors in the past are no longer reliable. John Eric (March 2008:154) noted: "In the past, after a long spell of cold weather, these mountains [appear to] rise and become steep. And these cliffs that are a great distance away are also clearly visible. Those people would say that there was a mirage effect and that the weather was about to change. Although that occurs these days, they no longer use it to predict weather change. And although Nelson Island rises because of the mirage effect, the weather never changes." Paul John (March 2008:595) added: "There isn't just one way that the weather turns out when a mirage occurs. Sometimes approaching warm weather causes it to occur, and sometimes calm weather causes it to occur. But sometimes these days, it doesn't materialize. After the weather seems as though it is about to turn out a particular way, it doesn't occur."

According to Camilius Tulik (March 2007:502): "The weather, they say, has become a liar. When it is supposed to be calm, big winds come. When it seems like it will not get calm, it does. [*chuckles*]" Paul John (March 2008:595) said: "I believe what the deceased shaman of the people of Nightmute said. He said that *ella* is becoming an incessant liar. He said that although it seemed the weather was going to turn out a particular way, these days it no longer materializes."

Paul John's statement confirms what many believe — that their own elders predicted this unpredictability. According to Paul Tunuchuk (March 2007:213): "The weather conditions aren't like they were when I first became aware. My grandfather told me many times that in the future there will no longer be a winter season." Martina Wasili (March 2007:218) added: "These days, how pitiful it is. There is no calm weather at all. I recognize their predictions nowadays and realize that those people talked about how it would become one day." John Eric (March 2008:156) concluded: "We have now reached the time that was predicted that weather changes would occur, and I have come upon the prediction that I used to hear."

John Eric (December 2007:333) explained the prediction that *ella* would return to its original state: "Concerning weather change, my deceased grandfather used to say that *ella* never had a winter season long ago. He said that when there is no longer a winter season in the future, *ella* will return to its original state. But he said that the Lower Forty-Eight states used to have a winter season and that the weather would switch. And since this area didn't have a winter season in the past, it will no longer have a winter season in the future." Peter John (March 2007:1198) was equally nonplussed: "My father and those first ones said that our land will no longer have a winter season in the future, and the area [to the south] will start to have winter seasons. Although they've announced that this [global warming] is occurring, I'm not surprised by it, since I already heard that prediction."

Qanikcarpaguatuuq-gguq | It Pretends to Have a Lot of Snow

John Phillip (October 2005:119) grew up in a land blanketed by snow:

> All winter long, we would shovel deep in places where we set our wooden conical fish traps under very deep snow. They wouldn't freeze, because the snow covered them. And these days, even though a lake is deep, it freezes all the way down. And the river no longer has fish, because it freezes, even though it has a current. Because the lakes are getting shallow, they have begun to freeze up. Those are the changes that I have observed on the land surrounding us.
>
> And all winter long, our animals would travel under the snow. The mink that we hunted always did that. Even though they declined in number, they wouldn't be completely depleted back when people used them.

Golga Effemka (January 2006:27) noted how much water came down from the mountains when the snow melted: "For those of us upriver back then, the water was even with the top of the bank before the river ice broke up. That doesn't occur these days, as we no longer have [a large amount of] snow." The same was true along the coast. According to Frank Andrew (June 2005:9): "There was a lot of snow here in the past, and it covered the houses. That's why there was a lot of water out in the wilderness." Times have changed: "*Qanikcarpaguatuuq-gguq* [It pretends to have a lot of snow]. Today when the snow melts, there won't be much water" (George Billy, February 2006:401).

Peter Elachik (October 2005:26) observed a change in the character as well as the quantity of snow: "There's a point in time, 1965, that I can relate to climate changes. I never did hear any more noise when people started walking, because I think there's moisture content in the snow when it fell down. In the 1940s it used to be really dry snow, but now it's damp when it first falls. And I say it's 1965 because I remember the year my daughter was born, my firstborn, Flora."

Peter Elachik (January 2007:22) also recalled four- to six-week changes in the timing of freeze-up at the mouth of the Yukon:

> I want to give one picture of the village [of Kotlik] in 1945 compared to November 10, 2005. Sixty years ago, I remember the cold around November 10 because that's when the fur-trapping season began. [At that time] a lot of traveling took place for hunting and visiting with little caution [because the ice was safe].
>
> Then they started using caution in the mid-1960s. We started noticing a slight warming, and then by the 1990s, it was dangerous. And last year, November 10, 2006, no travel was being done, and the first travel began a few days before Christmas. So there's a lot of difference between 1945 and 2005. Big change.

Not only is freeze-up later, but rain continues intermittently through fall. Peter (October 2005:27) continued: "[In 1965] after it snowed in early October, then later that month we experienced rain after the snow. Prior to that, it never got warm, never did rain [after freeze-up]. But from 1965, I noticed that it rains in October. Ever since that, we've had rainfall every month of the year, from January to December, at least once during that month. Last big rain that we had was around Christmas back in Kotlik. I thought the electric wires would break because there was so much ice."

Imarpiim Cikua Imarpigmiutaat-llu Cimirtut |
Sea Ice and Its Inhabitants Are Changing

The sea ice bordering Nelson Island and the lower Kuskokwim coast during winter and spring makes up the southern edge of an enormous ring of ice surrounding the North Pole. Because of the ability of this ice to modify the world around it, some view the changes in sea ice that are now occurring as perhaps the most far-reaching physical change of the earth in our lifetimes.[2] Under these circumstances, the experiences and knowledge of Bering Sea hunters traveling in ice near its southernmost limit may be quite valuable in helping those living farther north adapt to changing conditions.[3] Certain qualities of the sea ice cover and its seasonality exhibit strong gradients with latitude, and there is some indication that these different zones are now shifting north as ice conditions get milder in the Arctic itself. Thus, the types of deformation mechanisms and ice features that Bering Sea hunters know well may have special relevance for northern hunters in the future.

As temperatures warm, Yup'ik elders have observed corresponding changes in sea ice and, as a result, access to the sea mammals that call the ice home. John Phillip (October 2005:116) noted that recently south of Nelson Island shore-fast ice is both thinner and less extensive: "The *tuaq* [shore-fast ice] used to be very thick, and it froze as much as six miles from shore. Nowadays our ocean doesn't freeze far from shore, and our *tuaq* and rivers become unsuitable for hunting because they are too thin and dangerous. And last year, we really couldn't go out seal hunting in the area below [Kongiganak and Kwigillingok] because the shore-fast ice was too thin." John Eric (December 2007:334) compared past and present: "The changes in the weather and the ocean have occurred in my presence. Back when the weather was [severely] cold, the ocean down below our village froze a great distance away from shore. And the shore ice would be dry; there were no wet spots on the snow. There were also many birds and seals. That no longer occurs." Paul John (January 2007:9) declared: "When shore ice formed a good distance out toward the ocean, they caught many sea mammals and knew that they wouldn't be scarce. Nowadays, the shore ice no longer extends far out because the weather isn't as cold as it was in the past."

North of Nelson Island at the mouths of the Ningliq and Aprun Rivers, however, water is getting shallow and the current decreasing, allowing an increase in ice buildup along the shore. Peter John (April 2009:78) explained:

Since the current is getting weaker, ice tends to easily stick [on to the shore ice] down below [Newtok] when the wind is blowing from the west. Although it isn't extremely cold out, the shore ice becomes extensive. In the past, the shore ice hardly ever extended a great distance out from shore, back when the current was strong.

Aprun used to be dangerous when the tide was coming in. Back when I used to sit [in a kayak] and hunt, I could feel a jolt [as the ice hit and passed by]. Today it no longer does that.

Many note that fewer *evunret* (ice piles) form. As some bays and river mouths become increasingly shallow, ice is also piling in new places. Simeon Agnus (January 2007:127) observed: "*Evunret* are starting to grow in places where they never formed before. Although no *evunret* grew in front of Kangirrluar before, *evunret* are starting to grow there nowadays. When the current moves large ice floes and they pile into a shallow area, they quickly rise and grow." Stanley Anthony (January 2007:128) noted that the same is true of *etgalqitat* (ice beached in shallow areas): "These days, Kangirrluar and the areas below Umkumiut and Qurrlurta that once had no *etgalqitat* now have *etgalqitat* because the ocean is getting shallower. Since the ocean is changing, that's what's happening; some areas that were deep are now shallow, and sandbars that were never visible are starting to appear."

Warming temperatures also translate into later freeze-up and earlier breakup in coastal communities. Peter John (March 2007:1196) explained the trend he observed: "In the past when it started to get cold, the ice froze and wouldn't stop freezing. It used to get cold in October and November. These days sometimes it doesn't get cold for a long time. And although it does get cold, it doesn't reach the cold temperatures that it reached when we traveled by dog team [through the 1960s]."

Many observe that shore-fast ice does not stay as long as in years past. Stanley Anthony (January 2007:129) said: "Kangirrluar no longer has genuine shore-fast ice on it. In the past birds would arrive while ice was still there, and the ice was safe for a long time." Mark John (March 2007:1197) commented on earlier breakup in spring: "When we came home from St. Marys [boarding school] in 1968, we landed in Tununak and they brought us [to Toksook Bay] by boat. I think it was May 28. Kangirrluar still had shore ice at the time. These days, the shore ice sometimes melts completely by early May."

With the decrease in shore ice, hunters often lack a safe platform for butchering seals. Stanley Anthony continued: "These days, when catching an animal, one has to search for a place to butcher it. Back when I first started hunting, a suitable place was nearby, and one could just quickly climb onto the ice and butcher it." John Eric (March 2007:262) noted that seals also lack ice: "Those adult bearded seals and even walrus could lay on top of large, moving ice floes when they were far from shore. Even though there were ocean swells, the water would only reach their top edges. Those large floes were good because they were thick and not dangerous. But ice floes have gotten thinner, and some break to pieces when we go on top of them. They are mostly snow."

Cikullaq (newly frozen ice along open water) is also less extensive. According to David Jimmie (January 2007:133): "When *cikullaq* forms, it thickens rapidly as thin sheets of ice stack and layer over one another. When one ice sheet piles over another, a boat is unable to travel right away. That's what happened in the past. Since *cikullaq* no longer forms extensively, that no longer occurs."

As ice conditions change, so does the presence of seals that make it their home. Simeon Agnus (January 2007:118) observed: "These days the seals have been arriving earlier. Sometimes they pass during winter. In the past an abundance of seals were available closer to the summer season." Elders claim that seals were also more plentiful in the past. David Jimmie (January 2007:118) noted: "I first started seal hunting using a kayak. We just ignored these spotted seals but only tried to hunt ringed seals and young bearded seals. Spotted seals would swim alongside us, but we never hunted them. These days, they are starting to get frightened when [people] make noise from afar." According to John Eric (March 2007:200, 248), not only are bearded seals and spotted seals becoming scarce but beluga whales are also rarely seen and the ringed seals that were once plentiful around bays have moved farther down into the ocean. Lizzie Chimiugak (January

2007:428) noted that even when seals are available, lack of ice shortens the hunting season: "Some time ago, men who went seal hunting in spring didn't catch as many young bearded seals as usual because of the ice. These days, it seems as though sometimes their seal-hunting season is cut short. They lack ice to hunt on. I know these things since I depend on [seal hunters] for food."

Ocean fish are likewise changing in availability. Many Nelson Island elders speak of *manignaleryiit* (Pacific cod), which were harvested until the 1950s but are no longer seen.[4] Capelin and herring are also declining in numbers. Edward Hooper (March 2007:1396) reported: "I used to go by kayak to fish for *manignaleryiit*, and after loading my kayak, I'd return home. We can no longer do that. No codfish, nothing. And although capelin are around, they aren't as abundant. And herring also—for two years now, the shore hasn't turned white [from herring eggs on kelp]." Commercial fishing on the high seas is blamed for this decline. Edward Hooper continued: "Seiners cause the lack of fish. They take food that Yup'ik people eat and then discard [herring] in the ocean after removing their eggs. The large boats fish in a way that causes the fish to deplete." Paul John (January 2007:8) declared: "It is obvious that the large trawlers that are dragging the bottom [of the ocean] for our fish have hurt [fish stocks]. Many years ago, when they first trawled down below Nunivak Island, they didn't catch any herring on Nelson Island. Although they caught a few, [the herring] were small and escaped their nets, and they didn't catch any after the first large trawlers fished there." John Phillip (January 2007:6) noted that what they do with the catch is as harmful as their harvesting methods: "When trawlers started to turn up along the ocean, some of these species of fish diminished in number. I heard that they dump and discard fish that aren't part of their intended catch. Our elders used to say that if people wantonly waste fish, they will decrease in number. Indeed, since they began to trawl for fish below Nunivak Island, the tomcod which we consume have been unavailable."

New species are also seen today. Stanley Anthony (January 2007:116) reported: "In the early days, [salmon sharks] were never seen. Although they fished for halibut [in deep water], I never heard that they caught sharks. But recently in the 1990s, I started to hear of people who were catching sharks." James James added: "Three years ago I set a net right below Up'nerkillermiut, to get fish for bait. When we checked the net, we caught a salmon shark that was brownish in color, maybe seven or eight feet long. We couldn't pull it into the boat as it was dead. We took its tail and brought it home and sank [the body]. Sharks started to be seen around the late 1990s, and we caught one here."

Nunam Qainga Kuiget-llu Cimirtut | The Land and Rivers Are Changing

The broad, marshy lowland of the Yukon-Kuskokwim delta has always been subject to slow subsidence (as sediments compact steadily under their own weight) and erosion (as sea level rises and river courses shift). Recently, however, the rate of change has escalated.[5] The rise in sea level and related effects of increased fall storm surges associated with global warming are of particular concern. Assuming the predicted rise in global sea level of approximately half a meter by 2100, large portions of the coastal margin will be underwater during high tides.[6]

Elders are well aware of these changes. Many comment on how the land is sinking. Peter Matthew (March 2008:38) stated: "Up to this day the entire wilderness has changed. The steep places that we used to see have sunk; especially the areas along the edges of hills where they used to set metal traps and hunt for mink during fall have disappeared." Many attribute this sinking to melting permafrost, and in fact, scientists estimate that the area affected by permafrost degradation, due in part to thaw settlement resulting from climate warming, has increased 3.6 percent from 1951 to 2008.[7] Paul John (January 2006:17) explained: "They refer to it as *cikuq* [ice]. They say that the grassy knolls have sunk because the [ice] underneath that had always stayed frozen has melted." Paul John (January 2006:15) described what he has observed: "When it used to get cold, I'd accompany others traveling with dogs. Then when I went to that place twenty years later, the land appeared different. Those that we call *pengut* [hills] and *evinret* [small grassy knolls] had started to sink in our old hunting area. Many of those had disappeared because they had melted. I thought I learned that area in the past, but since it has transformed, it has become unfamiliar and one can easily get lost." Paul (March 2008:568) added: "The old villages of Cevv'arneq and Arayak-caarmiut that used to be situated on high ground have sunk down. And inland around the hills, many *allngignat* [tundra islands] have sunk and become scarce. The melting land is obvious."

Land is sinking all across the delta. David Jimmie (March 2008:75) spoke of the area around Chefornak: "Last year when I went to the large grassy knoll that was once located alongside that lake, it was gone." Peter Jacobs (January 2006:41) reported the same for the tundra region: "When I became aware, there were a number of grassy knolls beyond where we stayed, alongside sloughs. Those have disappeared. When it wanted to, one would sink." Nick Andrew (January 2007:16) asserted: "It is obvious through those [hills sinking] that the ice underground is melting." In the distant past, elders recall, the land was thin. Nick concluded: "One day, when [the hills] sink and the land becomes level, it will become thin again."

Sinking land affects the animals that live on it. John Phillip (January 2006:26) spoke of the lower Kuskokwim coast: "There were many mink [in the past]. Since their birthing dens are sinking into the ground, their numbers are declining." Muskrats have also declined: "Although there didn't seem to be many muskrats around in the past, people would catch lots. But now that people no longer hunt these muskrats, I wonder, 'Why haven't their numbers increased in lakes?' Out in the wilderness, I no longer see a muskrat swimming in the evening. Or is it because some entity made things available for people to use in the past [and no longer does so]?" (John Eric, March 2007:208).

Periodic storm surges, especially in fall, have always flooded coastal lowlands but have likewise become more frequent and severe. Frank Andrew (September 2005:162) said: "During this time there are many floods in my village. Tuntutuliak and Kipnuk are experiencing more and more floods these days, and these villages also have more erosion and more water." John Phillip (December 2005:109) reported: "Today it floods more often. In the past it flooded in fall only once in a while, covering our muddy lowland. This past fall it flooded and covered the land three times when it was windy down there." Like sinking ground, flooding affects the land's inhabitants. John continued: "There used to be plenty [of Arctic hare] in the summer downriver, but after it began to flood, they decreased in number. Voles also become few after floods in our village. The flood would kill them. In the past when it rarely flooded, there were plenty. At summer fish camps we had young Arctic hare as pets."

Land is also eroding faster. Newtok has lost over three thousand feet of shoreline in the last fifty years, and the community has begun a move to higher ground on the north side of Nelson Island. Although Newtok's location on the north side of the Ningliq River has made it especially vulnerable to stormy weather coming from the south, coastal erosion is regionwide. John Eric (December 2007:334) observed: "Indeed, concerning changes in *ella*, the land and our river are eroding before my eyes; the mouth of our river has gotten wide." Many old village sites have eroded, and in some cases their imminent demise led to their abandonment.

River channels are changing, new ones forming and old ones disappearing. Joe Asuluk (January 2006:31) of Toksook Bay shared an example: "I became aware in Kayalivik. And I would just step over the [rivers] where my father set his wooden fish traps. One cannot do that any longer, as they have become wide, and some are no longer rivers. The land has truly changed inland. Some rivers have formed mouths elsewhere, and they are flowing out into other sources." Scientists agree that channel migration, through a combination of erosion and deposition, is the most prevalent process on the delta today, affecting close to 5 percent of the land's surface.[8]

As the land melts and erodes and river currents slacken, bays and rivers fill with silt and mud. Paul John (March 2008:567) recalled a Nelson Island shaman's prediction that this was in their future: "Just as we moved to this village [of Toksook Bay], that shaman told Tim Akagtaq [what was to come]. Frank Amadeus said that he had dreamed of the village of Nunakauyaq [Toksook Bay], that Kangirrluar had gotten shallow and turned to land. He said that Toksook River was the only river reaching the area down below the village of Nunakauyaq. He said that there were automobiles lined up in the village. He said that he was watching [the village] from down below [along the shore]." Ruth Jimmie observed: "That's probably how [the village] will become when permafrost melts, and the people from the lowland areas start to move [to Toksook Bay]. [They will move] to places that are steep."

John Phillip (January 2006:26) noted that less snow today means less water, and currents have weakened in many coastal rivers and channels. Peter Matthew (March 2008:38) gave an example: "This river called Urrsukvaaq used to have a strong current. These days, it no longer has a [strong] current, and the river mouth to the ocean has become shallow." John Eric (March 2007:203) described changes at the river's mouth: "During eight years, eight feet of land was eroded [from the bank] and widened. When I went up from the ocean some time ago during an extreme low tide, the inside of our river downstream was covered by sandbars since it has become wide. These rivers change and fill with sediment, and the deep, navigable channels also change. And the sandbars are also changing down on the ocean and aren't like they were in the past."

Lakes are also drying up. Paul John (January 2006:16) said: "Some lakes that were once filled with water are now empty, as they have a means of draining, probably due to the melting. They begin to refer to them as *nanvallret* [dry lake-beds] when they empty and different grasses start to grow in them. The changes that have occurred are noticeable through those indicators. The warming of *ella* has caused that to occur." Paul Kiunya (October 2005:6) observed: "The land is drying, too. There isn't as much water as there was in spring. These old lakes we call *nanvallret*, the lakes that had emptied through channels cut through the land, would fill with water [from snowmelt], and it became possible to travel through them with kayaks, before the water drained."[9]

Most freshwater fish are still plentiful. Simeon Agnus (December 2007:340) said: "The fish that we eat haven't changed, but for a number of years now black-fish have decreased down in our village. [These days] they usually become plentiful just before freeze-up; then they become scarce when winter comes." John Eric (March 2007:207) agreed: "Although these blackfish are still available, their numbers are decreasing." John Phillip (February 2006:176) attributed this decline to increasingly shallow streams freezing to the bottom: "Some rivers have a run of blackfish all winter if they are deep. The [availability of blackfish] doesn't cease, and eventually, when the water where blackfish are located gets dark, they taste a little like feces when cooking them, but they are still good eating. But since there isn't a lot of snow nowadays, since [the streams] freeze [down to the bottom], it is no longer like that during our time."

Some freshwater fish have reportedly become less accessible because of the rapidly increasing beaver population all over southwest Alaska. Peter Matthew (March 2008:38) remarked: "Beavers are ruining rivers that they build their dams on. The places where they fished for blackfish aren't like they were in the past."

Simeon Agnus (December 2007:381) declared: "The two rivers downriver from [Cakcaaq], Talarun and Urumangnaq, once had [an abundance of] fish back when there were no beavers. But they no longer have that many [fish], since the upper parts of those rivers are no longer rivers." John Phillip (January 2006:26) said: "In the past no beavers were down on the coast whatsoever. Now that there are many [beavers] it has added to the decline [of fish], damming the rivers when they are trying to live off the fish." Michael John (June 2008:80) added: "In the past, rivers and sloughs existed around lakes, and there were many lakes that hadn't formed channels out to other sources. When beavers started to come around, they caused these [lakes] to form channels and empty into other sources, and some no longer have water in them." John Eric (January 2007:9) remarked: "These beavers are destroying these wonderful natural rivers, and they don't ask [permission], but here we mention that it is on federal land. We should write to those beavers [*laughter*]." Some non-Natives suggest that trapping beavers might alleviate the problem, but coastal people—who neither eat beaver nor use many beaver pelts—are reluctant to hunt what they do not use.

Generally, river and lake ice was thicker in the past: "It seems that ice [today] isn't as thick as it was back when we used to dipnet at Cevv'arneq" (John Walter, March 2007:1393). Tommy Hooper (March 2007:1391) agreed: "Looking at the ice, it isn't as thick as it was back in those days."

Vegetation is also changing. Tommy Hooper (March 2007:1391) observed: "Things that grow on the land are starting to grow earlier. These salmonberries and crowberries are starting to grow before the time that they grew in the past." Lizzie Chimiugak (January 2007:306, 426) stated: "The salmonberry growing season comes early sometimes, and they are ready [to pick] very early. It's probably because of the temperature increase." According to Ruth Jimmie (March 2008:587): "The crowberries turn brown quicker, and they taste different. These days, one cannot delay in picking them. Sometimes we wait to pick them until they taste better, and when we finally pick them, we'll see that they've turned brown."

Nick Andrew (January 2006:12) observed changes in tree growth: "These days, since it doesn't get cold although it's winter, trees are growing in the area behind [Marshall]. When I was traveling in that area, what had been small trees along the mountain about ten years ago had become full-grown spruce trees, big ones. The warming weather has caused those [trees] to grow. And when I traveled from [Bethel] to [Marshall] I couldn't find the trail through the trees." This increase in tree and shrub growth in turn creates ideal habitat for both beaver and moose. Nick Andrew continued: "There were no moose in our lands when I was a child. When I first saw a moose's tracks, I didn't know what they were. Evidently it was a moose. And since we listened intently to those men so that we would learn, they said that when an animal made tracks, it was like a dead animal. They said that when they want to hunt them, they follow them."

Ben Angaiak with a beaver hunted
on the Cakcaaq River, July 2007. *Ann
Fienup-Riordan*

Beavers and even moose are now seen along the Bering Sea coast, where they were not present thirty years ago. Golga Effemka (January 2006:30) commented on their westward migration: "Concerning these animals, I used to hear that they are going downriver to the ocean. They said that when [animals] reach the ocean and drink the ocean waters, they will become scarce; they will die from the saltwater. I am starting to witness what they said would occur." Such movement is associated with impending scarcity. Johnny Thompson (February 2006:141) declared: "These animals are obvious when they are about to get scarce. *Tua-i-gguq anelrarturluteng uatmun* [They say they begin to continually move downriver]. Even rabbits moved downriver in the past and otters also. Some of our elders said that they went somewhere to transform into other forms. And they would say that belugas would go upriver to transform into wolves."

Southwest Alaska's coastal wetlands support hundreds of thousands of migrating waterfowl each year, and most shorebirds and songbirds continue to be numerous. Yet these numbers do not compare to the millions seen in the past. Peter Elachik (January 2007:1) spoke of what he had observed near St. Michael: "During the 1940s there was an abundance of seabirds, cormorants, puffins, and eider ducks. And during the 1950s and 1960s we started to notice their decline. Then in the 1980s and 1990s, they're gone. So all the seabirds declined." Lizzie Chimiugak (January 2007:429) recalled migrating waterfowl around Nelson Island, especially snow geese: "These birds were the first to become scarce.

Those many snow geese didn't seem as though they could deplete in number, and when we were camping in tents, the area used to turn white from the many snow geese, and even the mouth of the Qalvinraaq River would be very white. The birds they called *kangut* [snow geese] are no longer around at all." John Eric (March 2007:200) noted the disappearance of snow geese as well as king eiders: "There were many birds when I first became aware of life. And large flocks of king eiders during spring would head our way looking like smoke down on the ocean. Indeed, I only grazed those things in the past. King eiders aren't that abundant nowadays, and other birds are also decreasing in numbers."

Ellam Cimillran Kangia | Why the World Is Changing

Elders offer their own interpretations of the underlying causes of these changes. Some say that changes began following the 1964 earthquake—felt as a rolling wave in coastal communities. Others mention the atomic bombs dropped on Hiroshima and Nagasaki in 1945 and the Exxon Valdez oil spill of 1989. These events likely stand as markers of change rather than causes. Paul John (December 2007:376) spoke at length on the impact one change in attitude has had on the abundance of animals. Animals, Paul said, remain plentiful where people share resources but no longer appear where squabbling over exclusive rights occurs:

> Our ancestors admonished us not to quarrel over food. They say when people have disputes over food, that it gradually becomes scarce.
>
> Things that people depend on for their livelihood are given to us by the One Who Is Unseen. The One who provided that resource will make it available after seeing that people need it; although people use it, he will make it available afterward.
>
> I use fur animals as an example. Back when we made efforts at catching mink during fall, after seeming as though they were all gone, there would be more the next year, since people depended on them and used them. And during spring, although we hunted muskrats to the best of our ability, after seeming as though they were all gone, some years they'd become available in greater numbers, back when [muskrats] were people's only source of income.
>
> Food is no different. Although a capable person hunts for food, it can be available again when its time comes around.

Simeon Agnus (December 2007:378) wholeheartedly agreed with his cousin and gave the Cakcaaq River as an example: "We welcome villages beyond ours to subsist at Cakcaaq. Since it belongs to the people of Nelson Island, it is ours. And we open subsistence [at Cakcaaq] for the villages beyond ours." Because Nelson Islanders share Cakcaaq's rich fishery, the river continues to provide abundant resources.

As these discussions make clear, the Yup'ik conception of *ella* includes both natural and social phenomena. How we treat our fellow humans directly affects our relations with the world around us. Thus, contemporary elders are as much concerned about changes in human relations as about changes in the so-called natural world. They recall with feeling how people treated each other in the past compared with the present day. Young people used to be shy and respectful toward their elders, whom they could address only with *tuqluutet* (relational terms), never their proper names.[10] Joe Asuluk (January 2006:31) recalled: "Our elders were respectable back in those days, and it was embarrassing to call them by their names. Also, it was very embarrassing to look directly at [elders'] faces back then. And those [elders], I cannot picture their faces now, but there were actually many of them." John Eric (December 2007:334) noted the respect he had for his older sister and the food she served him: "After playing outdoors, when I'd go inside, my older sister would give me food to eat, or even fry bread, or when there were a number of us, sometimes she'd split one small piece of candy into three pieces when I'd bring my friends inside. What a large amount that was to me. And we never said, 'I want more.' That was something difficult to say. During that time, before *ella* changed, people were much better. And the people who instructed them led respectable lives."

Paul John (March 2008:602) noted that although food was sometimes scarce in the past, people did not let their relatives starve:

These days when the weather stays cold past its usual time, our refrigerators and freezers don't allow us to experience starvation. If our situation was like it was in the past, we would probably die of starvation during this time.

Back when we first became aware, during this time of year, some people had no food whatsoever, back when there were no food stamps. But since we Yup'ik people had close family bonds, our relatives who had food to give would help their relatives and allow them to reach a time when food was available to eat, because it is a tradition of these poor Yup'ik people.

Peter Jacobs (January 2007:13) explained how past elders also tried to understand change:

Our elders always tried to gain an understanding of why something was occurring. Our elders' predictions are now coming true. These days we recognize their predictions when we encounter them.

And they mentioned the following. I used to see a chunk of land eroding from a steep place, and down below, we'd see tools made by past people, including wooden artifacts that they used long ago. The land had become thick. But they mentioned that when the land became thin, the weather would get warm.

They also said that while living life we would start to hear that [the land to the south] would begin to get cold, and this area would become warmer. These things that they predicted are now taking place.

In the past they mentioned that, even though something will eventually occur, they were trying to delay it, what they called *yaaveskaniqangnaqluku* [trying to delay an impending occurrence].

They constantly spoke [and gave instructions]. And they also said, "We are speaking of this out of experience." They'd mention that the weather wasn't like it was in the past. They tried to understand what caused it to become that way. *Qantullruut augkut, yun'i maliggluki-gguq ella waten ayuqliriuq* [People said in the past that the weather has become like this following its people].

Our elders feared Ellam Yua long ago, and I think Ellam Yua gave them the wisdom to know about things that would occur in the future. It's no different than when they convene in Washington, DC, to pass legislation. Since *ella* will continue to worsen, [Ellam Yua] can also give them wisdom.

Peter Jacobs's reference to the world following its people cuts to the heart of Yup'ik understandings of change. Dozens of elders repeated this phrase in many contexts. Many made a direct connection between disrespectful treatment of *ella* and subsequent changes. According to Paul Tunuchuk (March 2007:110): "During my lifetime, *ella* is worsening since we are no longer treating it with care and respect." John Phillip (December 2005:120) stated: "I hear the *qaneryaraq* [saying] that our land and weather will worsen and change along with the people. What our ancestors said stays in my mind. . . .The environment isn't like it was in the past because it is no longer treated with respect and care." Paul Kiunya (October 2005:1) noted: "I have reached the *qanruyun* [instruction] that we were given concerning the weather. They said that if people get bad, the weather will get bad following its people."

Martina Wasili (March 2007:209) spoke of changes she has seen in her long life: "My grandchild, concerning the question you asked about the weather, people in the past said, '*Ella* is getting worse because you aren't treating it with respect.' I'm starting to see signs of what they were talking about: 'It will deteriorate along with its people.' We have reached that prediction during my lifetime." Sophie Agimuk (January 2007:431) also spoke of *ella* changing with its people, relating these changes directly to people's failure to follow traditional *qanruyutet* (instructions) and *eyagyarat* (abstinence practices):

> They mentioned that long ago, people always treated *ella* with respect because all the customs they followed had a purpose. They said that since people have stopped treating *ella* with respect these days, its condition has deteriorated.

And people who carried out subsistence activities by traveling down to the ocean and out on the land would observe *eyagyarat*. And when they lost a family member, they always followed *eyagyarat*, and the bereaved hardly traveled out to the wilderness.

Again, when women had their first menses, they didn't allow them to walk around outdoors. It is said people in that situation followed those customs because of their possible effect on *ella*. These days, people don't know about those practices, and our poor world is deteriorating. It's as though we poor people have become thoughtless and irrational. These days, we no longer practice the ways that they followed when we first became aware of life.

Roland Phillip (November 2005:245) declared that *ella* was rarely windy in the past because there were no rule breakers: "They said that when some great hunters became too sleepy, they would wish that it would get windy [so they could stop hunting and rest]. And they attributed [the calm weather] to the fact that there were no rule breakers down on the ocean at that time." John Phillip (October 2005:23, 117) testified to the continued awareness of *ella* and its reaction to those who do not respect its rules:

> By following us people, our weather has changed; it has gotten bad. And they would say this about *ella*: those who had their first menses and those who had a miscarriage are told not to travel. Those people had a teaching like this also: if their relative died in the wilderness, they didn't bring him anywhere but buried him there. It is said the weather will know about that dead body and cause bad weather if they take a person's body elsewhere.
>
> Those people tried to live following their *inerquun* [cautionary rule]. Today those rules are no longer followed. And now, they say that people have turned into white people, so to speak. When people die nowadays, they travel by plane and return home. I have noticed after [the deceased] is transported, the weather gets bad sometimes. It knows [of the death]. *Ella* is aware. . . .
>
> And our young people no longer follow *inerquutet* [admonishments] down on the ocean, even if his spouse just had a child or a miscarriage. It is no longer how it was back then. *Ella* is changing because it is not treated with care. . . .
>
> And our water must be treated with care also. It is said that *mer'em makuara* [the water's *makuat* (particles)] have good eyesight. It is said that [*makuat*] will be aware of those in that circumstance. Not respecting our land has caused it to change.

John Phillip is not alone. David Jimmie (January 2007:137) attributed the early arrival of ocean swells and subsequent breakup of shore ice and loss of many snowmobiles off the coast of Chefornak to the fact that a man whose son had died traveled to the ocean to hunt: "The ocean is aware and knowing, and the ocean has good eyesight. That person who went seal hunting after [his son] died caused the ocean swells to form when he went down [to the ocean]. The customs that our ancestors followed won't be lost."

Elders south of Nelson Island emphasize the negative impact of failure to follow *eyagyarat* and the world's awareness of human transgressions. Nelson Islanders focus on changes in interpersonal relations rather than relations between humans and their environment as the primary causes of change. Simeon Agnus (December 2007:327) stated: "They said that the world is following us, its people, since we have become shameful, since we no longer live like they did in the past. Those people had great compassion for one another although they were poor."

Paul John (December 2007:323) shared his reflections on changes in the world around him, both natural and social:

> When I was a boy, one of the elders told us the following: "Since you young people are no longer making efforts at improving the world, storms have become more frequent today."
>
> This is how I understood the meaning behind that statement. We young people during that time did not work hard like they did, and we didn't lose sleep over efforts at leading proper lives. We did not experience what our ancestors experienced, what they call *cilkiaryaraq* [strenuous training to become good hunters], like they did.
>
> Since we weren't losing sleep over how to live proper lives, since we weren't working hard and suffering like our ancestors, that person told us that storms are becoming more frequent because we weren't improving it [through our efforts].
>
> It has become obvious today that *ella* has changed. Through statements made by white people, there are more believers [in climate change]; they mention that the hole in the [ozone] layer that blocks the heat from hitting the land has gotten large. And these days that [ozone depletion] has become obvious as the constant melting of the land has become noticeable.

Like many of his generation, Paul John believes that these changes were predetermined. Martina Wasili (March 2007:210, 218) agreed: "I recognize the prediction that they made that *ella* would deteriorate. How did those people see our future? There is nothing surprising about what's occurring these days. . . . [My father] always said, '*Ella ikiurciiquq yugtuumarmi. Tamana tekiskunegteggu naklegnarqeqatarpagtat* [The weather will become terrible along with its people. When they reach that time, how pitiful they will be].'"

Qanruyutet Cimirngaitut | Qanruyutet Will Not Change

I want us to give our future descendants these things to the best of our knowledge.
Or you can make a book out of [the instructions that we are sharing] and show them

to the people, because the qanruyutet [oral instructions] concerning our way of life won't change.

—John Eric, Chefornak

Although both *ella* and its inhabitants are changing, *qanruyutet* remain the same. Elders admonish us to look to the *qanruyutet* for a solution to climate change and global warming. If we correct our behavior toward one another, they say, the world will be a better place. Simeon Agnus (December 2007:134) reminded his listeners: "No matter which village you travel to, if you are among people from the coastal region, you will listen and recognize the things that we talked about. The *qaneryarat* [teachings] of the Yup'ik people, their *alerquutet* [admonishments] are the same. At this time, we are telling stories based on the few things that we've heard and aren't adding things [that are fabricated]. I wish that the people who have died were talking to you now in the authentic way!"

Peter John (March 2007:1164) spoke of his father's instructions, which he uses to this day: "Taking the oral instructions with you when traveling gives you good judgment. I continually take the instructions with me, and it is like I'm constantly accompanying my father by following his instructions. Indeed, when one makes an effort at following the instructions on the land and ocean and doesn't forget them, they are valid." John Eric (March 2007:251) likewise carries his instructions with him: "If we pay attention to the words spoken by those who brought us traveling, we will obtain information, and they will be like provisions to us when traveling." Paul Jenkins (March 2007:233) declared: "They said that all people on this earth who live by the *qanruyutet* and don't scatter their *qanruyutet* will be alive an extra day [live a long life]."

At the close of our Nelson Island gatherings, Paul John (December 2007:366–72) shared a long, eloquent account of some of the most important instructions that he was given while growing up, instructions that he continues to live by.

> I grew up near the village at Cevv'arneq. I listened to people who were speaking, and I was given instructions also. Although she was a woman, my father's deceased older sister left me instructions about the way of living, like she was filling my bag with advice that I could live by.
>
> When my father's older sister instructed me, she would constantly pretend as though other people were asking where I was as she spoke: "If your parents do not have wood, people will say, `Where is their son who should be getting wood for them?' If your parents do not have food, people will say, `Where is their son who should be getting food for them?'" I am grateful for that. Keeping in mind the idea that people would constantly ask where I was, from the time I became capable, I tried to make a livelihood to the best of my ability. . . .
>
> If we give these few traditional moral instructions to young people and not

try to keep them to ourselves, if we speak of them, we will be passing them down to our young people. And we must not say that a young person will not listen. Although they seem as though they aren't listening, if a young person hears people giving moral advice, he will be listening to things that he will think about during his entire life.

We who live to be elders are actually filled with large amounts of wisdom; we are holding on to the ways of leading good lives. If we do not speak of them, we just keep them stored away. But if we speak of them, we can encourage a person to live as though he can speak for himself as he is living.

There is the following instruction: "If you listen intently to *qanruyutet* during your short life, you will be your own instructor as you live."

You [Alice] are now starting this process yourself. If you carry out a task in the proper way following what you learned, you will work on it, instructing yourself. It is indeed true that if a person scrutinizes his actions, he will live by instructing himself. While he is living, when he comes upon something, he will say, "Oh yes, they've mentioned that engaging in something like this has a good outcome. Let me do it." The result is gratitude. And a person will also be grateful if he doesn't engage in something that they said would have a bad outcome.

They say our ancestors, after they experienced something and saw that it was good to engage in, spoke of it. That's evidently true.

Like Paul John, many remembered with gratitude their own elders sharing *qanruyutet* and the admonishment that they in turn share. David Jimmie (March 2007:212) recalled: "My grandfather told me, 'When I eventually stop advising you when I am gone, remember the instructions your father and mother gave you.' And he told me not to keep the instructions that my mother and father gave me but to constantly reveal them." David remembered these admonishments sounding like scolding: "Since my grandfather spoke to me critically, I asked my grandmother, 'Why does my grandfather speak to me as though he despises me?' Then she replied, 'Oh my, don't think that way; since he loves you, he's talking to you in that manner, advising you about what will come in your future.'"

Elders agreed on the need to continue to instruct young people today for the same reason—because they love them. Simeon Agnus (January 2007:3) placed particular importance on teaching about the ocean: "One has to use caution down on the ocean. I tell these young men not to say that they've learned the ocean, since I stopped going to the ocean before I learned it."

Knowing the land is also important. Simeon continued: "[Traveling] on the land isn't as worrisome [as traveling on the ocean]. A young man who is attentive will study the land well. That's the only thing that I mention to some young men. I tell them to look around and study surroundings in unfamiliar territory. 'Although you won't always travel in that area, one day when you arrive there, you will recognize it.'"

Paul John (December 2007:385) noted that only a person who instructs himself and takes responsibility for his actions can lead a good life: "We people are living by our own individual minds. A person can evidently live properly only through his consciousness. A person who is constantly instructing himself through his own consciousness to live in a way that will result in his own well-being, through his own efforts, is evidently following a course of life that will result in leading an honorable life through his mind." Simeon Agnus (December 2007:365) held a similar view of personal responsibility: "Nelson Island, in my thinking, is situated in the middle of fish. But although there is an abundance of fish, those of us who are lazy run out [of fish] when winter comes. We are responsible for our own livelihood."

Given their view of personal responsibility, it is no surprise that elders make a connection between human impacts on the environment and the "natural" effects of climate change. Throughout our discussions, they continually referred to the role of human action in the world when describing changes in the environment or species availability. Their insistence that "the world is changing following its people" logically flows from their view of the world as responsive to human thought and deed. *Ella* has always been understood as intensely social. The Western separation between natural and social phenomena sharply contrasts with the ideas expressed in our Yup'ik conversations, which eloquently focus on their connection.

Elders attribute environmental change not only to human action—wasteful fishing, burning fossil fuels—but to human interaction. To solve the problems of global warming elders maintain that we need not only to change our actions but to correct our fellow humans. They encourage youth today to attend to traditional *qanruyutet*, believing that if their values improve, correct actions will follow. Chris George (March 2007:796) concluded hopefully: "It is true that the weather is following its people by getting bad, but if we make efforts to improve ourselves, the weather will follow suit and become better."

Finally, elders do not dissociate themselves from observed changes in their homeland but accept personal responsibility. They relate the negative impacts of change they observe today to their failure to instruct their younger generation in proper behavior. Now, they say, is the time to reverse this trend. Observing uninstructed young men and women, John Eric (January 2007:26) remarked, "We must talk to them, to delay them from becoming like dogs." Elders warmly embraced our work together, which they view as much more than passive documentation of change, but as part of an active solution.

NOTES

Introduction

1 Andrew 2008; Fienup-Riordan 2005a, 2005b, 2007; Meade and Fienup-Riordan 2005; Rearden, Meade, and Fienup-Riordan 2005.

2 Fienup-Riordan 2005a:1–41.

3 US Fish and Wildlife Service 2002.

4 Shaw 1998.

5 Oswalt 1990:51.

6 Fienup-Riordan 1991, 1994; Oswalt 1990.

7 Drebert 1959:42; Society for Propagating the Gospel 1916:41.

8. Fienup-Riordan 1991.

9. Oswalt 1990: 145; Fortuine 2005.

10. See Fienup-Riordan 2012 for a detailed discussion of this period of rapid change.

Qanruyutet Anirturyugngaatgen/*Qanruyutet* Can Save Your Life

1 Observations of satellite-tracked drifters deployed just offshore of Quinhagak confirm what Atertayagaq learned from experience more than a century earlier—that coastal waters can be advected toward the southwest (Danielson et al. 2010).

2 Jacobson 1984:621.

3 Fienup-Riordan 2000:109–49.

4 Burch 1988; Prior and Graburn 1980, cited in Kishigami's excellent 2002 summary.

5   Feit 2002; Haraway 1989; Rose 2002.

6   Fienup-Riordan 1999.

7   Brightman 1983; Hallowell 1964; Scott 2002.

8   Alaska Injury Prevention Center 2007.

9   Cruikshank 2005:9.

Nuna-gguq Mamkitellruuq | They Say the Land Was Thin

1   McAtee 2008.

2   McAtee 2008.

3   James Kari, personal communication, February 2009, Fairbanks.

4   Young girls traditionally used carved and incised ivory or wooden story knives to draw
     designs in mud or snow as they told each other stories. In the 1970s and 1980s butter
     knives often sufficed.

5   Jacobsen 1977:187.

6   Boas 1901–7:149; Rasmussen 1921–25:11, 1938:200; Søby 1969:54.

7   Pratt and Shaw 1992:13.

8   Noatak, Amos, and Drozda 2007.

9   O'Leary 2008.

10   June McAtee, personal communication, March 26, 2009.

11   Guy 1988; Nicori 1988.

12   Andrew 1988; Guy 1988; Hutchinson 1991.

13   Mellick 1988a.

14   Nelson 1899:445–46; *Delta Discovery* 2002.

15   Pratt 1993.

16   Samuels 1986.

17   Lantis 1947:13–14; Pratt and Shaw 1992:8.

18   Mellick 1988b.

19   Pratt and Shaw 1992:6.

20   Pratt and Shaw 1992:13.

21   Pratt and Shaw 1992:8.

Ella Alerquutengqertuq | The World and Its Weather Have Teachings

1   Many have written about the relationship between human communities and the
     environment. Though the concept of environmental adaptation has a long history in
     anthropology, newer approaches are beginning to emerge that emphasize the cultural
     significance of weather and climate. See especially Basso 1996; Ingold 2000; Strauss and
     Orlove 2003.

2   Nelson 1899:481–82; Lantis 1946:268; Fienup-Riordan 1994:263.

3   Sienkiewicz 2003.

4   *Wikipedia*, s.v. "Halo (optical phenomenon)," accessed March 2011, http://en.wikipedia.
     org/Wiki/halo_(optical_phenomenon); James Brader, personal communication, May 4,
     2009, Fairbanks.

5   See Fabian 2001 for a good introduction to ethnoastronomy, and MacDonald 1998 and
    Knudsen 2008 for star lore of the circumpolar north.
6   Lantis 1946:197.
7   James Brader, personal communication, May 4, 2009, Fairbanks.
8   Knudsen 2008:3.
9   The Big Dipper is associated with caribou or reindeer throughout the Arctic.
10 Lummerzheim 2008.
11 James Brader, personal communication, May 4, 2009, Fairbanks.
12 Fienup-Riordan 1994:310–11, 315–16.
13 James Brader, personal communication, May 4, 2009, Fairbanks.
14 Fienup-Riordan 1994:161.
15 For a discussion of the origin of winds in Doll's ritual circuit of the earth-plane, see
    Nelson 1899:497–99; and Fienup-Riordan 1994:259.
16 See Fienup-Riordan 1996.

Nunavut | Our Land

1   Jorgenson and Ely 2001:129.
2   Shaw 1998.
3. For a detailed version of this tale, see Fienup-Riordan and Rearden 2011: 28–33.
4. US Geological Survey maps mistakenly refer to Ingriik as Tern Mountain.

Kuiget Nanvat-llu | Rivers and Lakes

1   Leary 2010.
2   June McAtee, personal communication, July 2010.

Yuilqumun Atalriit Qanruyutet | Instructions Concerning the Wilderness

1   The literature on place-names and their meanings is vast (e.g., Cruikshank 1990; Basso
    1996). See Fienup-Riordan and Rearden 2011 for a detailed discussion pertaining to
    southwest Alaska.
2   Matthew Sturm, personal communication, March 2010, Fairbanks.
3   Lipka 1998:187.

Qanikcaq | Snow

1   For those curious about the physics of snow and what it means to Native Alaskans,
    Matthew Sturm (2009) has written an excellent children's book and accompanying
    teachers' guide combining his expertise as a snow scientist with observations by elders in
    Barrow, Alaska.
2   Martin 1986.
3   Aporta 2010; Kaplan 2005; Pullum 1991:159–71.

4   Matthew Sturm, personal communication, August 2009, Fairbanks.

5   Sturm and Benson 1997; Matthew Sturm, personal communication, August 2009, Fairbanks.

6   Matthew Sturm, personal communication, March 2010, Fairbanks.

7   See Sturm 2009 for a clear discussion of the physics of the effects of wind on snow.

Imarpik Elitaituq | The Ocean Cannot Be Learned

1   See Frank Andrew's map (2008:36) showing the sandbars' locations.

2   Jorgenson and Ely 2001:134.

3   Fienup-Riordan 1999.

Ciku | Ice

1   In our discussion of Yup'ik sea ice we stand on the shoulders of classic studies, especially Nelson 1969, as well as extensive recent sea ice research in other parts of the Arctic, including Aporta 2002 and 2010; Druckenmiller et al. 2010; Eicken 2010; Eicken et al. 2009; Gearheard et al. 2006; Krupnik et al. 2010; Laidler et al. 2010; and Oozeva et al. 2004. See http://www.aspect.aq/ for a glossary and image library of scientific sea ice terms, e.g., nilas, frazil ice, pancake ice.

2   Hajo Eicken, personal communication, August 14, 2009, Fairbanks.

3   Hajo Eicken, personal communication, August 14, 2009, Fairbanks; Druckenmiller et al. 2010.

4   Hajo Eicken, personal communication, August 14, 2009, Fairbanks.

5   Hajo Eicken, personal communication, August 14, 2009, Fairbanks.

6   Hajo Eicken, personal communication, August 14, 2009, Fairbanks.

7   Hajo Eicken, personal communication, August 14, 2009, Fairbanks.

8   Hajo Eicken, personal communication, August 14, 2009, Fairbanks.

Yun'i Maliggluki Ella Ayuqucimitun Ayuqenrirtuq | The World Is Changing Following Its People

1   The literature on climate change across the Arctic is extensive, including Huntington and Fox 2005; Kolbert 2006; Krupnik and Jolly 2002; Laidler 2006; and Oakes and Riewe 2006.

2   Rozell 2009:xi.

3   Hajo Eicken, personal communication, August 14, 2009, Fairbanks.

4   See Drozda 2010 for a discussion of Pacific cod and other fish species near Nunivak Island.

5   Analyzing photographs taken from the air, Jorgenson and Dissing (2009) estimate that as much as 11 percent of the landscape has changed in some interpretable manner from 1951 to 2008.

6   Jorgenson and Ely 2001:134.

7   Jorgenson and Dissing 2009.

8   Jorgenson and Dissing 2009.

9   Torre Jorgenson (personal communication, May 2010, Anchorage) notes that lake drainage or tapping by channel migration is not common, affecting only 3 percent of the coastal landscape between 1951 and 2008. Where it does occur, however, such drainage has dramatic effects, as elders observe.

10 For a detailed discussion of *tuqluutet*, see Fienup-Riordan 2005a:209–53.

REFERENCES

Alaska Injury Prevention Center. 2007. "Suicide in Alaska." Juneau: Alaska State Department of Health and Social Services.

Andrew, Frank, Sr. 2008. *Paitarkiutenka/My Legacy to You*. Seattle: University of Washington Press.

Andrew, Wassillie. 1988. Oral history interview. Lisa Hutchinson, interviewer; John Andrew, interpreter. Kwethluk, AK. July 12. Tape 88CAL64. Bureau of Indian Affairs ANCSA Office.

Aporta, Claudio. 2002. "Life on the Ice: Understanding the Codes of a Changing Environment." *Polar Record* 38(207):341–54.

———. 2010. "The Sea, the Land, the Coast, and the Winds: Understanding Inuit Sea Ice Use in Context." In Krupnik et al. 2010:163–80.

Basso, Keith. 1996. *Wisdom Sits in Places: Landscape and Language among the Western Apache*. Albuquerque: University of New Mexico Press.

Boas, Franz. 1901–7. *The Eskimo of Baffin Land and Hudson Bay*. Bulletin of the American Museum of Natural History, vol. 15, pts. 1–2. New York: Trustees of the American Museum of Natural History.

Bobby, Pete 1987. Oral history interview. Margie Connolly and Robert Waterworth, interviewers. Alaska. June 25. Tape 87LMV019. Bureau of Indian Affairs ANCSA Office.

Brightman, Robert Alain. 1983. "Animal and Human in Rock Cree Religion and Subsistence." Ph.D. diss., University of Chicago.

Burch, Ernest, Jr. 1988. "Mode of Exchange in Northwest Alaska." In *Hunters and Gatherers: Property, Power and Ideology*, edited by Tim Ingold, D. Riches, and James Woodburn, 95–109. Oxford: Berg.

Cruikshank, Julie. 1990. "Getting the Words Right: Perspectives on Naming and Places in Athapaskan Oral History." *Arctic Anthropology* 27(1):52–65.

———. 2005. *Do Glaciers Listen? Local Knowledge, Colonial Encounters, and Social Imagination*. Vancouver, BC: UBC Press.

Danielson, S., L. Eisner, T. Weingartner, and K. Aagaard. 2010. "Thermal and Haline Variability over the Central Bering Sea Shelf: Seasonal and Interannual Perspectives." *Continental Shelf Research* 31:539–54.

Delta Discovery (Bethel, AK). 2002. "Is It a Bird, Is It a Plane? Huge Winged Creature Sighted off Nelson Island." March 13.

Drebert, Ferdinand. 1959. *Alaska Missionary*. Bethlehem, PA: Moravian Book Shop.

Drozda, Robert M. 2010. *Nunivak Island Subsistence Cod, Red Salmon and Grayling Fisheries, Past and Present*. US Fish and Wildlife Service, Office of Subsistence Management Fisheries Resource Monitoring Program, Final Report (Study 05–353). Fairbanks, AK.

Druckenmiller, Matthew L., Hajo Eicken, John C. George, and Lewis Brower. 2010. "Assessing the Shorefast Ice: Iñupiat Whaling Trails off Barrow, Alaska." In Krupnik et al. 2010: 203–28.

Eicken, Hajo. 2010. "Indigenous Knowledge and Sea Ice Science: What Can We Learn from Indigenous Ice Users?" In Krupnik et al. 2010:357–76.

Eicken, Hajo, Rolf Gradinger, A. Graves, A. Mahoney, and I. Ribor. 2005. "Sediment Transport by Sea Ice in the Chukchi and Beaufort Seas: Increasing Importance Due to Changing Ice Conditions." *Deep-Sea Research II* (52): 3281-3302.

Eicken, Hajo, Rolf Gradinger, Maya Salganek, Kunio Shirasawa, Don Perovich, and Matti Lepparanta, eds. 2009. *Field Techniques for Sea Ice Research*. Fairbanks: University of Alaska Press.

Fabian, Stephen M. 2001. *Patterns in the Sky: An Introduction to Ethnoastronomy*. Long Grove, IL: Waveland Press.

Feit, Harvey. 2002. "Animals as Political Partners: James Bay Cree on Reciprocity, Persons, Power and Resistance." Paper presented at the Ninth International Conference on Hunting and Gathering Societies, Edinburgh, Scotland.

Fienup-Riordan, Ann. 1991. *The Real People and the Children of Thunder: The Yup'ik Eskimo Encounter with Moravian Missionaries John and Edith Kilbuck*. Norman: University of Oklahoma Press.

———. 1994. *Boundaries and Passages: Rule and Ritual in Central Yup'ik Oral Tradition*. Norman: University of Oklahoma Press.

———. 1996. *The Living Tradition of Yup'ik Masks: Agayuliyararput/Our Way of Making Prayer*. Seattle: University of Washington Press.

———. 1999. "'*Yaqulget Qaillun Pilartat* (What the Birds Do)': Yup'ik Eskimo Understandings of Geese and Those Who Study Them." *Arctic* 52(1):1–22.

———. 2000. *Hunting Tradition in a Changing World: Yup'ik Lives in Alaska Today*. New Brunswick, NJ: Rutgers University Press.

———. 2005a. *Wise Words of the Yup'ik People: We Talk to You because We Love You*. Lincoln: University of Nebraska Press.

———. 2005b. *Yup'ik Elders at the Ethnologisches Museum Berlin: Fieldwork Turned on Its Head*. Seattle: University of Washington Press.

———. 2007. *Yuungnaqpiallerput/The Way We Genuinely Live: Masterworks of Yup'ik Science and Survival.* Seattle: University of Washington Press.

———. 2010a. "The Ice Is Always Changing: Yup'ik Understandings of Sea Ice, Past and Present." In Krupnik et al. 2010:303–28.

———. 2010b. "Yup'ik Perspectives on Climate Change: 'The World Is Following Its People.'" *Études/Inuit/Studies* 34(1):55–70.

———. 2012. *Mission of Change in Southwest Alaska: Conversations with Father René Astruc and Paul Dixon on Their Work with Yup'ik People, 1950-1988.* Fairbanks: University of Alaska Press.

Fienup-Riordan, Ann, and Alice Rearden. 2011. *Qaluyaarmiuni Nunamtenek Qanemciput/ Our Nelson Island Stories: Meanings of Place on the Bering Sea Coast.* Seattle: University of Washington Press.

Fortuine, Robert. 2005. *Must We All Die? Alaska's Enduring Struggle with Tuberculosis.* Fairbanks: University of Alaska Press.

Gearheard, S., W. Matumeak, I. Angutikjuaq, J. Maslanik, H. P. Huntington, J. Leavitt, D. Matumeak Kagak, G. Tigullaraq, and R. G. Barry. 2006. "'It's Not That Simple': A Collaborative Comparison of Sea Ice Environments, Their Uses, Observed Changes, and Adaptations in Barrow, Alaska, USA, and Clyde River, Nunavut, Canada." *Ambio* 35(4):203–11.

Guy, James. 1988. Oral history interview. Lisa Hutchinson, interviewer; John Andrew, interpreter. Kwethluk, AK. July 13. Tape 88CAL67. Bureau of Indian Affairs ANCSA Office.

Hallowell, A. Irving. 1964. "Ojibwa Ontology, Behavior and World View." In *Primitive Views of the World*, edited by Stanley Diamond, 45–62. New York: Columbia University Press.

Haraway, Donna. 1989. *Primate Visions: Gender, Race, and Nature in the World of Modern Science.* New York: Routledge.

Harkin, Michael E., and David Rich Lewis, eds. 2007. *Native Americans and the Environment: Perspectives on the Ecological Indian.* Lincoln: University of Nebraska Press.

Huntington, H. P., and S. Fox. 2005. "The Changing Arctic: Indigenous Perspectives." In *Arctic Climate Impact Assessment*, edited by C. Symon, L. Arris, and B. Heal, 61–98. New York: Cambridge University Press.

Hutchinson, Lisa. 1991. "Report of Investigation for Arnasagaq." BLM AA-9852, BLM AA-9853. Anchorage, AK.

Ingold, Tim. 2000. *The Perception of the Environment: Essays in Livelihood, Dwelling, and Skill.* New York: Routledge.

Jacobsen, Johan Adrian. 1977. *Alaskan Voyage, 1881–1883: An Expedition to the Northwest Coast of America.* Abridged translation by Erna Gunther from the German text of Adrian Woldt. Chicago: University of Chicago Press.

Jacobson, Steven A. 1984. *Yup'ik Eskimo Dictionary.* Fairbanks: Alaska Native Language Center.

Jorgenson, M. Torre, and Dorte Dissing. 2009. "Landscape Changes in Coastal Ecosystems, Yukon-Kuskokwim Delta." Report for the National Wetland Inventory, US Fish and Wildlife Service, Anchorage, AK.

Jorgenson, Torre, and Craig Ely. 2001. "Topography and Flooding of Coastal Ecosystems on the Yukon-Kuskokwim Delta, Alaska: Implications for Sea-Level Rise." *Journal of Coastal Research* 17(1):124–36.

Kaplan, Lawrence D. 2005. "Inuit Snow Terms: How Many and What Does It Mean?" In *Building Capacity in Arctic Societies: Dynamics and Shifting Perspectives*, edited by François Trudel, 263–69. Proceedings of the Second IPSSAS Seminar. Quebec: CIERA.

Kishigami, Nobuhiro. 2002. "A Typology of Food Sharing Practices among Hunter-Gatherers, with a Special Focus on Inuit Examples." Paper presented at the Ninth International Conference on Hunting and Gathering Societies, Edinburgh, Scotland.

Knudsen, Ole. 2008. "A Collection of Curricula for the STARLAB Inuit Star Lore Cylinder." Science First/STARLAB, Buffalo, NY. *www.starlab.com*.

Kolbert, Elizabeth. 2006. *Field Notes from a Catastrophe: Man, Nature, and Climate Change.* New York: Bloomsbury.

Krupnik, Igor, Claudio Aporta, Shari Gearheard, Gita Laidler, and Lene Klelsen Holm, eds. 2010. *SIKU: Knowing Our Ice; Documenting Inuit Sea Ice Knowledge and Use.* New York: Springer.

Krupnik, Igor, and D. Jolly, eds. 2002. *The Earth Is Faster Now: Indigenous Observations of Arctic Environmental Change.* Fairbanks, AK: ARCUS.

Laidler, Gita. 2006. "Inuit and Scientific Perspectives on the Relationship between Sea Ice and Climate Change: The Ideal Complement?" *Climate Change* 78(2–4):407–44.

Laidler, Gita J., Pootoogoo Elee, Theo Ikummaq, Eric Joamie, and Claudio Aporta. 2010. "Mapping Inuit Sea Ice Knowledge, Use, and Change in Nunavut, Canada (Cape Dorset, Igloolik, Pangnirtung)." In Krupnik et al. 2010:45–80.

Lantis, Margaret. 1946. "The Social Culture of the Nunivak Eskimo." *Transactions of the American Philosophical Society* (Philadelphia) 35:153–323.

———. 1947. *Alaskan Eskimo Ceremonialism.* American Ethnological Society, Monograph 11. Seattle: University of Washington Press.

Leary, Mark. 2010. "Iyana Gusty, Sr., Captain of the Kuskokwim." *Delta Discovery* (Bethel, AK), January 27, 20.

Lipka, Jerry. 1998. "Expanding Curricular and Pedagogical Possibilities: Yup'ik-Based Mathematics, Science, and Literacy." In *Transforming the Culture of Schools: Yup'ik Eskimo Examples*, by Jerry Lipka with Gerald V. Mohatt and the Ciulistet Group, 185–200. Mahwah, NJ: Lawrence Erlbaum.

Lummerzheim, Dirk. 2008. "Can You Hear the Aurora?" In *Frequently Asked Questions about Aurora and Answers.* http://odin.gi.alaska.edu/FAQ/.

MacDonald, John. 1998. *The Arctic Sky.* Toronto: Royal Ontario Museum.

Martin, Laura. 1986. "Eskimo Words for Snow: A Case Study in the Genesis and Decay of an Anthropological Example." *American Anthropologist* 88(2):418–23.

McAtee, June. 2008. "Traditional Use of Pigments and Mineral Material from Nelson Island." Paper presented at the Thirty-Fifth Annual Meeting of the Alaska Anthropological Association, February 27, Anchorage, AK.

Meade, Marie, and Ann Fienup-Riordan. 2005. *Ciuliamta Akluit: Things of Our Ancestors.* Seattle: University of Washington Press.

Mellick, Nick, Jr. 1988a. Interview with Alec Aloyouis of Kalskag, June 4. Mellick Tape 2:44, side A. National Park Service, Anchorage, AK.

———. 1988b. Interview with Peter Smith, September 16. Mellick Tape 29. National Park Service, Anchorage, AK.

Nelson, Edward William. 1899. *The Eskimo about Bering Strait*. Bureau of American Ethnology Annual Report for 1896–97, vol. 18, pt. 1. Washington, DC: Smithsonian Institution Press. (Reprinted 1983.)

Nelson, Richard K. 1969. *Hunters of the Northern Ice*. Chicago: University of Chicago Press.

Nicori, Fannie. 1988. Oral history interview. Lisa Hutchinson, interviewer; Ida Alexie, interpreter. Kwethluk, AK. June 29. Tape 88CAL044. Bureau of Indian Affairs ANCSA Office.

Noatak, Nuratar Andrew, Nakaar Howard Amos, and Robert Drozda. 2007. "Uraquralrig/Sibling Brothers." In *Words of the Real People*, edited by Ann Fienup-Riordan and Lawrence Kaplan, 102–21. Fairbanks: University of Alaska Press.

Oakes, Jill, and R. Riewe, eds. 2006. *Climate Change: Linking Traditional and Scientific Knowledge*. Winnipeg, MB: Aboriginal Issues Press.

O'Leary, Matt. 2008. "Report of Reinvestigation for Kegginaquq." US Bureau of Indian Affairs, ANCSA Office, Anchorage, AK.

Oozeva, Conrad, Chester Noongwook, George Noongwook, Christina Alowa, and Igor Krupnik. 2004. *Watching the Ice Our Way/Sikumengllu Eslamengllu Esghapalleghput*. Washington, DC: Arctic Studies Center, Smithsonian Institution.

Oswalt, Wendell. 1990. *Bashful No Longer: An Alaskan Eskimo Ethnohistory, 1778-1988*. Norman: University of Oklahoma Press.

Pratt, Kenneth L. 1993. "Legendary Birds in the Physical Landscape of the Yup'ik Eskimos." *Anthropology and Humanism* 18(1):13–20.

Pratt, Kenneth L., and Robert D. Shaw. 1992. "A Petroglyphic Sculpture from Nunivak Island, Alaska." *Anthropological Papers of the University of Alaska* 24(1–2):3–14.

Prior, F. L., and Nelson Graburn. 1980. "The Myth of Reciprocity." In *Social Exchange: Advances in Theory and Research*, edited by Kenneth J. Gergen, Martin S. Greenberg, and Richard H. Willis, 215–35. New York: Plenum.

Pullum, Geoffrey K. 1991. *The Great Eskimo Vocabulary Hoax*. Chicago: University of Chicago Press.

Rasmussen, Knud. 1921–25. *Myter og Sagn fra Grønland, I-III*. Copenhagen: Gyldendal.

———. 1938. "Knud Rasmussen's Posthumous Notes on the Life and Doings of the East Greenlanders in Olden Times." Edited by H. Osterman. *Meddr Grønland* 109(1).

Rearden, Alice, Marie Meade, and Ann Fienup-Riordan. 2005. *Yupiit Qanruyutait/Yup'ik Words of Wisdom*. Lincoln: University of Nebraska Press.

Rose, Deborah. 2002. "The Dead, the Missing, the Lost, and the Voiceless: Some Thoughts on Extinction from a Dingo Perspective." Paper presented at the Ninth International Conference on Hunting and Gathering Societies, Edinburgh, Scotland.

Rozell, Ned. 2009. "Encounters with Northern Sea Ice." In Eicken et al. 2009:xi–xx.

Samuels, Willie. 1986. Oral history interview. Harley Cochran and Phyllis Gilbert, interviewers; Pauline Samuels, interpreter. Platinum, AK. March 18. Tape 86PLA001. Bureau of Indian Affairs ANCSA Office.

Scott, Colin. 2002. "The Epistemology and Ethics of Interpersonal Agency in Wemindji Cree Hunting." Paper presented at the Ninth International Conference on Hunting and Gathering Societies, Edinburgh, Scotland.

Shaw, Robert. 1998. "An Archaeology of the Central Yupik: A Regional Overview for the Yukon-Kuskokwim Delta, Northern Bristol Bay, and Nunivak Island." *Arctic Anthropology* 35(1):234–46.

Sienkiewicz, Joe. 2003. "Is There Scientific Validity to the Saying 'Red Sky at Night, Sailors' Delight; Red Sky in the Morning Sailors Take Warning'?" http://www.superstringtheory.com/forum/philboard/messages21/326.html.

Søby, R. 1969. "The Eskimo Animal Cult." *Folk* 11–12:43–78.

Society for Propagating the Gospel. 1916. *Proceedings of the Society for Propagating the Gospel among the Heathen*. Moravian Archives, Bethlehem, PA.

Strauss, Sarah, and Benjamin S. Orlove. 2003. *Weather, Climate, Culture*. New York: Berg.

Sturm, Matthew. 2009. *Apun: The Arctic Snow*. Fairbanks: University of Alaska Press.

Sturm, Matthew, and Carl Benson. 1997. "Vapor transport, grain growth and depth-hoar development in the subarctic snow." *Journal of Glaciology* 43(143):42–59.

Sundown, Teddy. 1975. Oral history interview. Susan Hansen and William Schneider, interviewers; Myron Naneng, interpreter. Scammon Bay, AK. October. Tape 75CAL001. Bureau of Indian Affairs ANCSA Office.

US Fish and Wildlife Service. 2002. "Yukon Delta National Wildlife Refuge Fact Sheet." Yukon Delta National Wildlife Refuge, Bethel, AK.

INDEX

Abraham, Jobe, 232, 258

Abraham, Ryan, 8

Abraham, Theresa, 5; on clouds and wind, 79; on hunting, 222; on *ircenrraat*, 133; on the moon, 72; on personal possessions, 115; on story of Ingriik, 132; on the sun, 64; on teaching young people, 242

abstinence practices, 39; and birth and death, 267; and menstruation, 317; and miscarriage, 267, 317; and sandbars, 228. *See also eyagyarat*

accidents: avoiding, 140, 141, 142–43, 176, 177; and ice, 145, 290; and instruction, 165, 166, 167; and kayaks, 240; on the ocean, 240–42

Agimuk, Sophie, 78, 80, 316–17; on flooding, 86; on thunder, 91; on weather prediction, 94

Agnus, Anna, 6, 8, 14; on husband/wife relations, 38; on sharing instruction, 9; on teachings of elders, 31; on weather, 60

Agnus, Simeon, 5, 165, 179, 239; on animals, 314; on climate change, 42, 306, 307,

311, 312, 318; on current, 232, 234, 235; on danger, 177; on fear, 296; on gaffs, 292; on grass, 173, 181; on hunger, 301–2; on hunting, 168; on hypothermia, 171; on ice, 257, 271–73, 275, 282, 293; on the land, 9; on the ocean, 215, 216, 222–23, 228, 233, 261, 267, 268; on oral traditions, 5; on personal possessions, 115; on personal responsibility, 321; on place-names, 46; on *qanruyutet*, 319; on risk-taking, 222; on sandbars, 225, 227; on snow, 254; on stealing, 115–16; on the supernatural, 108; on teaching young people, 320–21; on travel, 282–83, 285–87, 288, 289–90, 290–91, 292–93; on *uiteraq*, 47; on waves, 231; on weather, 88, 93, 95, 96

Ahklun Mountains, 43

Akagtaq, Tim, 310

Akula, 156

Akulmiut, 73

Akuluraq. *See* Etolin Strait

akutaq (Eskimo ice cream), 15, 119, 190–91, 207, 212

Alaska, 14, 16–19, 244; and beaver population, 311–13; geology of, 43–45, 52–53, 123; statehood, 18; stone figures of, 48–51. *See also* Alaska Native Claims Settlement Act

Alaska Native Claims Settlement Act, 12, 19

alcoholism, 40

alerquutet (laws), 8, 29, 30, 319. *See also* instruction

Alexie, Sam, 224

Alexie, Theresa, 112–13

Alirkar, John, 60, 72, 279; on ice, 248–50, 255–58, 261, 280

Aloysius, Bob, 76, 140, 141, 144, 145; on cutting one's trail, 169; on fog, 209; on ice, 147, 149, 150; on snow, 186, 194, 195, 203, 205, 207–8; on stars, 66, 68; on weather prediction, 94; on winter, 77

Amadeus, Frank, 310

Amiigtalek (Door Mountains), 130–31

Amiik, 127

amikuk (legendary creature), 108

Anchorage, 11, 155

An'gaqtar (Stone Lady), 48–50, 58

Andreafski Hills, 43

Andrew, Elizabeth, 113

Andrew, Frank, 4, 56, 111, 218; on accidents, 240; on An'gaqtar, 48–49; on animals, 85; on awareness of *ella*, 203; on *ayaruq*, 178; on blinding the universe, 34, 218; on care of carcasses, 112; on climate change, 302; on danger, 175, 177; on *ellanguaq*, 99; on flooding, 310; on gratitude, 32, 36; on kayaks, 238, 239; on the land, 109, 110; on the Milky Way, 65; on months, 74–75; on the moon, 72, 73; on navigating, 174, 223, 224; on the northern lights, 70; on the ocean, 219–20, 221, 229, 242; on place-names, 166; on *qanruyutet*, 7, 28, 30; on *qasperrluk*, 178–79; on rivers, 112, 140; on rules, 30, 33; on sandbars, 225; on shamans, 75; on snow, 195, 200, 201, 304; on tides, 233, 235; on travel, 173, 179, 180; on weather, 101–2; on weather prediction, 35, 63, 77–78, 88, 94, 95–96, 97; on wind, 82, 84, 87; on wood, 186

Andrew, John, 54, 55

Andrew, Nick, 29, 45, 55, 56, 179; on awareness of *ella*, 203; on clouds, 78–79, 81; on fall, 164; on fish, 85; on hail,

191; on hoarfrost, 192; on ice, 145, 146, 147, 149, 151; on Ingriyagaq, 129–30; on instruction, 165; on *ircenrraat,* 129; on kindling, 179; on lakes, 156; on the land, 41, 122; on mountains, 133; on navigating, 174; on the northern lights, 68, 69; on refuse, 111, 112; on rivers, 137, 139–40, 144; on self-reliance, 183; on sharing knowledge, 96; on sleeping, 60; on snow, 173, 187, 189, 191, 193–95, 199, 200, 205, 207, 208; on stars, 66, 67; on staying warm, 172; on the sun, 64; on thunder, 91; on travel, 170, 201, 210–11; on tree growth, 312; on walking sticks, 150; on water, 145; on waves, 142; on weather prediction, 61, 63, 82, 89, 93, 188; on wind, 84, 141

Andrew, Paul, 147–48; on ducks, 223; on the ocean, 219; on restraint, 236; on the sun, 64; on weather, 101

Angaiak, Ben, 8, 313

Angaiak, Rita, 6, 102, 103, 114

Angaiak, Susie, 46, 210, 253

angalkut (shamans), 97, 98, 99, 121, 129. *See also* shamans

Aniak, 13

Aniak River, 54, 144, 186

animals, 123; abundance of, 314; availability of, 168; awareness of, 32, 38–39; and awareness of ocean, 220; and climate change, 313; fur, 205, 305, 314; in human form, 23, 127, 130; and ice, 272; newborn, 205–6; offerings to, 114; in other forms, 313; personhood of, 17, 18, 23, 220; relations with humans, 37–38; and sinking land, 309; and snow, 304; and tide, 233; tracks of, 205, 312; and wind, 85. *See also* sea mammals; names of individual animals

Anthony, Dick, 69–70

Anthony, Stanley, 232, 239–40; on climate change, 306–8; on ice, 275, 282, 286; on instructing the young, 42; on the ocean, 9

Anthony, Theresa, 180

Anuurarmiut, 98, 114

Aprun River, 306

Arayakcaaq, 5

Arayakcaarmiut, 87, 114, 309

Arctic: and ocean conditions, 244; peoples, 49; and sea ice, 245, 305; tales, 62

dance, 20, 68, 97–100, 104. *See also* Messenger Feast

dangers: behavior during, 176, 177, 295; of bluffs, 176; of cliffs, 176; and dogs, 182; of fall, 164–65; and gaining awareness, 202; of ice, 143–51, 152, 245, 246, 247, 249, 255, 261, 263–64, 269, 271, 273, 275, 279, 283–91; and importance of listening, 200–201; of lake edges, 147, 149, 176; and landmarks, 165–66; and learning from experience, 220–21, 240, 264; man-made, 290; and the ocean, 216–17, 221–22, 266; places of, 175–77; and *qanruyutet* regarding travel, 235–42, 293; of riverbanks, 146; of sandbars, 227; of shallow water, 225, 226; of sleeping, 60; of snow, 194, 195, 202, 206, 208–11, 250, 254; of spring, 143; of streams, 144; of tide, 234, 235, 306; of waves, 232; of whirlpools, 145

dawn, 60, 63–64, 82, 188

death: and abstinence practices, 267; and care of the body, 109; from hypothermia, 170, 171; from not listening, 201; on the ocean, 240; responsiveness of the land to, 109; responsiveness of the ocean to, 217, 218, 219, 220; rules regarding, 34, 317–18; signs of, 132; stone figures responsive to, 57, 58; in the wilderness, 177

dew, 107

Diomede Island, 245

dipnetting, 13, 15, 179, 312

dirt, 32, 121, 134, 161, 218

dogs, 3, 77, 175, 179, 184, 270, 307, 309; and cold weather, 301; on crusted snow, 204; feces of, 187; food for, 111; on ice, 149, 217, 274, 288, 294–95; intelligence of, 149, 181–82; sleeping by, 68; and surviving blizzards, 195, 196–97

dog sleds, 166, 171, 186, 203, 212

dolls, 59, 102–3

driftwood, 13, 233

drowning, 143, 147, 150, 249; and panicking, 176; and rivers, 159

drumming, 74, 99–101, 279

ducks, 99, 217; eider, 313; long-tailed, 223, 237

Dull, Albertina, 102, 158

Dull, Peter: cutting a channel, 153; on dogs, 182; on freeze-up, 247; on grass, 180–81; on ice, 258, 293, 296; on personal possessions, 116; tools for travel, 179

earthquake, 314

eclipse, 75

economy, 18, 19, 40

ecosystem, 59, 60

Ecuilnguq River, 144

education, 10, 19, 39. *See also* young people

Effemka, Golga, 152; on abundance of snow, 185; on Amiigtalek, 130–31; on climate change, 301, 304, 313; on loss of Yup'ik language, 40; on stone people, 53–54

eggs: gathering of, 14–15, 115, 151, 158, 160; on islands, 156; on sandbars, 225

Egnatty, Jack, 55

Eicken, Hajo, 244–45, 275, 283

eider ducks, 313

Elachik, Peter, 147, 304–5, 313

elders: gratitude of, 28, 36; as instructors, 4, 33, 42, 165–66, 167, 316; and prediction of climate change, 303–4, 318; preserving knowledge of, 3, 4, 5, 8, 20, 31; as scientists, 94, 96; and sharing knowledge, 26, 29, 31, 39, 40; sharing with, 35–36; and understanding change, 315–16; and weather prediction, 93, 96

ella (weather, world, universe): awareness of, 10, 203, 317–18; bringing bad weather, 89; and climate change, 167, 311; elders' knowledge of, 96; and erosion, 310; glossary for, 104–7; and importance of its teaching, 104; and loss of traditional knowledge, 10; making newborn bedding, 206; meanings of word, 8, 32, 59; and observing the sun, 62; original state of, 304; respect for, 171; as responsive to human behavior, 42, 97, 202, 242, 315, 319, 321; and weather prediction, 94; wisdom from, 316

ella maliggluku (following the direction of the universe), 49, 102

ellam iinga (eye of *ella*), 60, 97

Ellam Yua (Person of the Universe): and Christianity, 60; and following one's mind, 33; and masks, 97; and sharing, 36; as watching the world, 59, 60, 115; wisdom from, 316

ellanguaq (model universe), 98–101

Elqialek (stone figure), 50–51

offerings of, 113–15; quarreling over, 314; respect for, 10; restrictions on, 180; sharing of, 9, 20, 36–37, 315; traditional Yup'ik, 13–15; and travel, 179–80. *See also* starvation

Fox, Andy, 79

foxes, 123, 168, 186, 202, 205

freeze-up: and climate change, 41, 244, 251–52, 305, 306–7; and flooding, 86; and fresh water, 246; and ice, 154, 249–50; and lake edges, 147; and rivers, 146, 176; and tundra, 189; and use of walking stick, 177–78

frost, 189, 213; in the past, 300, 302, 303; and wild celery, 73, 192, 303; and wild rhubarb, 221

gaff, 287, 291–92. *See also negcik*

geese, 74, 82, 313–14

geology, 43–45, 52–53, 123

George, Chris, 321

George, Mary, 41

global warming, 42, 304, 309, 318, 319. *See also* climate change

glossaries: for ice, 296–99; for land, 121, 134–35; for the ocean, 243; for rivers and lakes, 161–63; for snow, 211–14; for weather, 104–7

Golden Gate Falls, 52

Goodnews Bay, 24, 57

GPS (Global Positioning System), 92, 124; and the changing ocean, 217; compared to *qanruyutet,* 174–75; and hunting in fog, 222; navigation before availability of, 245

grass: and dolls, 102–3; and fishing streams, 112; gathering and uses of, 15, 180–81; and navigation, 166, 167, 173, 198; rye, 15; tussocks of, 199, 122, 134, 135; and weather prediction, 35, 78, 122, 188; wheat, 175

grassy knolls, 309

gratitude: of An'gaqtar, 49; of elders, 28, 36; of Ellam Yua, 61; of fish, 111; of food, 111; of hunters, 33, 37; of *ircenrraat,* 132; of land, 123; of lenders, 116; of nonsentient things, 32; power of, 28, 33–34, 35–37, 38, 49, 53; for *qanruyutet,* 320; for safety, 295; of stones, 58

gray jays, 188

Great Death, 50, 55

greens, edible, 14, 115, 158, 178

gulls, 112, 223, 225

Guy, James, 53

hail, 190, 191–92, 212

halibut, 15, 226, 308

halo, solar, 64

harvesting, subsistence: and climate change, 312; and contemporary village economies, 19, 20; customary places of, 120–21, 134; and seasonal cycle, 13–16, 17, 18; and tides, 233

haze, 79, 92

heat waves, 79, 92

herring, 15; and climate change, 41, 301, 308; as a travel food, 180; and wind direction, 84

hoarfrost, 190, 192, 212

Hoffman, Chief Eddie, 75

Hoholitna River, 55, 130

holes, exit, 149, 161

Holitna River, 53

Holokuk River, 54

Holy Cross, 55

Hooper, Edward, 132; on climate change, 308; on clouds, 78, 80; on ice, 253, 274; on sharing instruction, 9; on rain, 90; on wind, 84

Hooper, Tommy: on climate change, 312; on eating snow, 210; on extraordinary beings, 108, 160; on ice, 272, 286, 295; on miscarriage, 218; on teaching young people, 8

Hooper Bay, 123, 124–25, 261

houses: and February thaw, 73; and hoarfrost, 192; and loss of *qanruyutet,* 39; and snow, 197; and sod, 17, 25, 187

hunters, 224, 238, 261, 312, 318; abandoning their catch, 236, 239, 286, 294, 295; announcing one's catch, 291–92; and offerings, 49, 54, 55, 113–14; partners of, 221–22, 273, 290–91, 293; and sleep, 168, 265; washed away on ice, 104, 240–42

hunting: chores, 27–28; and climate change, 244–45, 305; and compassion, 37, 39; customary places of, 165; after a death, 220; and fog, 91, 92; and gratitude, 33, 37; and ice, 270–71, 276, 278, 280, 281–82, 286, 287, 288, 293; and knowledge of

hunting (continued)
water, 136; and the ocean, 216, 217,
228, 229; and overhunting, 42; and
qanruyutet, 32; restrictions on, 218, 219;
rules for, 220–22; and snow, 202, 203,
254; in spring, 13, 152; and tide, 233; and
weather prediction, 80, 82, 95; and wind,
85, 87, 284
hypothermia, 143, 170, 171, 181, 201

ice: ability to part, 287–88; beached, 259–60,
281; belts, 277, 287, 297; blink, 261; brash,
276; broken pieces of, 245, 289, 296;
camping on, 253; and climate change,
250, 251–52, 257; colored, 145, 147, 151;
coves in, 262–63; in creation stories,
45–46, 48; crevices and cracks in, 257,
260–61, 263, 272, 273–74, 291, 296, 299;
crystals, 174, 223–24, 243, 297; dangerous,
144, 152, 178, 264, 289–91, 294; dark, 148,
149; and drinking water, 221, 265; falling
through, 143, 147–50, 247, 264, 291; fog,
64, 82, 107, 198, 213; freshwater, 154–55,
158, 249, 265; glossary for, 296–99;
hockey, 187; hole in, 161; jams, 152; as a
lookout, 257–58, 259; melting, 22, 263,
264, 291, 298; mixed with sand, 151;
needle, 148–49, 155, 162, 178; from the
north, 280; and ocean swells, 229–30;
overhanging edges of, 147, 278–79, 292,
297; pebbles, 276; people washed away
on, 104, 240–42, 272, 273–74, 290; pick,
248, 250, 292; ridges, 146, 147, 162; on
rivers and lakes, 152–53, 154; rolling,
274, 296; rough, 249–50, 251, 283, 288,
298; saltwater, 154–55, 248, 249, 265; and
sandbars, 258–59, 266, 269; scattered,
275–76; sediment laden, 256, 290,
292–93, 296, 299; shapes of, 260, 269;
shuga, 276; skating, 147, 155; sleeping
on, 265; slush, 253; smooth, 249–50, 251,
258, 264, 298; under snow, 143, 144, 150,
207; and solar halo, 64; sounds of, 148,
219–20, 267; in spring, 262–65, 276, 278;
strength of, 147, 155; surfacing, 255, 298,
299; tester, 150; thin, 288, 296, 298, 299;
trails in, 250; and travel, 147, 149, 292–96;
walking on, 248, 249, 251; and wind, 293;
winter, 255–61. *See also* floebergs; entries
immediately below

ice, floating, 221, 286, 297, 298; pieces of,
276–77; and safety, 288; and wind, 293
ice, newly frozen, 23, 41, 150, 154, 155, 246–
48, 270–72, 296, 297, 307; and climate
change, 307, 312; layered, 281; and
sandbars, 258; sheets of, 248–49, 289;
snow covered, 250; and travel, 287, 290;
and water, 304, 311
ice, packed, 270, 275–76, 298, 299; and
current, 277–78; and hunting, 288; ice-
free areas, 282–83; travel in, 286, 287, 293
ice, piled, 41, 154, 241, 245, 252, 255–58, 285,
288, 297, 299; and climate change, 306;
in creation stories, 45–46, 48; crevices
in, 291; dangers of, 289–90; and ice floes,
271, 272, 273; mixed with mud, 274–75;
named, 257; as shelter, 292, 293
ice, river, 143–49, 176, 249; and
environmental warming, 41; and
flooding, 86; in spring, 263–64, 266
ice, sea, 257, 259; and climate change, 41,
257, 260, 305–8; and erosion, 275; fall
formation of, 245–48; in spring, 262–65
ice, shore-fast, 221, 247–48, 249, 250–51,
261, 264–65; and climate change, 41,
251–53, 302, 306, 307; compared to seal
oil, 230; and ice floes, 270, 272, 273; and
ice pieces, 277; and newly frozen ice,
248–49; and ocean swells, 265, 267, 268;
and overhanging ice edges, 278–79; and
piled ice, 256–57, 258–59; sleeping on,
224; in spring, 253, 262, 263, 266, 288;
and travel, 286; and waves, 229; and
wind, 85
icebergs, 219
ice floes, 251, 256, 258, 296, 306, 307;
detaching from shore, 272, 273; drifting,
269–74; and foam, 263; and piled ice,
293; and spring melt, 265; and travel,
286, 293–94
igloos, 201. *See also* snow shelters
Ilkivik River, 91, 142
illness: caused by *ircenrraat,* 133; and eating
soil, 114–15; and epidemics, 16, 18, 19, 50,
55; and following admonitions, 32, 159;
and hoarfrost, 192; and the ocean, 217;
rules concerning, 34; and stone figures,
50, 55, 58
inerquun (cautionary rule; pl. *inerquutet*), 8,
29, 30, 317

story knife, 46

streams: dangers of, 144, 175; fish in, 148; mountain, 144–45, 161, 178

suicide, 40

summer, 75, 77, 84, 89, 92, 105, 226, 265; arrival of, 208, 221; and climate change, 301, 307, 310; clouds in, 78, 79, 81; and fish, 15, 17, 49, 62, 85, 110, 137; fog in, 198; and ocean swells, 229; rains of, 90, 91, 157; sun in, 65, 76; and thunder, 191; and waves, 139, 140; wind in, 83

sun, 104, 105, 142; and climate change, 301; finding direction of, 174, 209, 223–24; and fish breath, 78; and fog, 92; and ice, 223–24, 247, 262, 263, 266; origin of, 61–62; and snow, 205; in spring, 76, 205, 206; in summer, 65, 76; and weather prediction, 62–64, 88–89, 94, 95

sun column, 62, 104

sun dogs, 64

Sundown, Teddy, 48, 50–51

swamplands, 122, 134, 135, 176

swans, 172, 180

swells, ocean, 219, 220, 243, 265–69, 296, 297; and awareness of the ocean, 317–18; and breakup, 248; dangers of, 229–30; and ice, 251; navigating by, 223, 229

Talarun River, 312

Tall, Louise, 113

Taqukatuli Channel, 225

Tarunguaq, 49, 50

teasing, 99, 113, 290. *See also* cross-cousins

temperatures: and climate change, 41, 300, 301; and inversion, 88, 106; and markers of cold, 301; regional differences in, 301; and sea ice, 306; as warming, 252, 253, 306–7, 312. *See also* climate change

Tengmiarrluk (giant bird), 57

Tengutellret, 55

Therchik, Nick, Jr., 261, 274

Thompson, Johnny, 118, 120, 138, 313

thunder, 91, 107, 191

tide, 233–35, 243; high, 225, 226, 246, 309; and ice, 148, 152, 154, 251, 255, 259, 263, 276, 277, 284, 306; incoming, 248, 254, 269, 281, 286, 266; low, 225, 233, 244, 255, 283; outgoing, 285, 286; and snow, 253; in spring, 241; and travel, 226, 227; variation of, 244; and wind, 137, 141

Tirchik, Walter, 219, 220

Togiak, 48–49, 50

Togiak River, 49, 57

Toksook Bay, 6, 44, 153, 227, 280; and climate change, 306, 307, 310, 311; and ice, 252, 260, 261, 262, 289; and *ircenrraat,* 127; rock formations at, 46, 47; sandbars of, 225; and tide, 285; and travel, 232; and weather, 87, 96

Toksook River, 153, 228, 310

Tom, Mark, 7, 93, 248, 255

tomcod, 13, 86, 308

Tommy, Elsie, 55, 64, 133–34

Tommy, Nicholas, 167

trails: checking one's, 168; cutting with a knife, 169; and ice, 250, 270, 271, 284–86, 288, 289; and spring melt, 264, 266

trapping, 13, 309

travel: and climate change, 305; and dogs, 181–82, 183; and fog, 92, 216; on foot, 168; and freeze-up, 143–49; and grass, 180–81; and ice, 147, 217, 246–47, 250, 269, 275–76, 283–91; instruction for, 170, 319; and *ircenrraat,* 132, 133; and knowledge, 9; and landmarks, 84; and the ocean, 220–22, 231, 236–38; preparing for, 60–61; and *qanruyutet,* 235–42; and right conduct, 161; and rivers, 136, 139–44; rules regarding, 109, 159, 219, 317; and sandbars, 227–28; and snow, 194, 195, 196–97, 203, 204, 206, 208–11, 254; in spring, 262; and stars as a compass, 66–67; and staying quiet, 176; and staying warm, 171–72, 194; in stormy conditions, 209; and tide, 226–27; tools for, 149–51, 177–80, 201, 202, 247, 257, 260, 279, 291–92; and water, 225, 226, 254; and waves, 227–28, 231, 232; and weather, 90, 94, 102, 200; in wilderness, 164–80; in winter, 144, 186, 301. *See also* boats; kayaks

trees: and climate change, 312; hoarfrost on, 192; navigation by, 174; as protection, 170; and snow, 195, 196; spruce, 312; as weather indicators, 94

tuberculosis, 19

Tulik, Camilius, 117, 176, 177; on climate change, 303; on current, 235; on fearlessness, 296; on ice, 252, 286, 289; on instructions, 31; on knowledge, 5; on

227–28, 231; tidal, 295; as weather indicators, 230; and wind, 229–33, 267. *See also* swells, ocean

weather: after a funeral, 203; becoming a liar, 303–4; and climate change, 82, 300–305; following its people, 316, 317, 318, 321; influencing, 59, 97–103, 188; instructions regarding, 171; observation of, 60–61; people hunted by, 202–3; personhood of, 35, 202; power of, 171; and reasons for change, 314–18; responsiveness of, 242, 302; and travel, 168–69, 172–73, 175, 176, 179, 222; and trust of, 209. *See also ella;* weather prediction

weather prediction, 35, 60, 61, 87–88, 93–97; and the Bible, 63; and climate change, 303–4; and clouds, 78–82; and current, 159; and the moon, 72; and the northern lights, 68–69; and ocean swells, 229; and packed ice, 278; and plants, 77–78; and relationship between spring and fall, 77; and smelling, 97, 189; and snowflakes, 193; and spring melt, 151; and the stars, 65–66, 67; and the sun, 62–64; and trees, 94; and waves, 230; and wind, 65–66, 68, 89, 278, 293

wetlands: grassy, 144, 161, 175, 178; and migratory birds, 14; and waterfowl, 313

whales, 226; beluga, 13, 15, 42, 56, 307, 313; killer, 128

whirlpools, 145, 163

whitefish, 15, 42, 172

whiteout, 195, 199, 212

white people, 54; ocean becoming like, 220, 242; Yup'ik becoming like, 35, 93, 242, 317. *See also kass'aq*

wild celery, 14; and frost, 73, 192, 303; as a guide, 166, 173, 198

wilderness: and climate change, 309; dangers of, 150; death in, 177; and getting lost, 172–75; instructions regarding, 164–67; personal property left in, 115; staying quiet in, 176; travel in, 164–80, 183; as unowned, 119

wild parsnips, 14

wild rhubarb, 73, 83, 192, 221

wind, 83, 98, 106; awareness of, 317; and breakup, 266, 268; and climate change, 302, 303; and clouds, 78–81; and current, 140–41, 231, 232; directions of, 83–84,

85, 89; and drinking saltwater, 221; east, 198, 211; and *eyagyarat*, 219; and fish, 137, 138; and fog, 92, 93; as a gift, 84; and hoarfrost, 192; and ice, 147, 250, 251, 255, 270, 280, 283, 284, 288; and masks, 97; and the Milky Way, 65; and navigation, 170, 173, 174, 183; north, 98, 138, 172, 173, 189, 199, 200, 208, 229, 231, 270; northeast, 200, 281; and the northern lights, 70; northwest, 231, 280; and ocean swells, 229; offshore, 281; onshore, 284; prediction of, 65–66, 68, 278, 293; on rivers, 136–39; and snow, 172–73, 187–90, 194, 195, 199–200, 204, 208, 210, 214; south, 142, 187–89, 199–200, 204, 207, 208, 210, 214, 231–32, 251, 270, 302; and snowmobiles, 182–83; and stars, 66; strength of, 197; travel in, 211, 237–40; and water, 137, 138, 139–42; and waves, 139, 140, 142, 229–33, 267; and weather, 89–90, 302; west, 89–90, 231, 232, 280, 302

winter, 75, 76; and climate change, 301–2; and clouds, 81; early onset of, 77; and *ella*'s original state, 304; and ice, 248, 250, 255–61, 264; male and female, 77; and rain, 157; signs of, 189, 192, 204; travel in, 144; and wind, 83

Wise Words of the Yup'ik People: We Talk to You because We Love You, 8

wolverine, 129

wolves: belugas transformed into, 313; in human form, 127; *ircenrraat* as, 126, 127, 129–30; protection from, 181

wood: availability of, 300; and breakup, 152; gathering, 319; people acting like, 143, 236; and sandbars, 225; and snow, 186

world (*ella*): following its people, 38–42, 300–21; as responsive to humans, 10, 17, 18, 35, 38, 42; shrinking, 102. See also *ella*

worms, 59

young people: becoming elders, 31; contemporary, 318; following admonishments, 161, 317; and ice, 245, 289; instruction of, 7, 8–9, 177, 239, 242, 292, 319–20; and the land, 9; and and the ocean, 220–21; in the past, 315; and *qanruyutet*, 8, 10, 25, 26, 29; and spring, 262; teaching restraint to, 235–36; and traditional knowledge, 31, 39